CABLE–SUSPENDED ROOFS

CABLE–

McGRAW-HILL SERIES IN MODERN STRUCTURES:
Systems and Management
Thomas C. Kavanagh Consulting Editor

Baker, Kovalevsky, and Rish *Structural Analysis of Shells*
Coombs and Palmer *Construction Accounting and Financial Management*
Desai and Christian *Numerical Methods in Geotechnical Engineering*
Forsyth *Unified Design of Reinforced Concrete Members*
Foster *Construction Estimates from Take-Off to Bid*
Gill *Systems Management Techniques for Builders and Contractors*
Johnson *Deterioration, Maintenance and Repair of Structures*
Johnson and Kavanagh *The Design of Foundations for Buildings*
Kavanagh, Muller, and O'Brien *Construction Management*
Kreider and Kreith *Solar Heating and Cooling*
Krishna *Cable–Suspended Roofs*
O'Brien *Value Analysis in Design and Construction*
Oppenheimer *Directing Construction for Profit*
Parker *Planning and Estimating Dam Construction*
Parker *Planning and Estimating Underground Construction*
Pulver *Construction Estimates and Costs*
Ramaswamy *Design and Construction of Concrete Shell Roofs*
Schwartz *Civil Engineering for the Plant Engineer*
Tschebotarioff *Foundations, Retaining and Earth Structures*
Tsytovich *The Mechanics of Frozen Ground*
Woodward, Gardner, and Greer *Drilled Pier Foundations*
Yang *Design of Functional Pavements*
Yu *Cold-formed Steel Structures*

PREM KRISHNA
Professor of Civil Engineering
University of Roorkee
Roorkee, India

SUSPENDED ROOFS

McGRAW-HILL BOOK COMPANY
New York St. Louis San Francisco Auckland Bogotá
Düsseldorf Johannesburg London Madrid Mexico
Montreal New Delhi Panama Paris São Paulo
Singapore Sydney Tokyo Toronto

Library of Congress Cataloging in Publication Data

Krishna, Prem.
Cable–suspended roofs.

(McGraw–Hill series in modern structures)
Bibliography: p.
Includes index.
1. Roofs, Suspension. I. Title.
TH2417.K7 690′.1′5 77-22803
ISBN 0-07-035504-5

1234567890 KPKP 7654321098

The editors for this book were Jeremy Robinson and Joseph Williams, the designer was Elliot Epstein, and the production supervisor was Teresa F. Leaden. It was set in Times Roman by Computype, Inc.

It was printed and bound by The Kingsport Press.

Contents

Preface

The search for a means to make better and more efficient utilization of covered space has been one of the important objectives of engineers, architects, and planners. The demand for large-span roofs to provide unobstructed use of covered space has intensified the search for new solutions. Recent advances in the material and construction technology have contributed significantly in finding answers to meet this need. Recent developments in the construction of cable-suspended and pneumatic roofs have added a new and exciting dimension to large-span roofing. Some very efficient solutions have been achieved, and there is a great promise for the future.

During the last two to three decades, considerable research has been done in this area of rapid progress, and a fairly large volume of technical material has appeared. The dissemination of ideas on cable- and air-supported roofs has taken place largely through the proceedings of symposia and conferences held at regular intervals and through journals. The papers presented or published have given valuable—but isolated—information on different facets of such roofs. This book attempts to bring together and make a critical appraisal of ideas on the form, analysis, design, and construction of cable structures, with the primary purpose of serving those involved in their design. The coverage of structural analysis and response of cable roofs is fairly comprehensive, yet a balance is maintained between this aspect and design and construction. The areas of research are also identified. In preparing the manuscript the existing literature has been freely used and an extensive list of references is included.

Chapters 1 to 7 discuss various aspects of cable roofs and Chapter 8 has a brief coverage of air-supported roofs. Though the latter are not directly covered in the first seven chapters, the solutions to many related problems will be found therein. Chapter 1 is an introduction to forms and design ideas and includes a note on the development of cable roofs. Chapters 2 and 3 have a detailed coverage of the analysis and behavior of free form as well as pretensioned cable-roof systems. These two chapters also contain material to help the designer in carrying out approximate analyses and in making preliminary design estimates. The listing of a computer program for a discrete-element nonlinear analysis of cable systems is given in the Appendix. Chapter 4 deals with the dynamics of cable roofs. Chapter 5 discusses experimental procedures related to model studies of cable systems. Some case studies are also given. Chapter 6 has a discussion on design problems, loads, and cost comparisons, and a brief description of a few existing structures. Also included is an example illustrating the procedure to carry out a preliminary design and to obtain a cost estimate. The materials used in cable-roof construction and related structural details, with examples from existing structures, are discussed in Chapter 7.

This book does not cover analysis and design problems on related conventional construction in steel or concrete, which are duly covered in a large number of textbooks on these subjects. This does not indicate their lack of importance, but has been done just to save space, for a more detailed coverage of cable roofs on which information is not easily available elsewhere. While every care has been taken in choosing material useful for design and graduate work, comments or suggestions for future improvements in the manuscript will be most welcome.

The author is indebted to many persons for various reasons: the late S. R. Sparkes for introducing him to cable structures; E. H. Gaylord, Jr., for encouraging the writing this book, and his many friends, colleagues and students; in particular, Brijesh Chandra for help during different stages of the work; S. P. Sharma and T. N. Gupta for offering valuable criticism of the manuscript; V. K. Gupta, J. K. Jain, Banbir Das, and M. R. Sethia for helping with examples. To the various professional organizations and friends who have kindly made available technical material for inclusion in the book, the author is indeed grateful. The work of Ramesh Sharma in typing the manuscript so painstakingly, Narendra Arya in making the architectural sketches, and R. K. Tyagi in tracing the figures is thankfully acknowledged.

PREM KRISHNA

Notation

A	Cross-sectional area of cable
B_s	Coefficient of earth pressure
B_u	Nominal uplift coefficient
b	Typical dimension of footing, as defined where used
$\mathbf{C}$	Damping matrix
C_p	Wind pressure coefficient
C_m	Coefficient of an additional mass
C_d	Coefficient of damping
c	Unit cohesion
D	A length parameter, as defined where used
E	Modulus of Elasticity
E_t	Tangent Modulus
EA	Axial rigidity of the cable
e	Distance of cable below chord joining one end and central point
e_x, e_z	Errors of closure along x, z axes, respectively
F_i	Initial prestress force in a cable element i
ΔF_i	Change in force F_i
ΔF_i^t	Change in force due to temperature
F_i'	Final value of force, equal to $F_i + \Delta F_i$
f	Dip of cable
G	Gust factor
g	Acceleration due to gravity
H	Horizontal component of tension in cable
h	Change in H
h^t	Change in H due to temperature variation
H'	Final value of H, equal to $H + h$

$\mathbf{K}$	Stiffness matrix
$\mathbf{K}'$	Modified stiffness matrix
k_x, k_z	Stiffnesses of spring along x, z axes, respectively
k	Stiffness of cable
L	Cable span
l	Length of a cable arc
$(\Delta l)_e$	Elastic stretch of a cable arc of length l
M	Bending moment
M_e	Bending moment in an equivalent simply supported beam
$\mathbf{M}$	Mass matrix
m	Mass, as defined where used
N	Number of nodes in a cable system
P	Concentrated applied load
P_x, P_y, P_z	Applied loads at a node along x, y, z axes, respectively
$\mathbf{P}$	Load vector
$\mathbf{P}(\mathrm{t})$	Time dependent load vector
P_u	Ultimate pull-out load on a footing
p	Uniformly distributed applied load, as defined where used
p_w	Wind pressure
q	Uniformly distributed dead load, as defined where used
R_x, R_y, R_z	Force residuals at a node along the x, y, z axes, respectively
$\mathbf{R}$	Column vector of residual terms
S	Support reaction at a cable end
s_i	Initial length of a cable element i
Δs_i	Change in length s_i
s_i'	Final value of length s_i, equal to $s_i + \Delta s_i$
T	Tension in cable
ΔT	Change in tension T
$\overline{T}$	Horizontal force at junction of cables in a truss
t	Change in temperature, time period, as defined where used
$\mathbf{U}$	Column vector of unknowns u, v, w
u, v, w	Displacements along x, y, z axes, respectively
V	Velocity
V_m	Time averaged component of wind velocity
W	Energy, as defined, where used
X, Y, Z	Projections of cable segment on x, y, z axes, respectively
x, y, z	Coordinates along x, y, z axes, respectively

ψ Angle of internal friction for soil
γ Unit weight of soil
ξ Angle of wall friction
δ Logarithmic decrement
ε Strain
σ Stress
ϵ Slip of the cable at the support
Θ Curvature, or, a constant with like dimension
$\overline{\Theta}$ A constant having dimensions of curvature
λ Scale ratio
$\boldsymbol{\lambda}$ Characteristic eigenvalue scalar
ρ Mass density
ϕ Nondimensional modal displacement
$\phi(x)$ A load function
ω Circular frequency of vibration
η Frequency of vibration, hertz
α Coefficient of linear expansion
∇ Specified accuracy required in an iterative solution
χ Shape factor for a footing
Ψ Peak factor
Ω Excitation by background turbulence
ϑ Size reduction factor
ζ Critical damping ratio

CABLE–SUSPENDED ROOFS

CHAPTER ONE

Introduction

1-1 GENERAL

The animal skin, and its tension-membrane action; the spider's web; and the presence of other such structures in nature must, in all likelihood, have spurred man to develop tension systems for his use. The primitive roofing systems would perhaps have used animal skin to form tents, which even today are considered perfect examples of a membrane roof with predominantly tensile forces. Cable systems are tension structures as well and take an important place among this class of structures.

The idea of using a suspended cable was suggested, perhaps, also by nature in the tropical countries of Southeast Asia and Africa, where there is an abundance of ropes of vines and creepers. The available historical evidence suggests that natural ropes were the first materials used for constructing small suspension bridges. These bridges employed wooden platforms for a walkway. With the availability of iron and steel as structural materials, the first forms of the modern suspension bridge, in which high-tension steel cables are used, began to take shape; the analysis, design, and construction of suspension bridges has developed today to a fine and exciting art. Scores of outstanding bridges of this type now exist, or are being planned. Many of these, such as the Golden Gate bridge, the George Washington bridge, and the Verrazano Narrows bridge, were built in the United States during the first half of this century. The failure of the 2800-ft (853-m) Tacoma Narrows bridge in 1940, because of aerodynamic oscillations, showed that it was not adequate to design slender structures for static stability only. The research that has followed the Tacoma Narrows disaster has provided many useful ideas for the design of aerodynamically safe flexible structures.

Nets formed of natural or nonmetallic fabric as well as steel wire rope have been used in a variety of ways, such as in fishing, camouflage, and combat [6]; for supporting membranes in balloons, roofs, cooling towers, and sails for a ship; and for supporting towers and chimneys. The development of some structural uses of cables, such as in the construction of suspended cable roofs, suspended shells and buildings, and large cooling towers,[1] is very recent. A historical review of suspended roofs suggests that the earliest version of a tension roof was a tent. The principle of using tensioned cables in the construction of roofs was first illustrated, according to recorded history, in A.D. 70, when the 620 × 513 ft (189 × 156 m) Roman Colosseum was achieved. Although the system was used off and on, the construction of the Colosseum was not followed by any appreciable advance in the construction of cable roofs until the 1950s; thus, this is an extremely recent system of roof construction. Some suspended roofs were built in Russia at the end of the last century and ranged in span to about 200 ft (61 m). In these roofs, a crisscross lattice of steel strips was used in place of cables, to support the roof surface. The real developments in the analysis, design, and construction of cable roofs have occurred only after the Raleigh arena, designed by Nowicki, was built in the United States in 1953. This progress has indeed been rapid.

The impetus for the design of modern cable roofs could well be attributed to the success of suspension bridges. In suspension bridges, the "gravity" stiffness of the cable interacts with the flexural stiffness of the girder and may be adequate sometimes to enable the elimination of the girder. The main load-bearing element in suspended roofs is a steel cable, which is occasionally combined with struts or light flexural members. Usually, such roofs are also light in weight and would therefore be limited to their low gravity stiffness, unless the system is pretensioned. Cable-suspended roofs can be

[1] A cooling tower supported on a hyperbolic net formed with a triangular cable mesh was recently completed (1977) in Schmehausen, Germany. The structure has a diameter of 459 ft (140 m) at the base and is supported on a ring 288 ft (88 m) in diameter and 479 ft (146 m) above ground, which in turn is suspended from a mast 590 ft (180 m) high [84].

classed as *all-tension* structural systems, because, for reasons of its geometric proportions, a cable has negligible buckling strength or flexural stiffness, and as such can transmit or sustain loads by developing tensile forces only.

There are many similarities among cable structures, e.g., their nonlinear behavior, the high degree of indeterminacy to be tackled in analysis, the problem of anchoring cables, and the capability for large-span construction. Therefore, although cable structures other than cable roofs are outside the scope of this book, the material contained herein can lead to an understanding of cable behavior that could be helpful in the analysis and design of such structures.

1-2 DEVELOPMENTS IN ANALYSIS

Most of the developments in the analysis of cable-roof systems have taken place after 1950, though some work on the subject is reported before this date as well. Before we undertake a broad review of this aspect of the problem (see Chaps. 2 to 4 for more details), two points most relevant to theoretical analysis will be briefly discussed in the next paragraph.

A cable exhibits a marked nonlinear behavior when loaded, and the degree of nonlinearity varies with the type of cable structure as well as loading. It is necessary to distinguish this nonlinearity, which may be called *geometric* nonlinearity, from possible material nonlinearity. In order to sustain loads, a cable has to follow the funicular curve of the loads and so undergoes large geometric adjustments, particularly when the loading is concentrated or unsymmetric. Geometric nonlinearity will therefore occur in cable behavior irrespective of whether the cable material has a linear or a nonlinear stress-strain curve.[2] Figure 1-1 illustrates this feature of the behavior of a cable, which to a smaller or larger degree is applicable to cable-roof systems and poses a serious problem in analysis. Nonlinearity is more pronounced in roofs where cables are not pretensioned. The other problem that demands analytical skill is the generally high degree of indeterminacy of these systems.

The problem of analyzing cable systems can prove to be quite

[2] It is well worth noting that stranded cables do not have a linear stress-strain curve, unless prestretched, and have an average value of Young's modulus which is of the order of 75 percent of the value for steel.

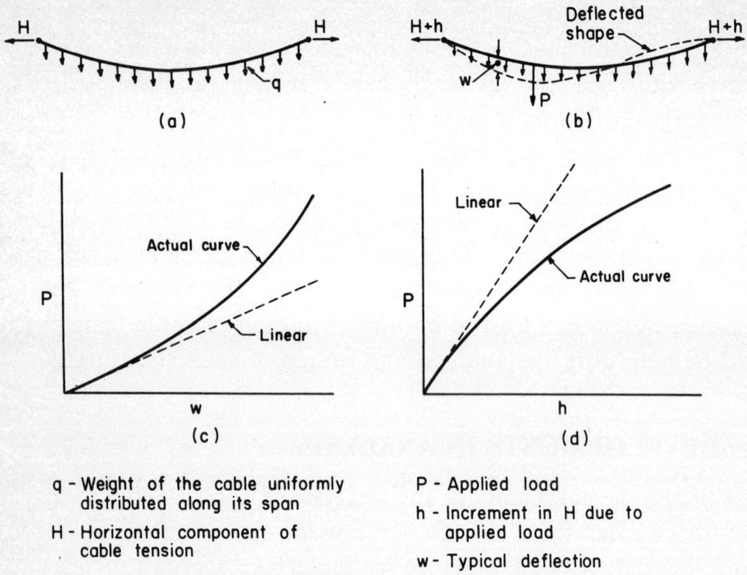

FIGURE 1-1 Nonlinearity of cable response.

complex in general, but simplifying assumptions [45] can be made to reduce the effort required, particularly for preliminary design work. There are two broad categories into which the development of cable-roof analysis can be placed, depending upon whether the network is treated as (1) a continuous membrane, or (2) a discrete system. In almost all analytical procedures published for cable roofs, the contribution of the cladding to the stiffness of the network is disregarded. That being the case, the bare network is in fact a discrete system and its treatment as a continuum is an approximation. Analytical procedures on the basis of the latter have been used to produce simplified approximate formulas for computations of cable systems (see Chap. 3).

If the structure is treated as a discrete system, any method of analysis must basically lend itself to the solution of *stiffness* or *flexibility* matrices set up for the structure. The use of iterative procedures becomes imperative for the solution of this nonlinear problem. Different iterative schemes [45] are possible for tackling the problem, such as the Newton-Raphson method, which has given good results. An alternative to stiffness- or flexibility-matrix analysis has been proposed in the use of the principle of minimization of total potential energy. This method of analysis uses the standard optimization techniques. The minimization procedure which uses *conjugate gradients* with step-length control and the Newton-Raphson procedure give good results in many cases.

Most analytical studies on cable systems have assumed the supporting structure of the roof to be rigid. However, evaluation of the effect of the support flexibility, which can be considerable, has engaged the attention of researchers recently [9, 68, 83]. A better understanding of the cable roof and the supporting structure, in relation to each other, is leading to a gradual departure from the early practice of making massive supporting beams.

It can be said in general that the problem of analyzing cable networks for static effects has been adequately met during the 25 years of their development after the 1950s, but there is only a meager amount of information available for direct and easy application in design office computation. This book attempts to fill that gap to the extent possible.

Cable roofs are flexible long-span structures and as such are prone to significant aerodynamic excitation. Some existing roofs of this type have shown a tendency to *flutter*, but there is little work reported [1] on the assessment of aerodynamic behavior of cable roofs. Considerable effort has, however, gone into producing information on the fundamental dynamic behavior of cable systems, by treating these both as *discrete* and as *continuum* systems.

1-3 FORMS AND CLASSIFICATION

A cable roof can be defined as one in which a cable or a system of cables is used as a load-carrying structural element. Such roofs can be placed in different categories depending upon the criterion used for classification. In accordance with the manner in which cables are used, they can be classified as

1. Cable-supported roofs
2. Cable-suspended roofs
3. Cable-cum–air-supported roofs

In cable-supported roofs, developed somewhat on the same lines as cable-supported bridges, the cables have only the auxiliary function of

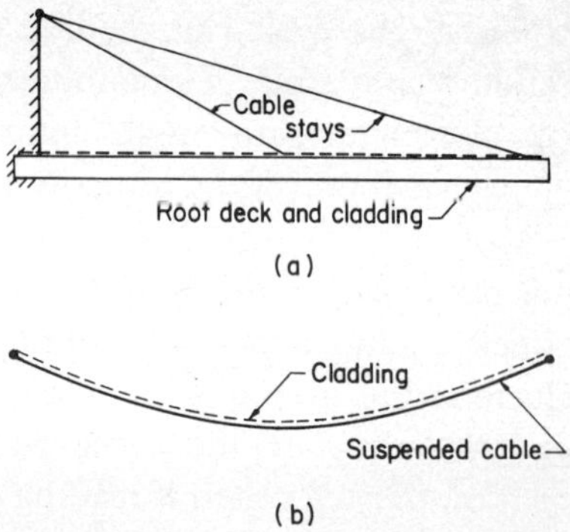

FIGURE 1-2 (*a*) Cable-supported roof. (*b*) Cable-suspended roof.

providing additional support for elements which are otherwise sufficiently strong to carry a major proportion of the loading. This type of roof has been built extensively in the United States and the European countries. In cable-suspended roofs, the system of cables carries the roof load directly and as such has a primary structural function. The cable system also serves as the falsework for the erection of the cladding. Another class of cable roofs, which is of comparatively recent development and has now gained considerable popularity, is the air-supported roof. These roofs are mostly tent- or balloon-type structures which are supported by a combination of cables and air inflation. The use of air-supported roofs was most extensively illustrated in the 1970 International Exposition in Osaka,

FIGURE 1-3 Roof of the Australian pavilion at the International Exposition, Osaka, Japan, 1970. (*Engineering News-Record—ASCE.*)

FIGURE 1-4 Roof for a sports stadium. (*British Constructional Steelwork Association, representing British Structural Steelwork Industry.*)

Japan. Chapter 8 is devoted to related aspects of air-supported roofs.

A classification of cable roofs is illustrated in Fig. 1-2 and examples of cable-supported roofs are shown in Figs. 1-3 to 1-6. Figure 1-3 shows the roof designed for the Australian pavilion for the 1970 International Exposition in Osaka, Japan. The 128-ft-high (39-m),

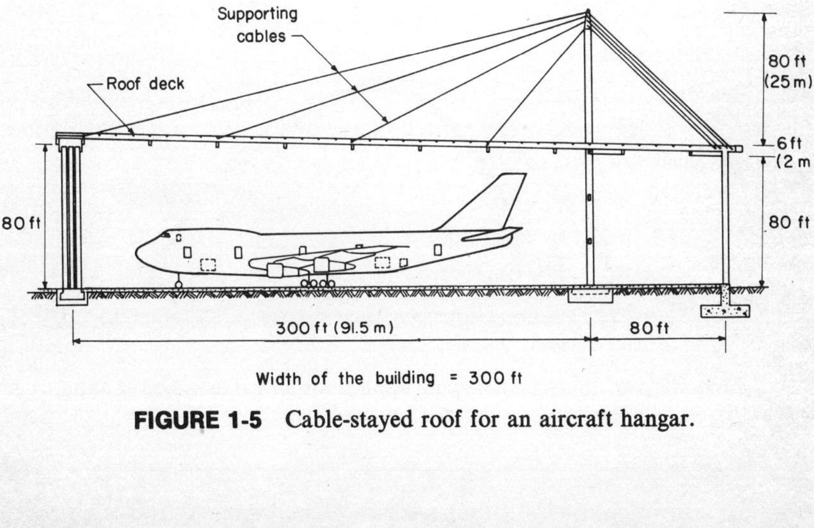

FIGURE 1-5 Cable-stayed roof for an aircraft hangar.

160-ft-diameter (48.8-m) roof for the 5-story pavilion is suspended from a hook through a cable network. Figure 1-4 shows a cable-supported roof in which a canopy is supported on a short strut, a long column, and the cable (marked 1, 2, and 3, respectively, on the photograph).

Using a design principle similar to that used for the canopy, several aircraft hangar roofs have been built, particularly in the United States, ranging in spans up to 300 ft (91.5 m); one such proposed design using a cable-supported roof is shown in Fig. 1-5. Another

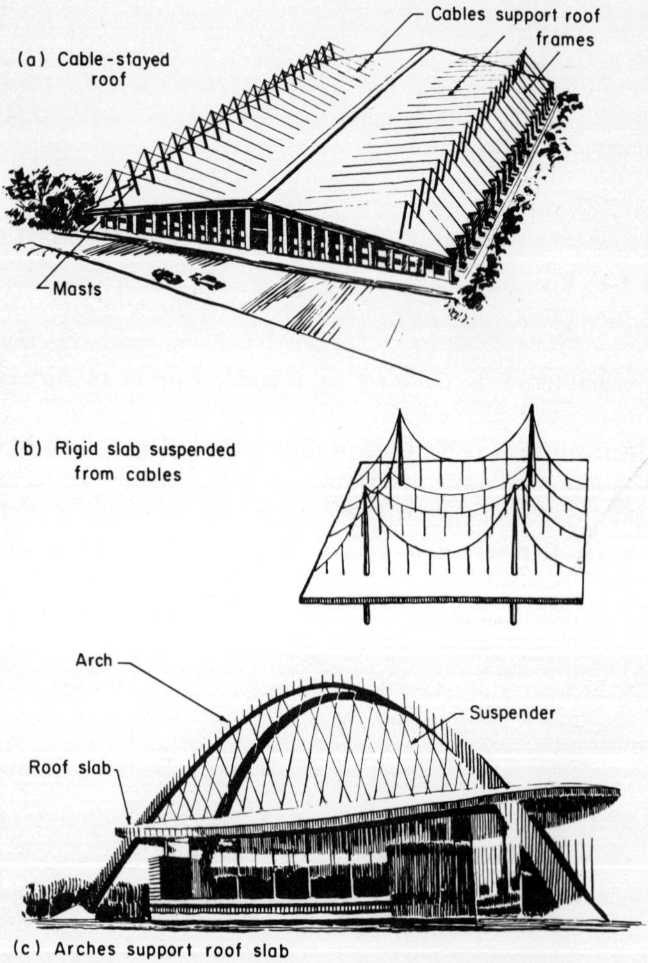

FIGURE 1-6 Sketches showing application of cable-stayed roofing.

example of a cable-supported roof is the 480 × 150 ft (146.5 × 45.7 m) roof for the Tilburg railway station in Holland. The roof is divided into four hyperbolic paraboloids which are supported on four piers at the ends of the building and also supported on cables suspended from six steel masts. Figure 1-6 shows sketches to indicate some other possibilities of employing the principle of cable-stayed roofing. Figure 1-6*a* depicts a general view of the 404 × 1200 ft (123 × 366 m) Exposition Center at Tulsa, Oklahoma. This building has a clear width of 300 ft (91.5 m). Figure 1-6*c* is an artist's impression of the Founders National Bank building in Oklahoma City, Oklahoma.

A further classification of cable-suspended roofs can be made, in accordance with their geometry, into three main types:

1. Single curvature
2. Double curvature (positive)
3. Double curvature (negative)

Roofs with single or positive double curvature either can be of the catenary form generated by a single layer of cables, or can be a structure employing *double-layer* cable trusses of various types. Roofs with negative double curvature, of which a hypar or a saddle shape are typical examples, consist of a two-way system of cables. The network thus formed gives the appearance of a grid. The gridwork can also be formed from other than quadrilateral units (e.g., tridiagonal or hexagonal). Figure 1-7 shows applications of the simply suspended (or single curvature) roofs, and Fig. 1-8 shows the inside view of a roof. The roof has a span of 80 ft (24.4 m) and employs $5 \times 4 \times \frac{1}{2}$ in thick (125 × 100 × 12 mm) tensile steel T sections in place of cables (see roof detail in Fig. 7-13).

The various types of roofs thus constructed can be further classified on the basis of whether or not they have to be pretensioned. It will be obvious that roofs with single curvature or positive double curvature, which consist of a single layer of cables, cannot be pretensioned except through preloading [91] with, for example, a concrete roof. These can also be stiffened by loading and then grouting the spaces between precast concrete slabs placed over the roof. When the grout hardens, the loads are removed, leaving a prestressed shell that is supported on and connected to the cables. An example of this is the roof over the passenger terminal at Dulles International Airport at

Chantilly, in which cables span between pairs of towers at 200-ft (61-m) spacing [76]. Some other roofs built on this principle are the municipal stadium roof at Montevideo, Uruguay, 300 ft (91.5 m) in diameter [85]; the roof of the Oakland–Alameda County Colosseum in Oakland, California, 420 ft (128 m) in diameter [92]; the 400 × 320 ft (122 × 97.5 m) elliptical paraboloid roof at Oklahoma City [56]; and a circular hyperbolic paraboloid roof in Phoenix, Arizona, 380 ft (116 m) in diameter.

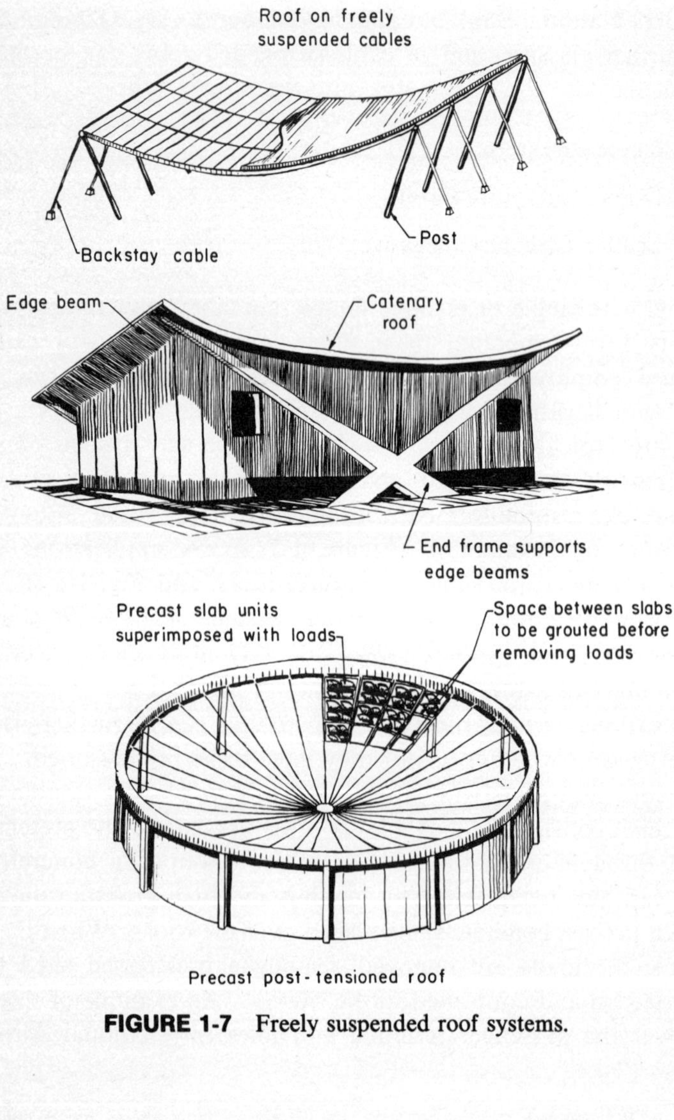

FIGURE 1-7 Freely suspended roof systems.

FIGURE 1-8 Inside view of a simply suspended gymnasium roof. (*Courtesy of A. R. Flint; Ref. 25.*)

As compared to simply suspended cable roofs, a much stiffer design for larger spans, initially developed in Europe, makes use of counterstressed cable trusses, as shown in Fig. 1-9*a*. Some of the possible variations in the design of such trusses are shown in Fig. 1-9*b* to *d*. Either of the two cables can be used to pretension the other through the web members. The cladding can be placed directly on the upper or the lower cable. The cable carrying the cladding is often called the *carrying* cable and the rest of the system is merely the pretensioning system. The structures in Fig. 1-9*b* and *c* are mechanisms and would become much more rigid, to resist unsymmetric loading, when the two cables are joined together at the center.

Cable trusses are perhaps the most popularly used among the different roofing systems, because these are easy to erect and can be employed to cover rectangular, circular, elliptical, or other plan areas quite conveniently. Some applications are illustrated in Figs. 1-10 and 1-11. Among the numerous examples of existing trussed roofs are the 240-ft-diameter (73.2-m) Utica Municipal Auditorium, Utica,

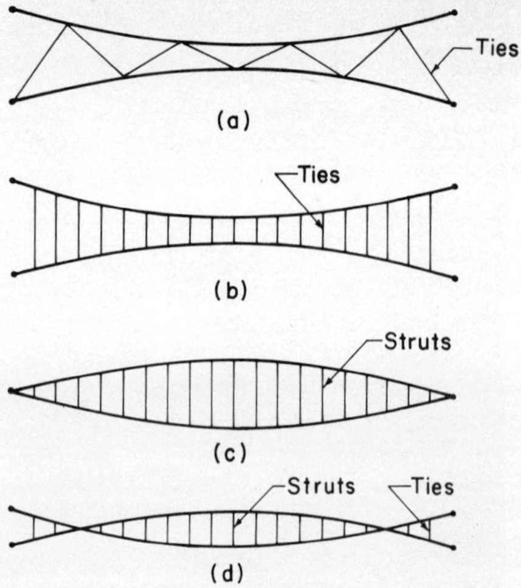

FIGURE 1-9 Various forms of cable trusses.

New York [14, 17]; the 262-ft-diameter (80-m) sports hall at Herning, Denmark [31]; the 185 × 123 ft (56.5 × 37.5 m) roof over the Whorley reservoir in England [10]; the 240-ft-diameter (73.2-m) roof of the forum sports arena at Inglewood, California; the 131 × 223 ft (40 × 68 m) sports hall at Billingham, England; the 252 × 419 ft (77 × 128 m) transit hall at Schiphol, Holland; the 240 to 272 × 393 ft (73 to 83 × 120 m) ice stadium-Johanneshov in Stockholm, Sweden; the 230-ft-diameter (70-m) Palasport at Genova, Italy [26]; and the 360-ft-diameter (109.7-m) roof of the Salt Lake County Civic Auditorium, Utah.

Doubly curved networks with sets of cables of opposite curvature forming the roof surface can be counterstressed just as trusses can be. This is another popular system for cable roofs and some ideas for its application are illustrated in Figs. 1-12 and 1-13. An example of a doubly curved roof is the Raleigh arena [62], which started a new era in cable-roof development. Another outstanding design is exemplified in the roof of the sports arena, constructed at Tokyo for the 1964 Olympic Games. (See Figs. 6-8 and 6-9.) However, perhaps the most talked about roof of the current times is the roof over the main stadium for the 1972 Munich Olympics (Fig. 1-14). Encouraged by the German pavilion at the 1967 Montreal Exposition [53], a similar type of structure, the German engineers and architects designed and achieved this outstanding structure. Hyparents, which make up the roof, are supported on edge cables and masts [53–55]. Besides these, there are numerous other examples of roofs using cable nets, such as

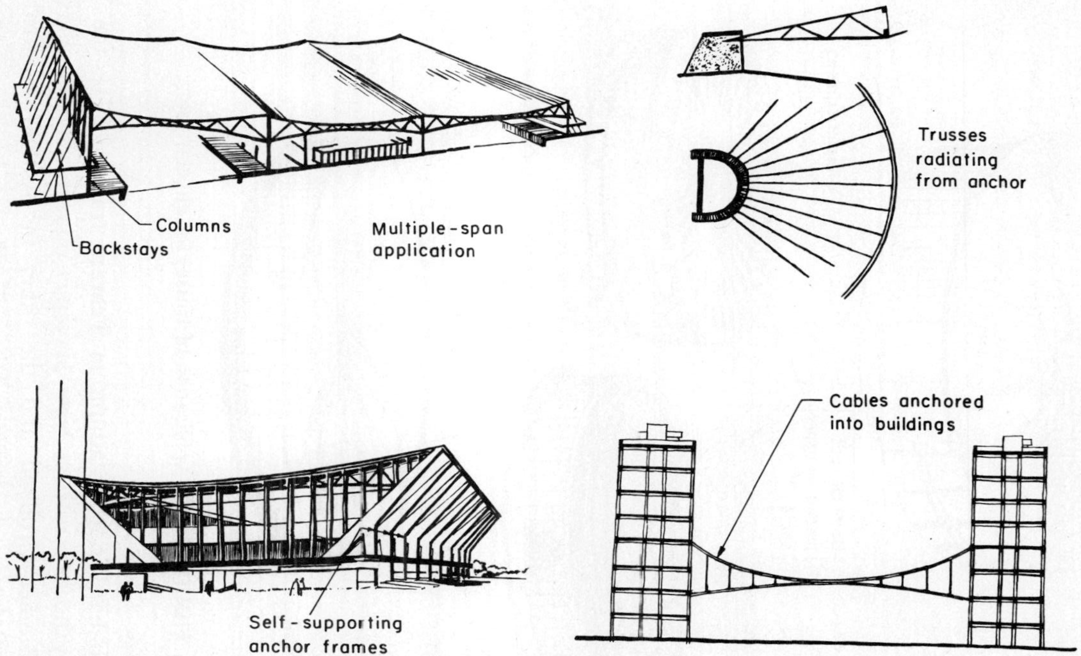

FIGURE 1-10 Cable truss applications.

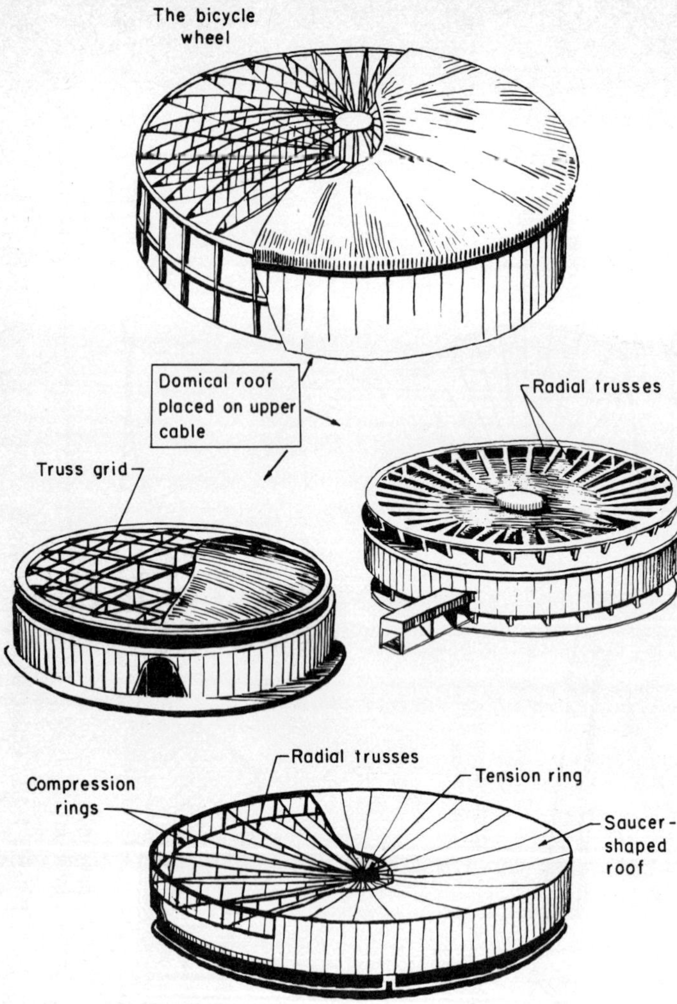

FIGURE 1-11 Application of trusses for circular roofs.

the Farahabad Sports Stadium in Tehran, Iran [59, 71]; the Yale University hockey rink at New Haven, Connecticut; two churches, one in Bremen-Grolland and the other in Stuttgart-Sonnenberg, Germany [6]. In doubly curved roofs the cladding is generally placed directly on the cable network formed by the two or more sets of cables. From the viewpoint of drainage of water or melted snow from the roof, surfaces with positive gaussian curvature (the shape of a saucer) may present some difficulty, while other systems should be amenable to a straightforward solution. It is necessary that in all cases adequate slopes are provided to avoid "ponding" of water.

In cable roofs the material of the cable is very efficiently utilized, and the result is a roof much lighter in weight than other comparable systems of roofing. However, the structural system to support or anchor the cables, if not carefully designed, can become quite heavy and expensive. As seen in Figs. 1-12 to 1-14, the cables can be supported on boundary cables, on flexural members, on arches, or on

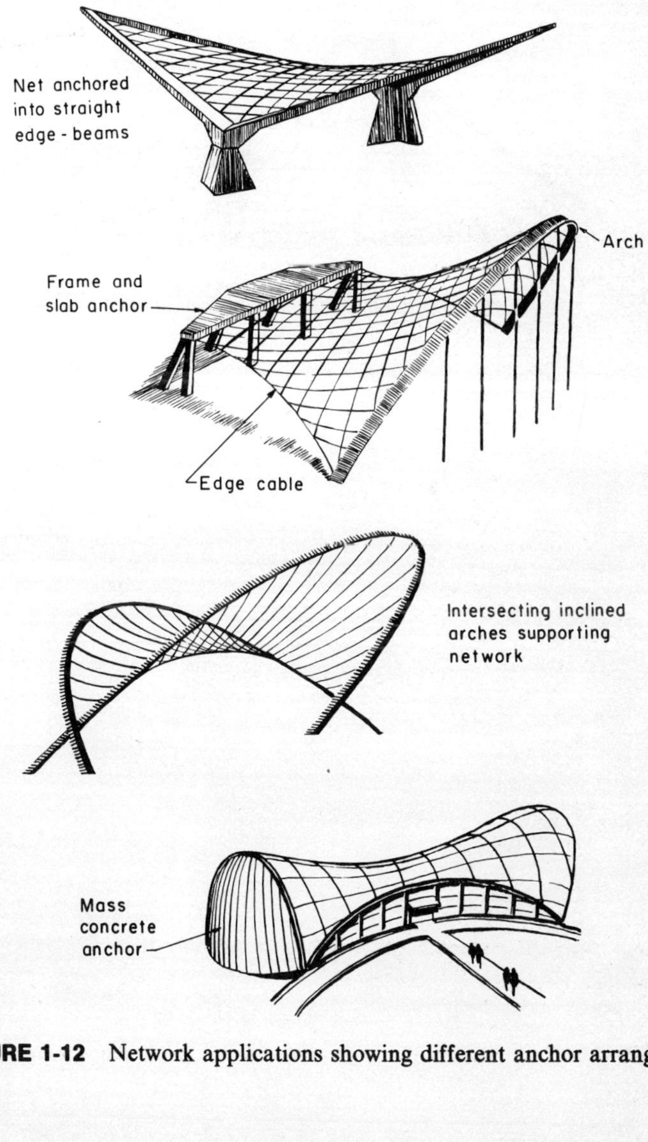

FIGURE 1-12 Network applications showing different anchor arrangements.

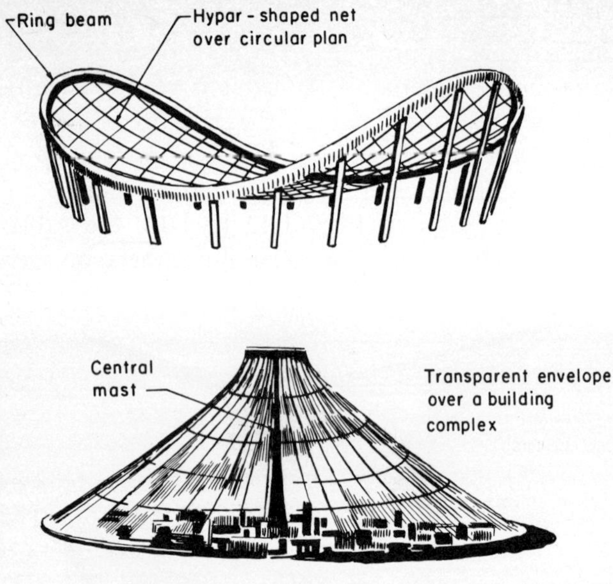

FIGURE 1-13 Networks for circular plans.

FIGURE 1-14 Suspended cable roofs over the 1972 Munich Olympic Stadium.

a combination of these. Perhaps the most efficient arrangement is to anchor cables into circular rings, although that places restrictions over the plan shape. Examples of trussed structures cited in the previous paragraphs will show that in most cases the cables are supported on towers and anchored into ground in the *deadman* anchor fashion, or they are anchored into circular rings. The Raleigh arena roof structure consists of a two-way system of cables, forming an anticlastic surface, which is anchored into two inclined arches. The arches cross each other at two points and counterbalance the horizontal component of their thrust, in the same manner as in a folding chair. The cables of the doubly curved roof for the Tokyo structure are supported on two curved beams and two main cables which are supported on towers.

Relevant details of some of the examples of cable roofs listed in the preceding paragraphs are given in Chaps. 5, 6, and 7.

1-4 GENERAL DESIGN CONSIDERATIONS

The problems usually encountered, or those which become more significant, in the design and construction of long-span roofs, such as the need for a more accurate assessment of the load distribution or a more critical study of structural details, are common to cable roofs too. In addition there are other problems, namely, the provision of adequate static stiffness, avoiding the occurrence of flutter; anchorage and pretensioning of cables and the design of the supporting structure; etc., which assume special importance for suspended roofs.

The need for large-span roofs is quite often governed by functional and aesthetic requirements, rather than by economic or structural considerations alone. In handling any large project such as of this nature, it is essential that both the architect and the structural engineer contribute to the development of the design concept and the construction schedule from the very initial stages of the project. Failure to achieve this may cause many difficulties in the execution of the project. A brief mention of some special considerations to be made in such structures now follows.

1-4.1 Planning

According to modern architectural concepts, functional flexibility and variety of purpose put new demands on space clearances. Quite often the geometry and in many cases the structural type will be governed by the variations in required clearances. A well-conceived cable structure serves such a function admirably.

If the building is to be put to varying uses, large-span cable roofs offer much flexibility in using the space enclosed, provided this is properly planned in the initial stages. The same flexibility cannot, however, be assumed for future expansions. Cable roofs are used in both temporary and permanent structures but are most ideally suited for temporary space systems of large spans, for use as exhibition halls or circus arenas. They can, with suitable design, be put to repetitive use most effectively.

Site conditions can have a strong influence on the basic structural form of the roof. For example, if there is a space restriction around the building, a self-supporting structure will have to be designed because cables cannot be anchored into the adjoining ground. If site conditions permit, a large-span suspended roof can be profitably planned between two buildings of adequate strength, which will serve as anchors for the cables. In planning a cable roof for a large hall next to an office building, the interaction of the two structures, and any possible sharp differences in their requirements, will have to be considered.

1-4.2 Materials

A cable-roof structure can be broadly subdivided into two parts—the supporting structure and the roof cladding. The supporting structure may be constructed of plain, reinforced, or prestressed concrete, steel, or a combination of the two. The final choice will be governed by aesthetic and structural considerations and economics. The cladding can be further subdivided into: (1) cables and their connections, (2) auxiliary framework that supports the decking and is placed over the cables, (3) roofing which is the external waterproofing skin, and (4) the insulating layer.

The cable is a strand or a rope made out of high-strength steel wire. The strand as well as the rope is protected with a uniform coating of pure zinc. Cables are available in standard sizes along with the appropriate fittings to facilitate cable connections with each other as well as with other structural members. The breaking strength of cables is in the region of 200 kips/in^2 (1380 N/mm^2) and the modulus of elasticity is of the order of 20,000 to 24,000 kips/in^2 (138 to 166 kN/mm^2). Further details are given in Secs. 7-2 and 7-3.

Auxiliary framing is required only if the decking cannot span the cable spacing. The decking may consist of a variety of materials, such as corrugated sheets of steel or aluminum, concrete, timber, gypsum, or rigid insulation boards. Steel and aluminum have the advantage of a high strength/weight ratio and will lead to a light roof. Concrete has a low strength/weight ratio and suffers from problems of creep and shrinkage but has the advantage of greater fire and corrosion resistance and adaptability to complex shapes. In nonpretensioned catenary roofs, which depend wholly or partly on *gravity* stiffness, it may be better to use concrete and take advantage of its greater weight. Wood and insulation boards, when used, will eliminate the requirement of an insulating layer. The use of glass sheets or synthetic material can be considered, in particular, for temporary roofs. Transparent or translucent sheets of such material, if placed over closely spaced cables, may eliminate the need for auxiliary framework or decking and can admit a large amount of natural light.

The roofing can be made of sheet metal, glass, plastics, or similar materials. Sheet metal used for roofing includes steel, aluminum, copper, stainless steel, etc., and the choice would depend upon economics and functional requirements. A control over transfer of heat through the roof (as well as the walls) is necessary to maintain comfortable temperatures inside the building and to economize on air-conditioning and heating costs. Transfer of heat by convection can be minimized by preventing the formation of air currents. Loss by radiation can be prevented by providing reflective surfaces, and loss by conduction can be controlled by using materials which are poor conductors of heat. Some possible ways of effecting proper insulation are the use of blanket insulation, structural insulation boards, or corrugated insulation, or by spraying on insulation material. The basic material in most cases is a mineral wool, such as rock glass or slag wool; vegetable or wood fiber; and expanded minerals such as vermiculite or perlite, animal hair, etc. The varieties available are numerous.

Section 7-4 gives more details regarding roof cladding and Figs. 7-10 to 7-13 show some practical details. In choosing the material for a large-span cable roof, connection procedures, details, and deformability are important considerations. The fireproofing qualities need not always be a dominant factor since such roofs will usually be quite high.

1-4.3 Economics

With increasing span, the cost per unit area of a roof structure also rises, and it will not be far from the truth to approximate the increase to be linearly related to the span. In terms of the direct structural costs, therefore, long-span roofs may not prove more economical than shorter spans, but their construction may well affect the usefulness of the covered space. For example, in countries with severely cold conditions uncovered sports stadia cannot be used during the winter season. If these stadia are covered, it could lead to greater interest in sports as well as increased economic gain. It is worth mentioning that a large-span roof will perhaps not be built where a small-span structure will satisfy the users' needs. But among the many structural systems in use for long spans, cable roofs claim several attributes through which greater structural efficiency and economy can be achieved. Further, the length of spans possible with the use of cables may not be physically or economically viable with other structural forms.

It is a well-established fact that a structural member, which is subject to axial forces only, utilizes the material more effectively than a member in bending. It is also true that steel is the most suitable structural material where tensile forces are to be sustained. In carrying applied loads, by developing purely tensile stresses, a cable, which is made of high-strength steel wire, satisfies both the conditions. In a structural sense, therefore, the cable is a very efficient structural system. Suspended roofs also eliminate problems of instability and the resulting reductions in allowable stress, whereas for trusses, shells, or domes of latticed steel construction, or for concrete shells and domes, this can be a predominant design consideration. In constructing cable roofs, little or no formwork is necessary. Even when concrete is used as a cladding material, the usual practice is to make precast slabs rather than laying the concrete in situ. In the construction of concrete shells, shuttering is an expensive item. Construction and erection time for cable roofs can be considerably less than for many other forms and results in a saving in financing costs and quicker returns from the structure. There are many reasons, therefore, for cable roofs to be more economical than other systems, such as trussworks, shells, and domes. Direct cost comparisons are, however, possible only for particular cases and generalizations are not appropriate. Cost data and comparisons with other systems, where possible, are given in Chap. 6. There are some cases where cable roofs are quoted to be several times cheaper than comparable structures.

As mentioned earlier a major proportion of the cost of a cable-roof structure is taken up by the supporting system—abutments, anchorages, etc.—and the cladding. The cost of cables and their fittings is comparatively small (see examples in Chap. 6). A cable system may be more economical when covering larger areas or spans, for the following reasons:

1. It is cheaper to buy larger quantities of cable.
2. The number and cost of fittings do not increase proportionately to the span.
3. The erection of a larger-span cable is as easy as for a smaller span.

The economy of cable-roof systems is greatly affected by the mode and degree of prestressing. As mentioned earlier, a cable system can be stiffened through *gravity* loading or prestressing. Gravity loading may lead to a heavy supporting structure and foundations, and prestressing appears to be a preferable way, wherever possible. In deciding the degree of prestressing, two schools of thought are prevalent.

1. Prestressing should be done to such an extent that no cable element becomes slack under any loading condition.
2. A few cable elements could be permitted to become slack under adverse loads, which may be expected to occur only rarely. This may, however, be done after making certain that the overall stability of the system will not be affected. If it occurs, a local minor damage in the roofing may have to be tolerated.

The second method of design may lead to considerable economy, since it is possible that the degree of prestress may have to be increased by a substantial amount to avoid the slackening of only a few members.

1-4.4 Durability and Fire Protection

Cable roofs are designed for both permanent and temporary use, and a study of their record shows that a number of them have been built for sports halls, aircraft hangars, and exhibitions. Some roofs are constructed to last many years, while others may be designed for not more than 1 or 2 years of life. In some seasonal structures, the roof is removable and thus it may be used only for small periods each year, but for many years. The material used and the design of details depend also upon the expected life of the structure. For semi-permanent or temporary suspended structures, canvas or suitable synthetic materials with shorter life may be used as cladding. It may also be possible to permit higher permissible stresses in the material, reduce minimum thicknesses, and make the design of the details less elaborate. The advantage of using cheaper forms of cladding, as mentioned above, cannot be had to the same extent for conventional temporary structures because of the limitation on the choice of materials in such cases.

Insofar as fire protection is concerned, there are no building codes specifically applicable to cable roofs. It may be difficult to interpret existing codes for conventional buildings for application to these structures because of their long spans and sometimes unusual geometry. Therefore, the designer will be particularly required to use judgment in this aspect of the problem.

Higher temperatures due to fire cause the cable material to *creep*. There are a decrease in the modulus and also a relaxation of pretension and, consequently, an overall decrease in the stiffness of the system, leading to greater deformations. Also, in redundant cable assemblies there is a redistribution of stresses. A qualitative study of the problem yields the following points:

1. At increased temperatures the cable sag increases, thus affecting a decrease in cable tension due to superimposed loads. Further, as already mentioned, there is a relaxation of pretension. Thus an increase in temperature affords a relief in tension and the stresses are reduced.

2. It is unlikely that in a large-span roof the full design loads will occur together with a full fire.

3. Therefore, although the strength of a cable is reduced at high temperatures, this may not be a major problem owing to

points 1 and 2 above. Reduction of tension may, however, be so large that pretensioning members in counterstressed systems may become slack, reducing them to simply suspended systems, with all the disadvantages of unconstrained cable systems.

4. Large deformations will usually occur at high temperatures and may cause damage to the cladding or the details.
5. Roofs of longer spans will usually be high, and fireproofing of the roof in itself may not be a dominant consideration.
6. In steel tension systems there is a better possibility of redistribution among the members than in rigid framed structures, where local failures due to buckling may occur at higher temperatures. A reasonable margin of time before collapse occurs should normally be expected. Jawerth has shown through tests on some of his structures that they remain stable for 30 to 60 min after a fire has started.

Pearson and Dickson [77] have carried out theoretical studies for a cable-roof structure, and reach the important conclusion that it would be quite safe if the normal precautions of maintaining free exits and providing sprinklers, vents, and alarms are observed for such structures.

1-4.5 Deflections

Cable roofs are generally classed among flexible structures. The deflection restrictions on these systems cannot always be as stringent as for conventional slab-and-beam structures. There are, however, no building codes dealing with this specific question for suspended roofs, and the designer would do well to use discretion where possible to arrive at reasonable values. It is necessary that not only deflections but also slopes be limited. Sharp changes in slope can damage the cladding, even if the maximum deflection is within the permissible limits. This situation may arise particularly in cases where heavy concentrated loads are suspended from the roof. Local damage has been reported in the suspended roof of a gymnasium building; the gymnasts used the roof to suspend themselves. In longer spans the effect of movements at the support points will be exaggerated and may need to be studied. It may be necessary to provide special details to accommodate these movements.

1-4.6 Safety

For cable roofs which are normally large in span, and may also be housing expensive equipment or machinery, the safety of the roof must be ensured through conservative design, if necessary. In terms of structural response to loads, cable roofs enjoy several advantages over other systems, such as shells or latticed domes, used for large spans. The latter carry the load in compression and can fail through instability. In latticed domes a local buckle or a *snap-through* could result in the overall failure of the structure. In a cable structure, on the other hand, a local buckle will not spread because of its being a tension system. It is well known that a concentrated load cannot be supported by a shell, since it could cause the local failure of the structure and finally lead to overall instability. There is no such problem in cable roofs. Further, it is worth noting that cable systems, networks in particular, are highly indeterminate structures, and being made from steel and as such ductile members, they are likely to offer considerable *overload* capacity beyond the working loads. Calculations on sample nets have shown that the ultimate load is as much as 60 to 170 percent higher than the nominal load [32]. The deformations will, of course, be large at or before the ultimate stage, and the cladding may not always be able to tolerate the same.

If the main cable of a truss begins to yield, it will lead to increasing deformations and collapse. In this respect, trusses differ from nets, in which the capability of redistribution of forces between cables is distinctly better. Further, if one of the carrying cables in a tensioned truss system snaps or slips out of the anchorage, the damage can indeed be serious and has to be guarded against. It is also most important to prevent ponding of water in saucerlike roofs, by providing an efficient drainage system.

1-5 AREAS OF RESEARCH

Architects and engineers have designed and achieved many outstanding cable roofs in recent years, but there are facets of analysis, behavior, loading, design, and construction which are not fully understood. Many problems that arise are solved in reference to particular projects but no general solutions are available. Indeed it is perhaps ambitious to expect general solutions to problems concerning a structural form with possibilities of such extensive variation, yet there is a need to seek greater information so that designs can be carried out with greater economy and certainty. Areas in which there appears to be a general lack of knowledge, and thus a need for investigation, are identified and discussed in the following paragraphs.

Methods of static analysis which have been a topic of considerable research are well developed, and structural behavior under static loading of commonly used roof systems is reasonably well understood. However, in the literature reporting attempts on nonlinear matrix analysis of cable systems, little attention seems to have been given to the effect of support flexibility on the structure response. It is obvious that this can be an important problem for some structural forms, particularly during the stages of erection, and needs a more thorough effort in investigation.

There has likewise been some effort in investigating the fundamental dynamic behavior of lumped-mass models for cable-roof systems. The information on expected dynamic response under time-dependent loads is, however, quite limited. Further, the absence of knowledge of realistic values of damping reduces the credibility of theoretical computations for the determination of response. There appears to be a need for carrying out experimental studies on both models and prototypes to obtain more information on damping and dynamic response. Another important factor which needs consideration is the nonlinearity of the system.

Both static and dynamic analysis seeking rigorously accurate results for these highly indeterminate systems can, in many cases, be rather expensive in terms of computer storage space as well as execution time. This is a general drawback in large structural-analysis problems using the matrix approach and is a subject of wide study in order that the required effort can be minimized. Further, it must be understood that an electronic digital computer is not available at the beck and call of a designer, at least not for computations connected with the assessment of design ideas and the preliminary design effort. Therefore, it is necessary to develop *desk-calculation* methods. Some work on these lines is reported in Chaps. 2 to 4, but more effort is necessary to improve upon available formulas and to cover more cases of structural form and loading.

Estimates for conventional static loads such as dead, live, and snow loads for roof surfaces are possible with good accuracy. Determination of wind loads, both static and dynamic, particularly for curved surfaces met with in cable roofs, is on the other hand quite difficult. It is also well understood that the best way to determine these loads is to make measurements on appropriately designed wind-tunnel models. Extensive research is required in this area. Studies on existing structures may also prove to be useful.

In cable roofs, the design of the cable cross section for given loads does not present much of a problem. Research is required to find more economical protective measures for cables. Other important areas for detailed investigation are the determination of fatigue strength of cables, the effect of clamp pressure on cables, and the behavior of fittings and clamps under load in order that these may be more economically designed.

Generally, ideas and innovation in design and construction precede research on related technology and analysis. This is the case with cable roofs. As new design ideas develop, they open up new areas for research. With wide interest in this form of construction, extensive research is presently in progress. If those involved in the design and construction of cable roofs would give their work greater publicity, in the form of published work, it would be a contribution of great utility.

CHAPTER TWO

A Freely Suspended Cable

2-1 GENERAL

Cables are the basic structural elements in suspended roofs. Whereas the analysis of a freely hanging cable is important from the point of view of understanding such structures, it is directly applicable to roofs in which cables are not prestressed—many air-supported and simply suspended roofs come under this classification. The analysis and behavior of a cable have been treated in detail in the existing literature mostly in connection with the study of suspension bridges. This chapter contains a review and extension of the existing material for better understanding and application.

Since its transverse dimension is very much smaller than its longitudinal dimension, a cable has negligible buckling strength. For the same reason, a cable cannot sustain any bending or torsional moments and can transmit loads only by developing direct tension. Ideally, the tensile stress across the cable cross section will be uniform and the entire cross section will therefore be fully utilized. For a cable with a sag/span ratio within the practical range, the variation of tension along the length is small. Thus, the entire cable material is utilized to almost 100 percent efficiency.

An important fundamental aspect of cable mechanics can be illustrated by the treatment of a weightless string supported at A and B and loaded as shown in Fig. 2-1a. The moment at a point X in the cable is given by

$$M_X = M_{e_x} - Hz$$

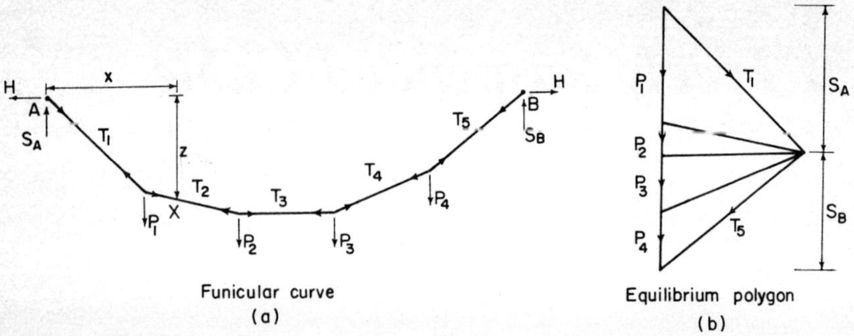

FIGURE 2-1 Funicular curve of a cable loaded with point loads.

but since the moment at any point on the cable must be zero, it follows that

$$z = \frac{M_{ex}}{H} \tag{2-1}$$

where M_{ex} is the simple beam bending moment at X.

Equation 2-1 indicates that the cable curve is the bending moment diagram to some scale for a beam supported at A and B and loaded in the same manner as the cable. The complete geometry of the string can be obtained if the value of H or the ordinate of any one point on the curve is known.

It is easy to see that the cable follows the funicular curve of the loads and that the segments are parallel to the corresponding rays in the equilibrium polygon in Fig. 2-1b. If the cable is subjected to further loading, it will have to undergo adjustment in its funicular shape. For this reason and also because of its inability to develop other than tensile stresses, a cable is a flexible structure and the problem of its analysis involves the consideration of large deflections.

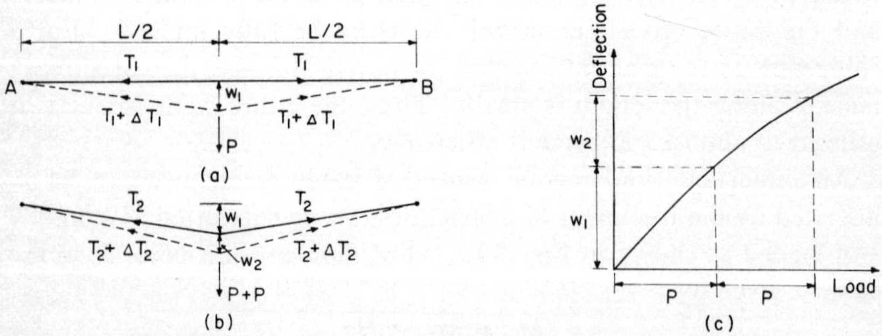

FIGURE 2-2 Load vs. deflections for a taut string.

Another characteristic of cable behavior is its nonlinearity. This is common to any structural system with a predominance of direct stresses in which large deflections occur. Consider a string stretched between A and B (Fig. 2-2*a*). When subjected to a central load P, it deflects by an amount w_1, and the value of the tension in it changes from T_1 to $T_1 + \Delta T_1$. For the equilibrium of the deflected string,

$$P = 4(T_1 + \Delta T_1)\frac{w_1}{L}$$

or

$$w_1 = \frac{PL}{4T_2} \tag{2-2a}$$

where $T_2 = T_1 + \Delta T_1$.

If the same string is further loaded with an additional load P, also applied at its center point, it undergoes a change in tension equal to ΔT_2 and a deflection w_2 as shown in Fig. 2-2*b*. For the equilibrium of the final deflected shape,

$$2P = 4(T_2 + \Delta T_2)(w_1 + w_2)\frac{1}{L}$$

or, on substitution of the value of w_1 from Eq. 2-2*a* and rearranging,

$$w_2 = \frac{PL - 4\Delta T_2 w_1}{4(T_2 + \Delta T_2)} \tag{2-2b}$$

A comparison of expressions for deflections[1] w_1 and w_2, both caused by application of loads P, shows that $w_1 > w_2$. It is obvious, therefore, that the load-deflection relationship for the string is nonlinear (see Fig. 2-2*c*). If the applied load P is small in comparison to the initial tension T_1, then the terms ΔT_2 and $4\Delta T_2 w_1$ could be neglected in comparison to T_2 and PL, respectively. The expression for w_2 then reduces to that for w_1. This indicates that whereas the response of a cable is approximately linear for small applied loads, it is essentially nonlinear for large loads.

2-2 CABLE UNDER ITS OWN WEIGHT

The complete analysis of a cable in any of its structural applications can be divided into two steps: the establishment of the cable geometry under its own weight and the analysis of the cable under applied loads. This section is concerned with the first of these two objectives. The weight of the cable is here assumed to include any

[1] The extension of the string has been neglected to simplify the expressions, but its inclusion will not alter the conclusion.

permanent uniform load superimposed on the cable. Rigorously there is only one form a cable segment can take: a catenary. For analysis it may be approximated to a parabola or a straight line.[2] The three cases are given below in detail.

2-2.1 Catenary

Consider a cable segment as shown in Fig. 2-3, hanging under its own weight, or a superimposed load q, distributed uniformly along the length of the arc. The governing equation of the arc is

$$H \frac{d^2z}{dx^2} + q \frac{dl}{dx} = 0 \tag{2-3}$$

Integrating the equation twice and substituting the boundary conditions, $z = 0$ at $x = 0$ and $z = Z$ at $x = X$, we get

$$z = \frac{H}{q} \left[\cosh \alpha - \cosh\left(\frac{2\beta x}{X} - \alpha \right) \right] \tag{2-4}$$

where $\alpha = \sinh^{-1} \left[\dfrac{\beta(Z/X)}{\sinh \beta} \right] + \beta$

$$\beta = \frac{qX}{2H}$$

From Eq. 2-4 it can be seen that the complete geometry of the cable is established if the value of H or the ordinate of any one point on the cable (for example, the cable sag at the center) is known. The

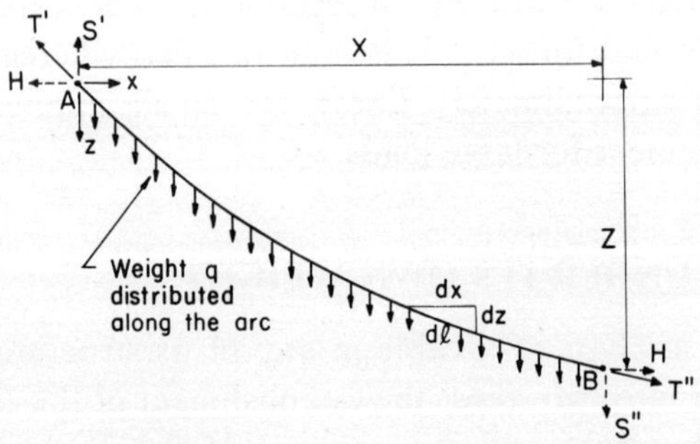

FIGURE 2-3 Segment of a catenary.

[2] Some other forms, which are perhaps only of academic interest, such as a heterogeneous cable or a catenary of uniform strength, have been treated in Ref. 80, chap. 1.

expressions for the end tensions T' and T'', reactions S' and S'', the length of the arc l, and the projections[3] X and Z in terms of these different parameters are as follows:

$$T' = H \cosh \alpha \tag{2-5}$$

$$T'' = \frac{q}{2}(l \coth \alpha - Z) \tag{2-6}$$

$$S' = H \sinh \alpha \tag{2-7}$$

$$S'' = \frac{q}{2}(Z \coth \beta - 1) \tag{2-8}$$

$$l = \frac{2H}{q} \sinh \beta \cosh (\alpha - \beta) \tag{2-9}$$

$$X = \left(\frac{H}{q}\right) \log_e \left(\frac{T' + T'' - ql}{T' + T'' + ql}\right) \tag{2-10}$$

$$Z = \left(\frac{S' + S''}{T' + T''}\right) l \tag{2-11}$$

$$(\Delta l)_e = \left(\frac{Hl}{X}\right)^2 \frac{1}{EA} \tag{2-12}$$

where $(\Delta l)_e$ = elastic stretch of the cable due to increased tension
EA = extensional rigidity of the cable

For the standard case of a cable, as shown in Fig. 2-4, with the supports A and B at the same level ($Z = 0$) and central sag f, the values of α and β reduce to

$$\alpha = \beta = \frac{qL}{2H} \tag{2-13}$$

and the expression for z (Eq. 2-4) becomes

$$z = \frac{H}{q}\left[\cosh \alpha - \cosh\left(\frac{qx}{H} - \alpha\right)\right] \tag{2-14}$$

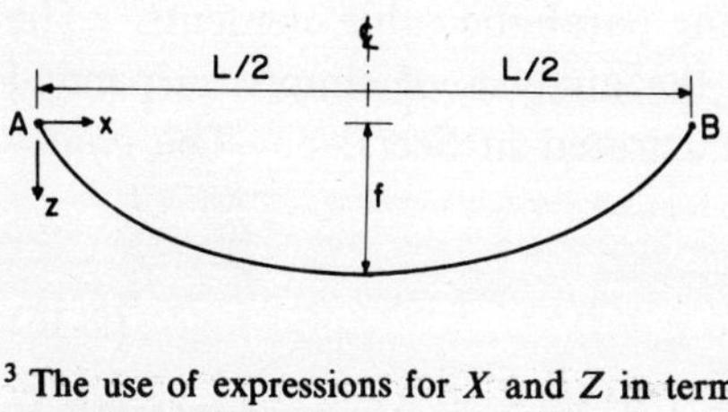

FIGURE 2-4 Simply suspended cable.

[3] The use of expressions for X and Z in terms of other properties of the cable segment as given here will be clear when studying Sec. 2-4.

Using the condition $z = f$ at $x = L/2$, Eq. 2-14 reduces to

$$f = \frac{H}{q}\left[\cosh \alpha - 1\right] \tag{2-15}$$

The value of H can now be computed from Eq. 2-15, and the geometry of the cable can be computed from Eq. 2-14. The parameters in Eqs. 2-5 to 2-12 can be derived for this case by substituting the values of α and β from Eq. 2-13.

2-2.2 Parabola

A cable segment whose weight q is assumed to be distributed uniformly along the horizontal length of span is shown in Fig. 2-5. The weight of a freely hanging cable of constant cross section is in fact distributed uniformly along its length, as assumed in the analysis of the catenary. However, the approximation in assuming the weight to be distributed along the cable span can be justified in the following cases:

1. When the sag/span ratio of the cable is small, the difference between the arc length and the span is very small, and the approximation makes almost negligible difference in the geometry obtained from the two analyses (see Fig. 2-7).
2. If the cable has a permanent superimposed load which is uniformly distributed along its span and is much heavier than the cable itself (as in the case of .a suspension bridge), the approximation also has justification.

It will be seen from the analysis to follow that the equations for the resulting curve, which is parabolic, are very much simpler to use than the equations for the catenary. A method for approximate analysis of pretensioned cable trusses and networks is developed in Sec. 3-5 by treating these structures as comprising parabolic cable elements. The procedure is equally applicable to the analysis of simply suspended cables under applied loading, as illustrated in Sec. 3-5. The condition of equilibrium for the curve is

$$H\frac{d^2z}{dx^2} + q = 0 \tag{2-16}$$

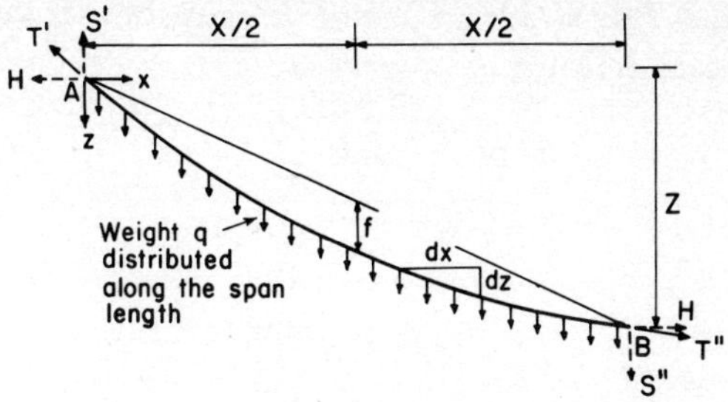

FIGURE 2-5 Segment of a parabolic cable arc.

On integrating the equation twice and using the condition $z = Z$ at $x = X$, we get

$$z = \frac{qx}{2H}(X - x) + \frac{Z}{X}x \tag{2-17}$$

As before, the complete geometry of the cable can be defined if the value of H is specified or the ordinate of any one point on the curve is known. If the sag of the cable at the center of the span is specified to be f below the chord AB, from Eq. 2-17 we get

$$\frac{Z}{2} + f = \frac{qX}{4H}\frac{X}{2} + \frac{Z}{2}$$

or

$$H = \frac{qX^2}{8f} \tag{2-18}$$

Substitution of this value of H in Eq. 2-17 gives

$$z = \frac{4fx(X - x)}{X^2} + \frac{Z}{X}x \tag{2-19}$$

The length of a cable can be expressed as

$$l = \int_0^X \left[1 + \left(\frac{dz}{dx}\right)^2\right]^{1/2} dx$$

Substituting the value of z from Eq. 2-19, expanding the expression binomially, and integrating the expression for the length l, we obtain[4]

$$l = X\left(1 + \frac{8f^2}{3X^2} + \frac{1}{2}\frac{Z^2}{X^2}\right) \tag{2-20}$$

[4] Higher-order terms of the sag/span ratio (f/X) are ignored in arriving at this expression.

The expressions for S', S'', T', and T'' can be derived from simple statics and are as follows:

$$S' = \frac{qx}{2} + H\frac{Z}{X} \tag{2-21}$$

$$S'' = \frac{qx}{2} - H\frac{Z}{X} \tag{2-22}$$

$$T' = (H^2 + S'^2)^{1/2} \tag{2-23}$$

$$T'' = (H^2 + S''^2)^{1/2} \tag{2-24}$$

$$T_x = H\left(1 + \frac{64f^2x^2}{L^4}\right)^{1/2} \tag{2-25}$$

For the standard case when the supports A and B are at the same level ($Z = 0$), with span L and central sag f,

$$H = \frac{qL^2}{8f} \tag{2-26}$$

$$z = \frac{4fx(L - x)}{L^2} \tag{2-27}$$

$$T_{\max} = H\left[1 + 16\left(\frac{f}{L}\right)^2\right]^{1/2} \tag{2-28}$$

$$l = L\left[1 + \frac{8}{3}\left(\frac{f}{L}\right)^2\right] \tag{2-29a}$$

If another term in the binomial expansion used to arrive at Eq. 2-20 is included, then to a greater accuracy Eq. 2.29a becomes

$$l = L\left[1 + \frac{8}{3}\left(\frac{f}{L}\right)^2 - \frac{32}{5}\left(\frac{f}{L}\right)^4\right] \tag{2-29b}$$

2-2.3 Straight-Line Elements

A cable segment under its own weight cannot remain straight. This idealization would be possible, strictly speaking, only if the cable were weightless along its length (or assumed to be so), or, approximately, if the ends of the segment were subjected to forces that are predominantly larger than the effects of the segment's own weight. This approximation, however, simplifies the analysis considerably, offering

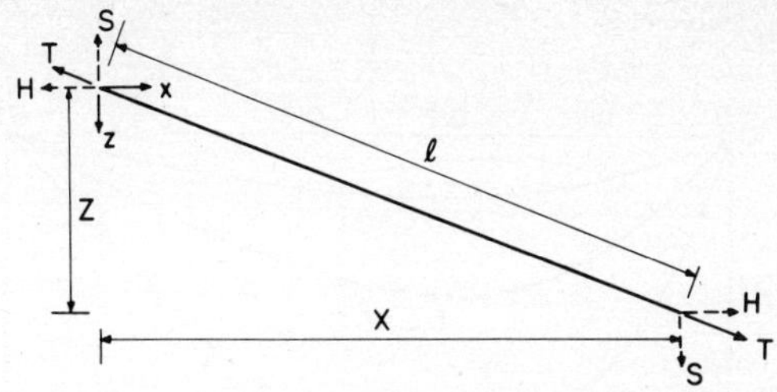

FIGURE 2-6 Straight weightless cable element.

a generalized procedure for the analysis of a freely hanging cable (see Sec. 2-4), and leads to procedures for the analysis of more complex cable-roof networks. Such procedures are discussed in Chap. 3.

Expressions for the different parameters governing a straight segment (Fig. 2-6) are easily derived from consideration of statics. These are summarized below.

$$z = \frac{Z}{X} x \tag{2-30}$$

$$l = (X^2 + Z^2)^{1/2} \tag{2-31}$$

$$T = (H^2 + S^2)^{1/2} \tag{2-32}$$

$$X = l \frac{H}{T} \tag{2-33}$$

$$Z = l \frac{S}{T} \tag{2-34}$$

$$(\Delta l)_e = \frac{Tl}{EA} \tag{2-35}$$

where $(\Delta l)_e$ = increment in the cable length due to tension T
EA = extensional rigidity of the cable

2-2.4 Comparison of Cable Shapes

A study of the existing literature on cable analysis shows that the parabolic cable has received greater attention than has the catenary. Apart from the simplicity offered by a parabolic cable in its analysis, the following two points are responsible for this:

1. The cable in a suspension bridge, which has been the subject of these studies, has the load more uniformly distributed along its span length than along the length of the arc.

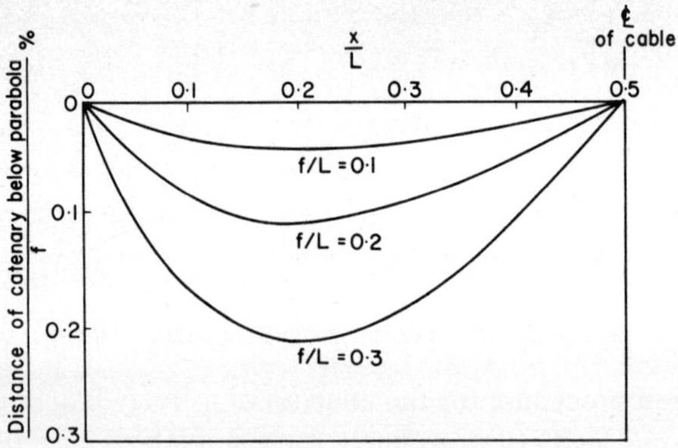

FIGURE 2-7 Difference between the ordinates of a parabolic cable arc and a catenary having the same span as well as central sag and supported on level supports.

2. The difference in shape between a parabola and a catenary, with the same span and central sag, is so small that it has to be plotted on an exaggerated scale to be perceived at all.

Figure 2-7 shows the difference between the two curves for different values of f/L. It will be seen that the difference in geometry is almost negligible for the f/L ratios within the practical range. It is sufficiently accurate, therefore, to treat a freely hanging cable as parabolic in order to ascertain its geometry.

2-2.5 Cable Deformations

If the two ends of a cable are fixed in space, the cable length between these points can change for the following reasons:

1. Cable extension due to its tension
2. Temperature change
3. Slip of the cable at a support

In any case, if the change in the sag of the cable due to change in the length can be computed, it is possible to set up the new geometry of the cable. The following procedure illustrates how this can be done for a parabolic cable with level supports. Similar treatment is possible for a cable hanging in the form of a catenary.

Differentiating Eq. 2-29b gives

$$\Delta l = \frac{16}{15}\frac{f}{L}\left(5 - 24\frac{f^2}{L^2}\right)\Delta f \tag{2-36}$$

Equation 2-36 relates the change of cable length to the change in sag, provided the span remains constant. If Δl is due to elastic stretching, it is given as

$$\Delta l = \int_0^l \frac{T\,dl}{EA}$$

where T is the cable tension at any point x in the cable (Eq. 2-25). On integration this gives

$$\Delta l = \frac{Hl}{EA}\left(1 + \frac{16}{3}\frac{f^2}{L^2}\right) \tag{2-37}$$

If Δl results from a change in temperature, it is given as

$$\Delta l = \alpha t l \tag{2-38}$$

where α = coefficient of expansion
t = change in temperature

If the cable slips by an amount ϵ at a support,

$$\Delta l = \epsilon \tag{2-39}$$

Inserting the value of Δl from either Eq. 2-37, 2-38, or 2-39, Δf can be computed from Eq. 2-36.

In the same way, if the span of the cable, L, changes with the movement of the supports while the length remains constant,

$$\Delta L = \frac{(16f/L)(5 - 24f^2/L^2)}{15 - 40f^2/L^2 + 288f^4/L^4}\Delta f \tag{2-40}$$

Once the value of Δf is obtained from either Eq. 2-36 or 2-40, the change in cable tension H can be obtained by partially differentiating Eq. 2-26,

$$\frac{\partial H}{\partial f} = -\frac{wL^2}{8f^2} = -\frac{H}{f}$$

$$\Delta H = -\frac{H}{f}\Delta f \tag{2-41}$$

Sections 2-3 and 2-4 will show how some basic results derived in this section for a cable segment in the form of a catenary, a parabola, or a straight line can be used to analyze a cable under applied loads.

2-3 ANALYSIS OF A CABLE FOR VERTICAL LOADING

The problem of analyzing a freely suspended cable with level supports has been treated for three cases of vertical loading:

1. A concentrated load
2. A short uniformly distributed load applied centrally
3. A short uniformly distributed load applied at an end

The procedure used to analyze these cases leads to closed-form solutions which are useful in obtaining results that are approximate within the bounds of the assumptions made in deriving the expressions. These results are not only helpful in understanding the behavior of a cable but can be used for preliminary design computations. It is possible to apply the procedure used here to derive expressions for more general combinations of horizontal and vertical loading, but the results may become far too involved. It is perhaps more expeditious to resort to numerical procedures as described in Sec. 2-4.

2-3.1 Case I: A Concentrated Load

The cable *AB*, shown in Fig. 2-8, has a dead weight of *q* per unit length of span and hangs initially in the form of a parabola. The geometry of the cable is described by Eq. 2-27, and the horizontal component of cable tension, *H*, is given by Eq. 2-26.

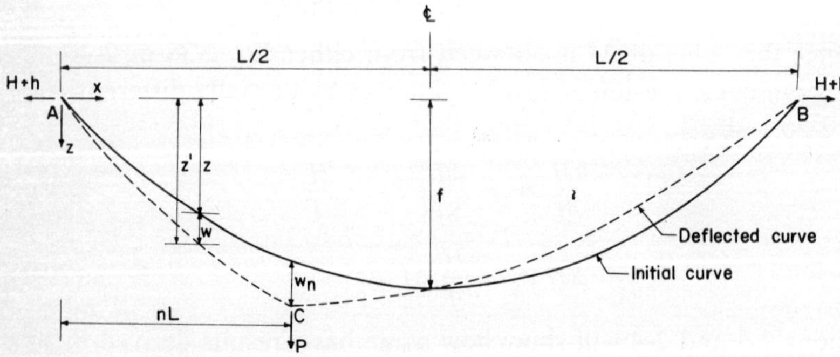

FIGURE 2-8 Cable subjected to a concentrated load.

When subjected to a load P, at a distance nL from A, the cable deflects to a new geometry given by z'. The cable tension H increases by an amount h. In the deflected position the cable consists of two parabolic curves AC and CB, with a discontinuity at C (Fig. 2-8).

The equations for the curves AC and CB can be written from considerations of statics as follows. For arc AC,

$$z' = \frac{qx(L - x) + 2P(1 - n)x}{2H'} \tag{2-42a}$$

and for arc CB,

$$z' = \frac{(qx + 2Pn)(L - x)}{2H'} \tag{2-42b}$$

where $H' = H + h$. On substituting $\theta = x/L$ and $\mu = P/qL$, we get

$$z' = \frac{qL^2}{2H'}\left[\theta(1 - \theta) + 2\mu\theta(1 - n)\right] \qquad \theta \leqslant n \tag{2-43a}$$

and

$$z' = \frac{qL^2}{2H'}\left[(\theta + 2\mu n)(1 - \theta)\right] \qquad \theta \geqslant n \tag{2-43b}$$

If the cable is assumed to be inextensible, then the length of the initial curve can be equated to the lengths of the curves AC and CB. Therefore,

$$l = l_{AC} + l_{CB} \tag{2-44}$$

Obtaining the value of l from Eq. 2-29a and the value of l_{AC} and l_{CB} by combining Eqs. 2-20, 2-43a, and 2-43b and inserting these into Eq. 2-44 yields the value of H' as

$$H' = \frac{qL^2}{8f}\left[1 - 3n(1 - n) + 3n(1 - n)(1 + 2\mu)^2\right]^{1/2} \tag{2-45}$$

The value of w can be derived by using the relationship

$$w = z' - z \tag{2-46}$$

and substituting the value of H' into Eqs. 2-43a and 2-43b:

$$w = 4f\frac{\theta(1 - \theta) + 2\mu\theta(1 - n)}{\left[1 - 3n(1 - n) + 3n(1 - n)(1 + 2\mu)^2\right]^{1/2}} - 4f\theta(1 - \theta)$$

$$\theta \leqslant n \tag{2-47a}$$

and

$$w = 4f\frac{(\theta + 2\mu n)(1 - \theta)}{\left[1 - 3n(1 - n) + 3n(1 - n)(1 + 2\mu)^2\right]^{1/2}} - 4f\theta(1 - \theta)$$

$$\theta \geqslant n \qquad (2\text{-}47b)$$

The maximum deflection in this case will occur directly under the loaded point, that is, when $\theta = n$, and its value is

$$w_n = 4f\frac{n(1 - n)(1 + 2\mu)}{\left[1 - 3n(1 - n) + 3n(1 - n)(1 + 2\mu)^2\right]^{1/2}} - 4fn(1 - n) \qquad (2\text{-}47c)$$

The envelopes for the vertical as well as horizontal deflections occurring under the load point are plotted in Fig. 2-9*a* for $\mu = 0.1$. The values of deflections are obtained by the general procedure of analysis explained later in Sec. 3-2, treating the cable as inextensible and comprising 10 straight segments. The vertical deflections have also been computed from Eq. 2-47*c* for comparison. It will be seen that the horizontal deflections are much smaller in comparison to the vertical deflections. This partly justifies the assumption of ignoring horizontal deflections in approximate analyses of cable systems. It may be noted, however, that vertical deflections computed from Eq. 2-47*c* by neglecting horizontal deflections are overestimated, particularly in the end quarter of the span.

It may be noted, too, that the maximum deflections occur near the quarter-span point rather than in the center, as may usually be expected for a slab or a beam. The influence line for h/H for a unit load of $\mu = 0.1$ computed from Eq. 2-45 is also drawn in Fig. 2-9*a*, and it may be noted that the maxima occur at the center.

Figure 2-9*b*, in which values of w_n/f and h/H have been plotted against the value of μ, illustrates the nonlinear response of the cable. In general, therefore, the principle of superposition and thus the use of influence lines are not applicable to cables. It is worth noting, however, that for small values of μ, nonlinearity is rather less predominant than for larger values of μ. Therefore, if the applied load P is small in comparison to the weight of the cable qL (that is, small μ), the use of superposition can be admitted as an approximate measure. This leads to the idea of using *deflection coefficients*. For the purpose of computing the deflection coefficients the cable may be

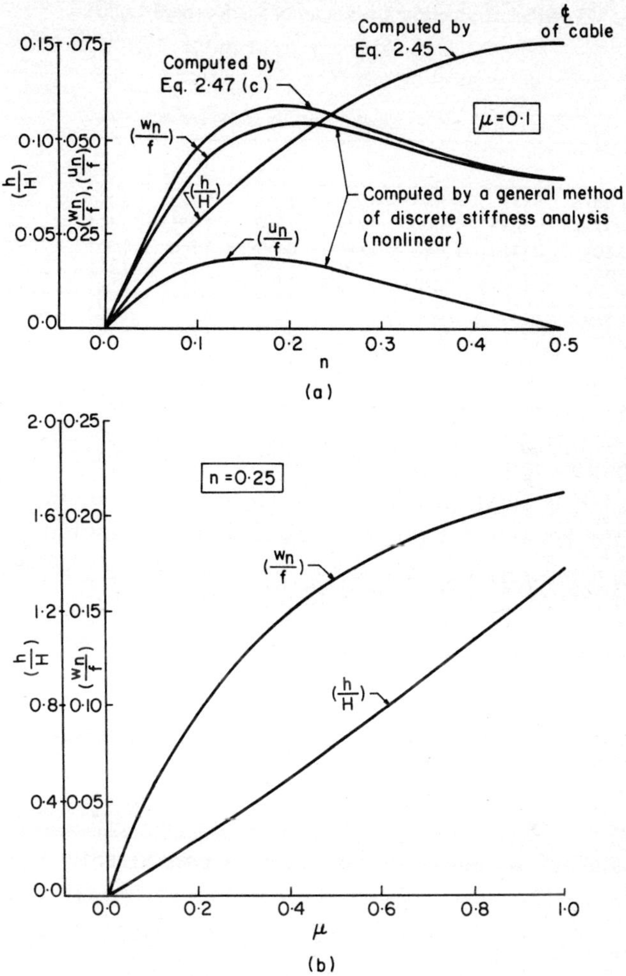

FIGURE 2-9 Static response characteristics of a cable when loaded as in Fig. 2-8.

divided into a number of small segments (say, ten for convenience) and deflections may be computed at each of the node points (nine in this case) by placing a small unit load at each of the nodes separately. A set of coefficients computed in this manner for a unit load of $\mu = 0.1$ is tabulated in Table 2-1. Deflection coefficients can be used to construct influence lines in the same way as for any other structure for analyzing a case of vertical loading.

TABLE 2-1 Vertical-Deflection Coefficients for Vertical Loads

Unit load: $\mu = P/qL = 0.1$

Load at node no.	*Deflection as percentage of sag f at node no.*								
	1	2	3	4	5	6	7	8	9
1	4.842	2.557	0.709	− 0.702	− 1.676	− 2.215	− 2.316	− 1.981	− 1.209
2	2.526	5.784	2.503	0.048	− 1.867	− 2.955	− 3.313	− 2.939	− 1.835
3	0.809	2.541	5.193	1.688	− 0.897	− 2.560	− 3.302	− 3.122	− 2.022
4	− 0.445	0.138	1.750	4.391	1.088	− 1.187	− 2.433	− 2.650	− 1.839
5	− 1.316	− 1.568	− 0.757	1.116	4.053	1.116	− 0.757	− 1.568	− 1.316
6	− 1.839	− 2.650	− 2.433	− 1.187	1.088	4.391	1.750	0.138	− 0.445
7	− 2.022	− 3.122	− 3.302	− 2.560	− 0.897	1.688	5.193	2.541	0.809
8	− 1.835	− 2.939	− 3.313	− 2.955	− 1.867	0.048	2.503	5.784	2.526
9	− 1.209	− 1.981	− 2.316	− 2.215	− 1.676	− 0.702	0.709	2.557	4.842

A simplification of Eqs. 2-45 and 2-47 is possible when the applied load P is small in comparison to the weight of the cable qL. Substituting back $H + h$ for H' in Eq. 2-45 and simplifying,

$$\left(\frac{H+h}{H}\right)^2 = 1 + 12n(1-n)\mu(1+\mu)$$

or

$$1 + 2\left(\frac{h}{H}\right) + \left(\frac{h}{H}\right)^2 = 1 + 12n(1-n)\mu(1+\mu)$$

Neglecting $(h/H)^2$ and μ^2 in this expression and simplifying,

$$H' = H + h = \frac{qL^2}{8f}\left[1 + 6n(1-n)\mu\right] \tag{2-48}$$

The value of w_n given by Eq. 2-47c is consequently modified as follows:

$$w_n = 8f\frac{n(1-n)(1-3n+3n^2)\mu}{1+6n(1-n)\mu} \tag{2-49}$$

Correction for Extensibility. In deriving all the results for an applied concentrated load, the extension of the cable on account of the tension increment h has been neglected. This can be justified as long as the applied load is small, and thus the tension increment is small. However, if the loads are larger, cable extensibility has to be considered.

Extension of the cable, Δl, is obtained from Eq. 2-37 as

$$\Delta l = \frac{hl}{EA}\left(1 + \frac{16}{3}\frac{f^2}{L^2}\right)$$

and inserting this into Eq. 2-44, we get, as a condition of compatibility,

$$l + \Delta l = l_{AC} + l_{CB} \tag{2-50}$$

The values of l, l_{AC}, and l_{CB} can be obtained from Eqs. 2-20, 2-29a, 2-43a, and 2-43b. Inserting these into Eq. 2-50 and simplifying, we get

$$L\left(1 + \frac{8}{3}\frac{f^2}{L^2}\right)\left[1 + \frac{H' - H}{EA}\left(1 + \frac{16}{3}\frac{f^2}{L^2}\right)\right]$$
$$= L\left[1 + \frac{q^2L^2}{24H'^2}(1 - 3n + 3n^2) + \frac{q^2L^2}{8H'^2}n(1 - n)(1 + 2\mu)^2\right]$$

This reduces to a cubic equation in H' of the form

$$aH'^3 + bH'^2 - c = 0 \tag{2-51}$$

where $a = \dfrac{d}{EA}$

$$b = \left(\frac{8}{3}\frac{f^2}{L^2} - \frac{d}{EA}H\right)$$

$$c = \frac{q^2L^2}{24}\left[(1 - 3n + 3n^2) + 3n(1 - n)(1 + 2\mu)^2\right]$$

$$d = \left(1 + \frac{8}{3}\frac{f^2}{L^2}\right)\left(1 + \frac{16}{3}\frac{f^2}{L^2}\right)$$

After H' has been computed from Eq. 2-51, the value of deflection w can be computed by combining Eqs. 2-43 and 2-46.

EXAMPLE 2-1

The following example illustrates the effect of extensibility of the cable. The cable shown in Fig. 2-10 has a weight of 3.16 lb/ft (45.6 N/m) of its length.[5]

$$l = 1000\left(1 + \frac{8}{3}\frac{100^2}{1000^2}\right)$$
$$= 1026.67 \text{ ft } (312.6 \text{ m})$$

The analysis in this section treats the cable weight as distributed

[5] The example has been solved by other general methods of cable analysis, and the values thus obtained are given for comparison in Sec. 2-4.

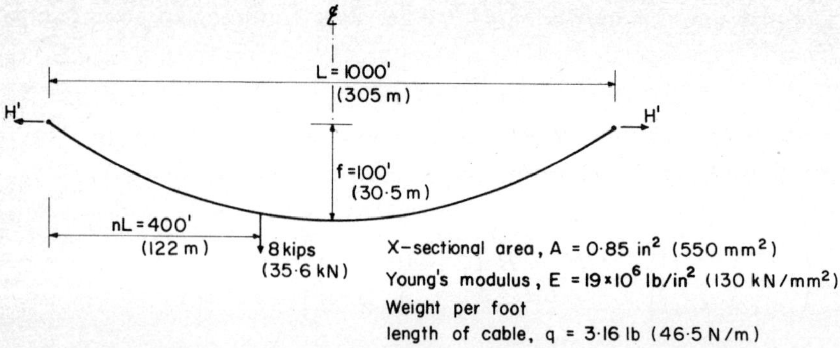

FIGURE 2-10 Simply suspended cable analyzed in Example 2-1.

uniformly along its span length, and consistent with this assumption we have

$$q = 3.16 \times \frac{1026.67}{1000} = 3.244 \text{ lb/ft } (47.8 \text{ N/m})$$

$$\mu = \frac{P}{qL} = \frac{8000}{3.244 \times 1000} = 2.465$$

$$n = \frac{400}{1000} = 0.4$$

$$H = \frac{qL^2}{8f} = 4.055 \text{ kips } (18.22 \text{ kN})$$

(1) If the cable extensibility is neglected, then the values of H' and w_n from Eqs. 2-45 and 2-47c work out as follows:

$$H' = 5.059 \times H = 5.059 \times 4.055 = 20.514 \text{ kips } (92.1 \text{ kN})$$

$$w_n = 16.52 \text{ ft } (5.04 \text{ m})$$

(2) If extensibility of the cable is considered, then H' from Eq. 2-50 works out to be

$$H' = 4.961 \times H = 4.961 \times 4.055 = 20.109 \text{ kips } (90.4 \text{ kN})$$

and w_n as computed by combining Eqs. 2-27, 2-43, and 2-46 is

$$w_n = 18.72 \text{ ft } (5.71 \text{ m})$$

This shows that an error of −11.75 percent in the deflections is caused by ignoring the effect of extensibility. It must be pointed out, however, that in this example the value of μ is far higher than what could be expected in practice when dealing with a cable roof. The errors due to neglecting extensibility will be much smaller for values of μ in the practical range. For example, if the same exercise is repeated for $\mu = 0.2465$, the deflections corresponding to (1) and (2) above are 8.40 ft (2.56 m) and 8.58 ft (2.62 m), respectively. The error in this case reduces to -2.1 percent.

2-3.2 Case II: Distributed Load Placed Centrally

The cable in Fig. 2-11 is loaded with a uniformly distributed load of length $(1-2n)L$ and intensity p per unit length. The loading is symmetrical with respect to the center of the span. The deflected cable will in this case consist of three parabolic arcs AC, CD, and DB, arcs AC and DB being similar.

The equations describing the geometry of the arcs can be obtained from consideration of statics as in case I. For arc AC,

$$z' = \frac{qL^2}{2H'}\left[\theta(1-\theta) + r\theta(1-2n)\right] \qquad \theta \leqslant n \qquad \text{(2-52}a\text{)}$$

and for arc CD,

$$z' = \frac{qL^2}{2H'}\left[(1+r)(1-\theta)\theta - rn^2\right] \qquad (1-2n) \leqslant \theta \geqslant n \quad \text{(2-52}b\text{)}$$

where r is the load ratio p/q. On application of the condition

$$l = l_{AC} + l_{CD} + l_{DB}$$

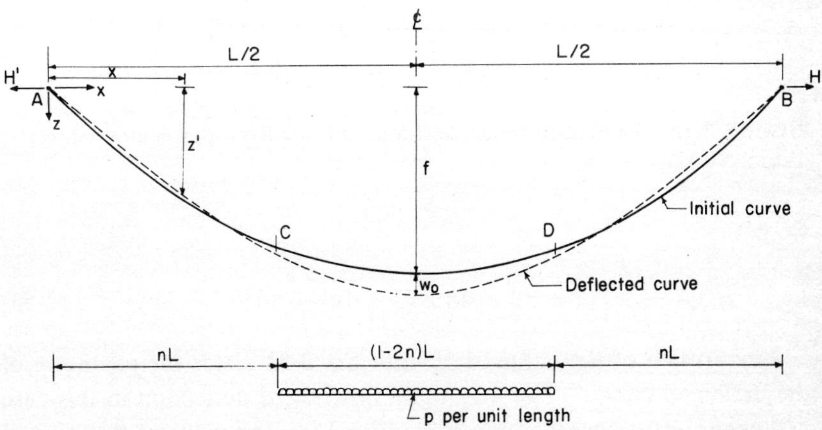

FIGURE 2-11 Cable subjected to a distributed load placed centrally.

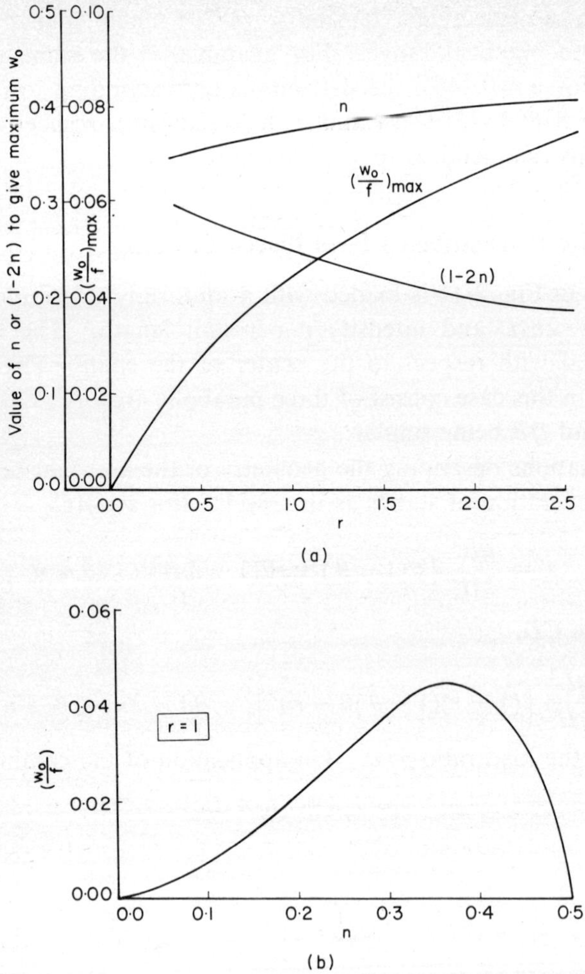

FIGURE 2-12 Static response curves for a cable loaded with a central uniformly distributed lag.

we get

$$H' = \frac{qL^2}{8f}\left[(1+r)^2 - 4n^2r^2(3-4n) - 4n^2r(3-2n)\right]^{1/2} \quad (2\text{-}53)$$

Substitution of the value of H' into Eq. 2-52 yields the geometry of the deflected curve. The maximum downward deflection in this case will occur at the center of the cable ($\theta = \frac{1}{2}$). The value of this central deflection w_0 as obtained from Eqs. 2-52*b* and 2-27 is

$$w_0 = f\left\{1 - \frac{(1+r) - 4n^2r}{\left[(1+r)^2 - 4n^2r^2(3-4n) - 4n^2r(3-2n)\right]^{1/2}}\right\} \tag{2-54}$$

Using the condition $\partial w_0/\partial n = 0$ (r constant), the optimum load disposition n to give the maximum value of w_0 can be computed. By this procedure the relation between the optimum value of n and the load ratio r is plotted in Fig. 2-12*a*. Also given is the corresponding maximum value of w_0/f. It may be noted that the value of n to give the maximum value of w_0/f does not vary much with the load ratio.

Figure 2-12*b* gives the value of w_0/f for varying lengths of the load (for $r = 1$). It is worth noting that for $n = 0$, that is, when the full span is loaded, the deflection is zero. It can be checked from Eq. 2-52 that the deflection anywhere in the span will be zero for this loading condition, irrespective of the intensity of load. This implies that the cable retains its initial shape, unless the extension of the cable is taken into account. The results for case II can be corrected for the effect of cable extensibility in the manner illustrated for case I.

2-3.3 Case III: Distributed Load at One End of the Span

When a short load of intensity p and length nL is applied at one of the ends of the span (Fig. 2-13), the deflected curve of the cable consists of two parabolic arcs AC and CB. From considerations of

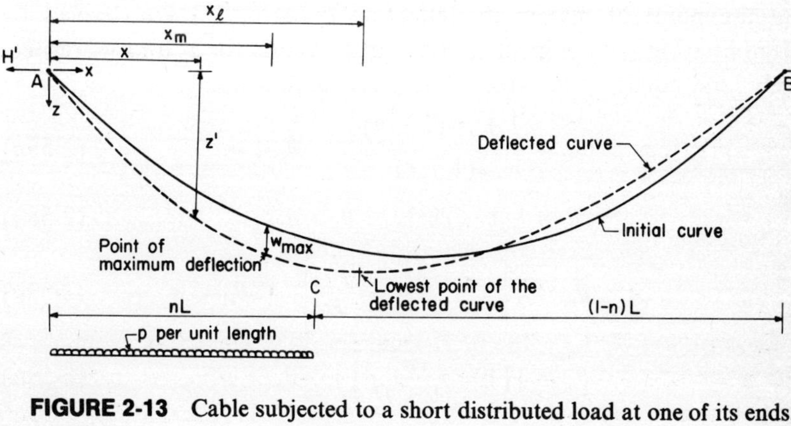

FIGURE 2-13 Cable subjected to a short distributed load at one of its ends.

statics as in the previous two cases, we get, for arc AC,

$$z' = \frac{qL^2}{2H'}\left[(1+r)\theta(1-\theta) - r\theta(1-n)^2\right] \qquad \theta \leqslant n \qquad (2\text{-}55a)$$

and for arc CB,

$$z' = \frac{qL^2}{2H'}\left[(\theta + rn^2)(1-\theta)\right] \qquad \theta \geqslant n \qquad (2\text{-}55b)$$

Substitution of the required values in Eq. 2-20 gives the lengths l_{AC} and l_{CB}, and applying the condition that

$$l = l_{AC} + l_{CB}$$

we get

$$H' = \frac{qL^2}{8f}\left[1 + 2r(3-2n)n^2 + r^2n^3(4-3n)\right]^{1/2} \qquad (2\text{-}56)$$

The value of w is now obtained as

$$w = 4f\frac{(1+r)\theta(1-\theta) - r\theta(1-n)^2}{\left[1 + 2r(3-2n)n^2 + r^2n^3(4-3n)\right]^{1/2}} - 4f\theta(1-\theta) \qquad \theta \leqslant n \qquad (2\text{-}57a)$$

$$w = 4f\frac{(\theta + rn^2)(1-\theta)}{\left[1 + 2r(3-2n)n^2 + r^2n^3(4-3n)\right]^{1/2}} - 4f\theta(1-\theta) \qquad \theta \geqslant n \qquad (2\text{-}57b)$$

The position of the lowest point on the deflected cable, $\theta_l(= x_l/L)$, and the point of maximum deflection $\theta_m(= x_m/L)$ are derived by differentiating with respect to θ the expressions for z' and w, respectively, and equating to zero. These are as follows:

$$\theta_l = \frac{1 + rn(2-n)}{2(1+r)} \qquad \theta_l \leqslant n \qquad (2\text{-}58a)$$

$$\theta_l = \tfrac{1}{2}(1 - rn^2) \qquad \theta_l \geqslant n \qquad (2\text{-}58b)$$

$$\theta_m = \tfrac{1}{2}\left[1 - \frac{r(1-n)^2}{1 + r - D}\right] \qquad \theta_m \leqslant n \qquad (2\text{-}59a)$$

$$\theta_m = \tfrac{1}{2}\left(1 - \frac{rn^2}{1-D}\right) \qquad \theta_m \geqslant n \qquad (2\text{-}59b)$$

where D equals the expression within the brackets in Eq. 2-56.

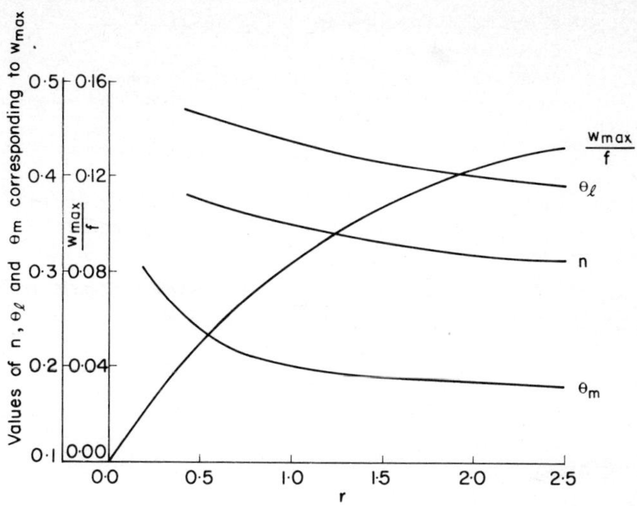

FIGURE 2-14 Cable response to load as stipulated for the case in Fig. 2-13.

Figure 2-14 shows the plot for the maximum deflection[6] w_{max}, the value of the loaded length nL to cause the maximum deflection, and the corresponding value of θ_l and θ_m for the varying value of the load parameter r. It will be seen that the deflection varies nonlinearly with load. The values of n, θ_l, and θ_m do not show much variation with r and lie within a fairly narrow band.

EXAMPLE 2-2

Example 2-2 employs the procedure for analysis discussed in Sec. 3-2 for comparing the deflection results of three loading cases shown in Fig. 2-15, to bring out a behavioral aspect of the cable. Extension of the cable is taken into account, and it is subjected to the same magnitude of load disposed in different manner. A comparison of results in Fig. 2-15 shows how the deflection values reduce from a maximum in case I, to a very small value in case III. The results emphasize the importance of considering concentrated loads or loads of short length when computing maximum deflections in simply suspended systems.

[6] The optimum value of the loaded length for any particular load-intensity ratio r to give the maximum deflection is obtained by using the condition $dw/dn = 0$ in Eq. 2-57.

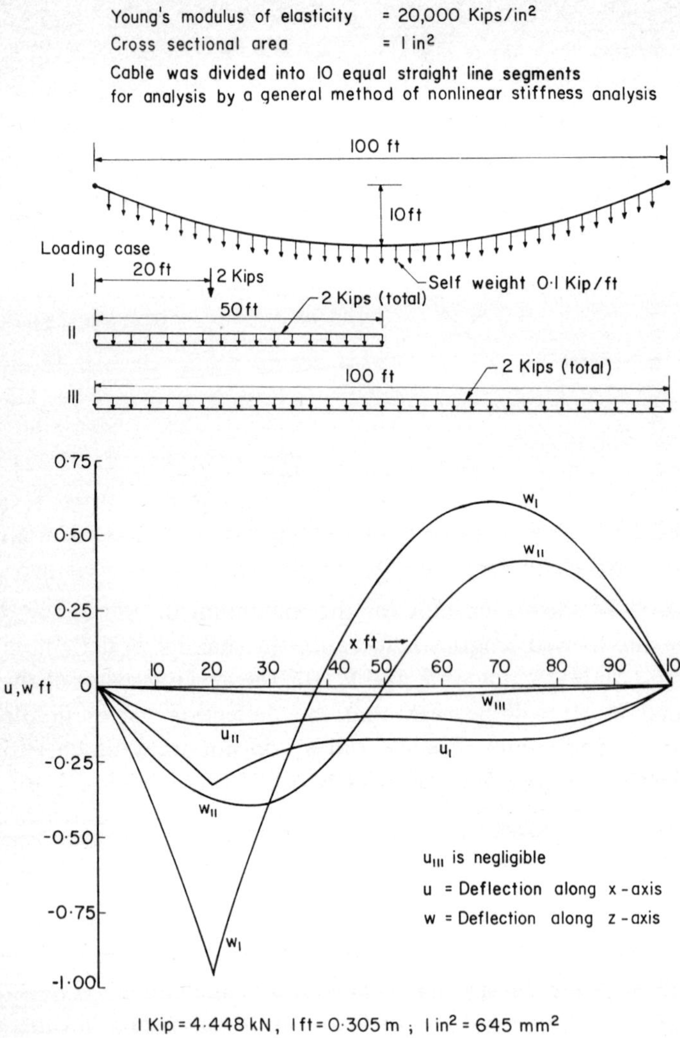

FIGURE 2-15 Comparison of cable deflections when the same load is disposed in different ways (Example 2-2).

2-4 METHODS FOR GENERALIZED ANALYSIS

The procedure used in Sec. 2-3 to obtain solutions for the different loading cases of a parabolic cable supported at two level supports can be extended to more general cases of cable geometry and loading. The scope for such analysis can be widened to include the effects of cable extensibility (as illustrated in case I of Sec. 2-3.1), temperature changes, or support movements. It is possible also to obtain horizontal deflections using these principles. However, an attempt to obtain closed-form mathematical expressions for these general cases could lead to algebraic complexity, and it is perhaps simpler and faster to resort to numerical procedures. The application of numerical procedures is facilitated if digital computers are used.

In a more general way there are two types of problems that can occur in analyzing the cable. The first type, which is encountered more frequently in practice, refers to problems in which the horizontal position of the applied loads is specified. The other type comprises problems in which the horizontal position of the point of application of the load changes on application of the load, and it is this final deflected position of the load that is specified. A study of the latter type is of some importance in the erection of suspension bridges, but not so important in cable roofs. However, the procedure described in the following paragraphs for the analysis of the first type of problem can be extended [36, 73] to solve the second type also.

The procedure of analysis described here is a numerical iterative scheme in which the satisfaction of geometrical compatibility is used as a point of termination of computations. The procedure is completely general, in that it may be used to analyze a cable with any geometry and loading. Second-order effects such as cable extensibility, temperature changes, and support movements can be included without much extra effort.

2-4.1 Cable Segment Treated as a Catenary

The steps in the method are explained by considering the cable segment to be a catenary [73]. This has been established in Sec. 2-2 to be the most exact treatment of a freely suspended cable element. For the cable shown in Fig. 2-16*a*, whose supports are spaced at a distance L horizontally and $\overline{Z}$ vertically, there are i points at which loads are applied, and the cable is thus divided into $i + 1$ segments, each hanging in the form of a catenary between the load points. The jth segment, after it has been isolated from the rest of the cable by cutting it immediately next to the loaded points j and $j + 1$, is shown in Fig. 2-16*b*. If the reactions H_j and S_j' for the segment are known, the values S_j'', X_j, Z_j, and Δl_j can be computed by the application of Eqs. 2-5 to 2-12. The values of H_{j+1} and S_{j+1}' are obtained by considering the static equilibrium of the node $j + 1$, and segment $j + 1$

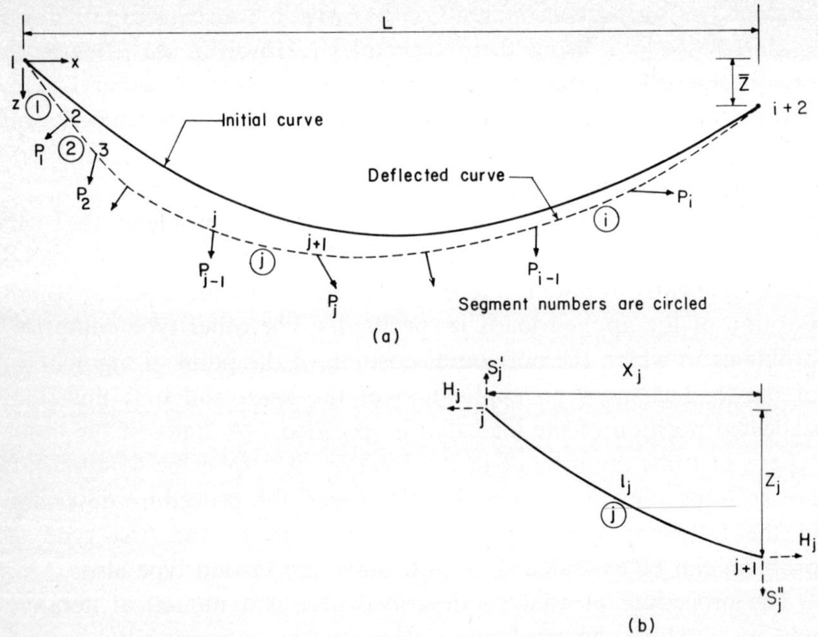

FIGURE 2-16 Discrete-element model for analysis.

can be analyzed in the same manner as segment j. It is obvious, therefore, that all the $i + 1$ segments can be analyzed, if the reactions H_1 and S_1' at the left-hand support are known or assumed.

If the cable supports are unyielding,[7] then for compatibility the summation of horizontal projections of the segments must equal the span L and the vertical projections must equal $\bar{Z}$. Mathematically,[8]

$$\sum X_j = L \tag{2-60a}$$

$$\sum Z_j = \bar{Z} \tag{2-60b}$$

Equations 2-60 will be satisfied if the assumed values of H_1 and S_1' are the correct ones. In the more common case, however, there will be an error of closure, e_x and e_z in x and z directions, respectively, given as

$$e_x = \sum X_j - L \tag{2-61a}$$

$$e_z = \sum Z_j - \bar{Z} \tag{2-61b}$$

[7] The effect of elastic supports can be taken into account and is illustrated later in this section.

[8] Summation in any of the equations in this section is implied over all the segments, that is, for $j = 1$ to $j = i + 1$, unless otherwise stated.

Therefore, the values of H_1 and S_1' have to be corrected and the computations described above have to be repeated until the errors e_x and e_z are within a tolerable range.

In order to evaluate the corrections required to be applied to H and S (exponents are dropped for simplicity) the following partial differential equation can be used:

$$\Delta X = \frac{\partial X}{\partial H}\,\Delta H + \frac{\partial X}{\partial S}\,\Delta S \tag{2-62a}$$

$$\Delta Z = \frac{\partial Z}{\partial H}\,\Delta H + \frac{\partial Z}{\partial S}\,\Delta S \tag{2-62b}$$

where ΔX, ΔZ, ΔH, and ΔS are the increments in X, Z, H, and S, respectively. If it is assumed that ΔH and ΔS have the same value for all the segments, Eq. 2-62 can be rewritten as

$$e_x = \sum \Delta X = \Delta H \sum \frac{\partial X}{\partial H} + \Delta S \sum \frac{\partial X}{\partial S} \tag{2-63a}$$

and

$$e_z = \sum \Delta Z = \Delta H \sum \frac{\partial Z}{\partial H} + \Delta S \sum \frac{\partial Z}{\partial S} \tag{2-63b}$$

Simultaneous solution of Eqs. 2-63*a* and 2-63*b* yields the value of ΔH and ΔS, which for a catenary is given as

$$\Delta H = \frac{A_1 e_x + A_2 e_z}{A_1 A_3 - A_2^2} \tag{2-64a}$$

$$\Delta S = \frac{A_2 e_x + A_3 e_z}{A_1 A_3 - A_2^2} \tag{2-64b}$$

where $A_1 = \sum \frac{X}{H}\left(\frac{X}{l}\right)^2$

$$A_2 = \sum \frac{Z}{H}\left(\frac{X}{l}\right)^2$$

$$A_3 = \sum \left(\frac{X}{H}\right) - A_1$$

The various steps in the procedure are summarized in the following list and are illustrated for a numerical example in Table 2-2.

1. Assume values of H_1 and S_1' for segment 1, that is, the left-hand support reactions.
2. Compute H and S'' for all the segments from considerations of statics.

TABLE 2-2 Results for Example 2-3 with the Segments Treated as Catenary

		Variable										
Trial	*Segment*	H, lb	S', lb (Eq. 2-7)	S'', lb (Eq. 2-8)	$(\Delta l)_e$, ft (Eq. 2-12)	l, ft	X, ft (Eq. 2-9)	Z, ft (Eq. 2-10)	e_x, ft (Eq. 2-61*a*)	e_z, ft (Eq. 2-61*b*)	ΔH, lb (Eq. 2-64*a*)	ΔS, lb (Eq. 2-64*b*)
1	1	19,500	6421.0	5116.0	0.515	413.399	396.365	117.213	− 1.685	− 1.617	+ 598	+ 29.3
	2	19,500	− 2884.0	− 4821.0	0.757	613.799	601.950	− 118.830				
2	1	20,098	6450.3	5145.3	0.537	413.420	397.173	114.548	− 0.021	− 0.047	+ 8.1	+0.96
	2	20,098	− 2854.7	− 4791.7	0.779	613.821	602.806	− 114.585				
3	1	20,106.1	6451.26	5146.26	0.536	413.419	397.179	114.512	− 0.002		+ 0.6	
	2	20,106.1	− 2853.74	− 4790.74	0.778	613.820	602.819	− 114.512				
4	1	20,106.7	6451.26	5146.26	0.536	413.419	397.180	114.509				
	2	20,106.7	− 2853.74	− 4790.74	0.778	613.820	602.820	− 114.509				

1 kip = 4.448 kN

1 ft = 0.305 m

SOURCE: Based on Ref. 73, Table 2, pp. 100–101.

Resulting deflections:

$u = 397.180 - 400.0 = -2.82$ ft

$w = 114.51 - 96.05 = +18.46$ ft

3. Compute for all segments the values of X, Z, and Δl by using Eqs. 2-5 to 2-12.
4. Evaluate values of corrections ΔH and ΔS from Eq. 2-64.
5. Repeat steps 2 to 4 until the corrections are small enough to satisfy a given tolerance.

There are two points which need explanation with regard to these five steps:

a. The speed of convergence of the procedure is not greatly dependent upon the accuracy of the initial guess for H_1. It can therefore be recommended that initially H_1 may simply be computed from consideration of statics on the basis of the configuration of the unloaded cable.

b. In executing step 3 in any particular cycle of iteration, the value of l for the segment is required, but this itself is dependent upon the values computed in the step. It is a good approximation, however, to use the value of l computed in the previous cycle of iteration.

After the correct values of H and S and hence X and Z have been computed, the calculation of deflections is simple. For example, the horizontal and vertical deflection at a node k are

$$u_k = \sum X_j - x_k \tag{2-65a}$$

$$w_k = \sum Z_j - z_k \tag{2-65b}$$

where the summation is over the segments $j = 1$ to $j = k - 1$, and

where u_k = horizontal deflection of node k
w_k = vertical deflection of node k
x_k, z_k = coordinates of node k

A freely hanging cable segment of uniform cross section takes the form of a catenary (see Sec. 2-2) but can also be treated approximately as a parabola or a straight line. Obviously the two extreme cases are those when a segment is treated as a catenary or a straight line. It is of interest therefore to compare the latter with the former case. The case of a parabolic segment will give results which are likely to be closer to those obtained for the catenary as compared with the ones computed with a straight line treatment.

The differences that arise in the general procedure, if the cable curve is approximated by a number of straight-line segments, are the following:

1. It will generally not be adequate, from the point of view of accuracy, to divide the cable into less than 10 segments, regardless of the number of load points i. This may mean that computations will need to be done for a greater number of segments than may actually be necessary when the segments are treated as suspended in the form of a catenary. The expressions for a straight line are certainly much simpler than those for a catenary, and the increased number of computational operations are to be weighed against this simplification.
2. Loads will need to be transferred to the node points. A certain amount of inaccuracy will doubtlessly be introduced because of the idealization of applied loads onto the node points.
3. In executing steps 3 and 4 of the iterative procedure, Eqs. 2-30 to 2-35 will have to be used. Equation 2-63 is still applicable if the appropriate values of X, Z, H, and l are used.

2-4.2 Sign Convention

A sign convention, which has been found convenient for use in this procedure [73], is to treat the vertical end reaction S' or S'' as positive when it is trying to rotate clockwise the end at which it acts. The vertical drop Z is positive when the left-hand end of the segment is higher.

EXAMPLE 2-3

If the problem in Example 2-1 were solved by the general procedure treating the two arcs AC and CB as catenary (see Fig. 2-17*a*), the value of load-point deflection obtained at the end of 4 cycles of iterations would be 18.46 ft (5.62 m). The computed results are given in Table 2-2. Alternatively, if the cable is divided into 10 straight segments as shown in Fig. 2-17*b*, the result at the end of 4 cycles is a deflection of 17.95 ft (5.47 m). The error thus caused is -2.75 percent. It is of interest to note that the deflection value obtained (in one cycle of computation) in Example 2-1, by using the procedure laid out in Sec. 2-3, case I, is 18.72 ft (5.71 m).

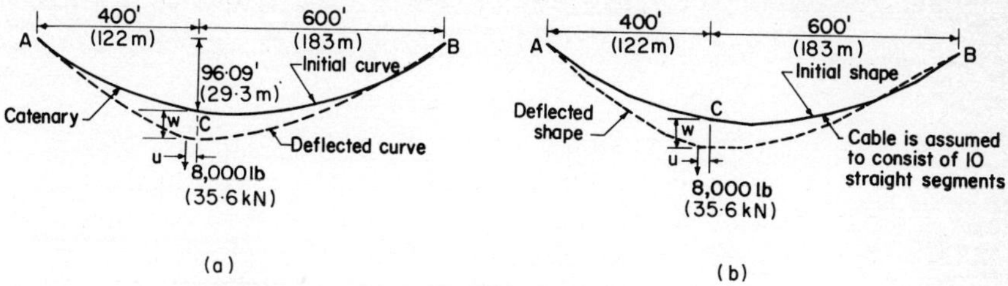

FIGURE 2-17 Analytical models for Example 2-3.

2-4.3 Corrections for Secondary Effects

The effect of temperature change or movements at the supports can be taken into account in the aforementioned analysis in a very simple way as described below.

1. *Temperature change*. This will introduce a change in length given by

$$(\Delta l)_t = \alpha t l \tag{2-66}$$

where α = coefficient of linear expansion
t = temperature change

In order to account for temperature changes in the analysis, $(\Delta l)_t$ can be added directly to the elastic stretch $(\Delta l)_e$ as given by Eq. 2-12.

2. *Support movements*. The stiffness of the cable supports can be expressed in terms of spring constants k_x and k_z

where k_x = stiffness of the spring along x axis
k_z = stiffness of the spring along z axis

In any iterative cycle, therefore,

$$\Delta L = [H_1 - (H_1) + H_{i+1} - (H_{i+1})]k_x \tag{2-67}$$

and

$$\Delta \bar{Z} = \{[S_1' - (S_1')] - [S_{i+1}'' - (S_{i+1}'')]\}k_z \tag{2-68}$$

where terms in parentheses ( ) are the values of these parameters in the unloaded state of the cable. Equations 2-61 thus become

$$e_x = \sum X - (L - \Delta L) \tag{2-69a}$$

and

$$e_z = \sum Z - (\bar{Z} - \Delta \bar{Z}) \tag{2-69b}$$

CHAPTER THREE

Pretensioned Cable Systems

3-1 GENERAL

The problem of theoretical analysis of pretensioned cable systems is discussed in this chapter. The factors simplifying as well as those complicating the problem, the assumptions and approximations usually made in analysis, and the numerical procedures required are treated in detail. This chapter also includes an exhaustive study of the structural characteristics of several typical cable-roof systems with a view to help the designer in his choice of form. Curves presenting relationships among nondimensional design parameters and formulas for approximate analysis are also included. As illustrated in Chap. 6, these can offer valuable help in the preliminary design work.

It has been mentioned earlier that cable systems, by virtue of the geometric proportions of cables, are incapable of carrying compressive stresses. If subjected to a compressive force, a cable element will become slack. This may cause large local deformations or sharp changes in slope, affecting the roof cladding adversely. Therefore, cable systems are generally designed, through the introduction of prestressing, to ensure that the entire system remains in tension under the worst conditions of loading. Cables also cannot carry bending or twisting moments. As a result, the only type of force to be considered in the analysis of cable systems is axial tension. This certainly simplifies the analysis considerably.

It has also been indicated (in Chap. 1) that the response of cable systems to loads is nonlinear in general. Thus, the principle of superposition cannot be considered as valid in the analysis of these structures, although its use as a measure of approximation can be admitted. In principle, therefore, an important tool of structural analysis—the use of influence lines—becomes inapplicable.

The analysis of large redundant structures like the cable beams or networks can be accomplished by treating them as discrete element systems in one of two basically different ways, namely the stiffness (displacement or equilibrium) method or the flexibility (force or compatibility) method. Any generalized formulation for the application of either of the methods will lead to solutions which will require considerable skill in numerical analysis. Most development in this direction has taken place in the application of the stiffness method and of the numerical techniques required to tackle the resulting nonlinear equations [4, 24, 42, 44, 58, 66, 69, 95]. Although Siev and Eidelman [89] have developed an analysis procedure for flat networks based on the flexibility approach, and the method is amenable to a generalized formulation, the flexibility approach has not found such extensive application as the stiffness method. In analyses considering the flexibility of the supporting structure, a *mixed* approach [68] has been attempted but does not appear to offer much advantage. A discussion on the flexibility or the mixed approach is not included herein, but a *tension-coefficient* type of approach leading to a stiffness-matrix analysis is formulated, although many of the problems and ideas discussed are applicable to the former two approaches of analysis as well. The method described has found wide coverage and application in recent literature, on account of its simplicity and versatility. Certain special problems in the analysis of tension systems that arise generally, and particularly when using this method, are given in detail in Sec. 3-2. The method described is strictly applicable to cable-suspended-roof systems only.

A somewhat more complex problem is met in analyzing networks or trusses, wherein cables are combined with flexural members. These structures can be analyzed [98, 102, 103] by an appropriate method which duly considers kinematic nonlinearity as well as other problems associated with cable elements. Bending stresses will be small if the configuration of the flexural members is kept as near as possible to the funicular curve of the loads. As for a cable-stayed roof, structurally it can be classified as a rigid flexural member, such as a beam or a plate, propped by elastic supports. The analysis of these structures may be carried out by the conventional procedures of structural analysis. *Cable* action and nonlinearity are unlikely to be important in cable-stayed roofs.

An entirely different approach (as also mentioned in Sec. 1-2) will be to approximate cable structures to membranes [96, 97]. A rigorous analysis of membranes can be mathematically quite complex, whereas a simplified analysis will have its own limitations. Perhaps for this reason there has been comparatively little effort to formulate a rigorous nonlinear analysis for membranes as applicable to cable systems. Continuum analysis with the intent of developing approximate relations has, however, been attempted [47, 66, 67, 75] and is dealt with in this chapter. Recent developments in the matrix analysis of cable systems have quite rightly laid stress on accounting for the flexibility of the supporting structure [9, 68, 83]. The flexibility of the supporting structure can have a marked effect on the behavior of the cable system particularly during erection. However, at the time of finalizing this text, experience in this context was too limited to draw any conclusion.

3-2 STIFFNESS ANALYSIS

A general method of analysis for cable systems, based on the well-known *tension coefficient* approach and leading to a stiffness-matrix solution, is described in this section. In carrying out the analysis, certain assumptions are made to simplify the computational work. These are enumerated below along with a discussion of their validity.

1. *The cable is treated as being completely flexible; i.e., it cannot sustain any bending moment.* The assumption is largely justifiable, although it has been shown that a certain degree of fixity is introduced in the cable at the supports [101]. It is also to be expected that there will be some degree of fixity at the points of connections between the cables. This causes secondary bending stresses, although their occurrence will be local.

2. *The cable is considered incapable of taking any compressive forces.* This assumption is certainly valid, as the cable has a negligible buckling strength on account of its geometrical proportions.

3. *The cladding does not contribute to the stiffness of the roof structure.* This is assumed generally for the purposes of analysis but is not correct. The assumption can be justified on three grounds:

   a. The contribution of the cladding is an uncertain factor and its neglect leads to greater surety.

   b. The error caused is on the safer side.

   c. It simplifies the analysis considerably.

4. *The intersection of two or more cables is treated as a joint.* In most practical cases this is true because the cables are adequately clamped.

5. *The cable elements lie along straight lines between the joints.* This implicitly assumes that all the loads including the self-weight of the cable are acting at the nodes only and that the cable element itself is weightless. The assumption should be considered quite fair at the prestress stage, when the pretension is far larger than the tension due to the self-weight of the cable.

The cladding is often connected to intermediate points on the cable between the joints, in a manner such that the dead weight of the cladding, as well as of the superimposed loads, will be transferred first to these intermediate points. Therefore, in the context of the loaded structure, this assumption must be considered as an approximation to simplify the analysis. A matrix formulation to account for loads distributed on the cable has been developed [5] to avoid this approximation, and it has been shown that the distribution of loading can affect the response of the structure. For the examples chosen, it is shown that the difference in results for nodal or distributed loading remains within 3 percent. In another work [12], however, it has been shown that differences in results due to concentrating the loads at nodes can be substantially more.

3-2.1 Formulation of Equations

The method can be best explained by considering a node q in any space cable network connected to adjoining nodes p through elements

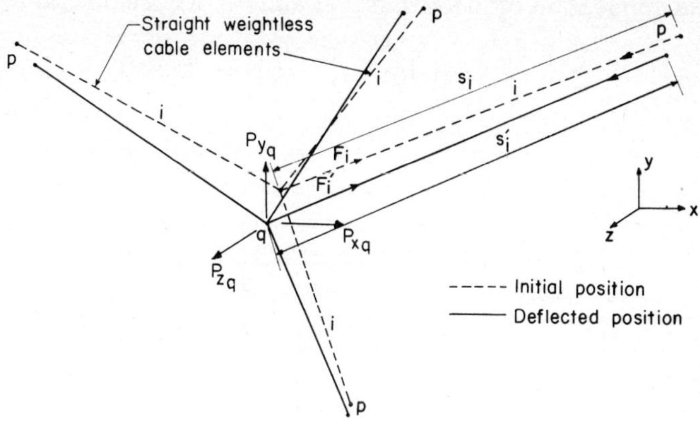

FIGURE 3-1 Elements in space meeting in node q.

i. The force of pretension in a typical element is F_i and its length is s_i. The initial coordinates of the nodes p and q are x_p, y_p, z_p and x_q, y_q, z_q, respectively. As shown in Fig. 3-1, a system of forces P_{x_q}, P_{y_q}, and P_{z_q} is applied at node q. The nodes p and q deflect through u_p, v_p, w_p and u_q, v_q, w_q, respectively, along the x, y, and z axes. The force F_i and the length s_i change by amounts ΔF_i and Δs_i and become F_i' and s_i', respectively.

The equations of equilibrium at node q, before the application of loads, can be written as

$$\sum^{i} \frac{F_i}{s_i}(x_p - x_q) = 0 \tag{3-1a}$$

$$\sum^{i} \frac{F_i}{s_i}(y_p - y_q) = 0 \tag{3-1b}$$

$$\sum^{i} \frac{F_i}{s_i}(z_p - z_q) = 0 \tag{3-1c}$$

This implicity assumes that the given geometry of the structure is in equilibrium with the applied forces. Cases where this assumption does not hold good, e.g., where the initial geometry itself needs to be determined, are treated in Sec. 3-2.2. The length of the element, s_i, can be expressed as

$$s_i = \left[(x_p - x_q)^2 + (y_p - y_q)^2 + (z_p - z_q)^2\right]^{1/2} \tag{3-2}$$

On the application of loads, Eqs. 3-1 and 3-2 will change as follows:

$$\sum^{i}\left[\frac{F_i'}{s_i'}(x_p + u_p - x_q - u_q)\right] + P_{x_q} = 0 \tag{3-3a}$$

$$\sum^{i}\left[\frac{F_i'}{s_i'}(y_p + v_p - y_q - v_q)\right] + P_{y_q} = 0 \tag{3-3b}$$

$$\sum^{i}\left[\frac{F_i'}{s_i'}(z_p + w_p - z_q - w_q)\right] + P_{z_q} = 0 \tag{3-3c}$$

$$s_i' = \left[(x_p + u_p - x_q - u_q)^2 + (y_p + v_p - y_q - v_q)^2 + (z_p + w_p - z_q - w_q)^2\right]^{1/2} \tag{3-4}$$

Also, $F_i' = F_i + \Delta F_i$

and $$\Delta F_i = EA_i\left(\frac{s_i' - s_i}{s_i}\right) = EA_i\left(\frac{s_i'}{s_i} - 1\right) \tag{3-5}$$

where EA_i is the extensional stiffness of the ith element.

Equations 3-2 and 3-4 can be combined to give

$$s_i' = s_i(1 + 2a_i + b_i)^{1/2} \tag{3-6}$$

where $a_i = \frac{1}{s_i^2}\left[(x_p - x_q)(u_p - u_q) + (y_p - y_q)(v_p - v_q) + (z_p - z_q)(w_p - w_q)\right]$

$$b_i = \frac{1}{s_i^2}\left[(u_p - u_q)^2 + (v_p - v_q)^2 + (w_p - w_q)^2\right]$$

Expanding the right-hand side of Eq. 3-6 and making the appropriate substitution in Eq. 3-5, we get

$$\Delta F_i = EA_i(a_i + \tfrac{1}{2}b_i - \tfrac{1}{2}a_i^2 - \tfrac{1}{2}a_ib_i + \tfrac{1}{2}a_i^3 + \cdots) \tag{3-7}$$

Similarly expressing Eq. 3-6 as

$$\frac{1}{s_i'} = \frac{1}{s_i}(1 + 2a_i + b_i)^{-1/2}$$

and expanding, we get

$$\frac{1}{s_i'} = \frac{1}{s_i}(1 - a_i - \tfrac{1}{2}b_i + \tfrac{3}{2}a_i^2 + \tfrac{3}{2}a_ib_i - \tfrac{5}{2}a_i^3 \cdots) \tag{3-8}$$

Making substitutions from Eqs. 3-1, 3-7, and 3-8 into Eq. 3-3, and rearranging terms such that only the first-order terms in u, v, w

remain on the left-hand side, we get

$$\sum^{i}\left[F_i\frac{u_p-u_q}{s_i}+(EA_i-F_i)\frac{x_p-x_q}{s_i}a_i\right]=-P_{x_q}+R_{x_q} \quad (3\text{-}9a)$$

$$\sum^{i}\left[F_i\frac{v_p-v_q}{s_i}+(EA_i-F_i)\frac{y_p-y_q}{s_i}a_i\right]=-P_{y_q}+R_{y_q} \quad (3\text{-}9b)$$

$$\sum^{i}\left[F_i\frac{w_p-w_q}{s_i}+(EA_i-F_i)\frac{z_p-z_q}{s_i}a_i\right]=-P_{z_q}+R_{z_q} \quad (3\text{-}9c)$$

where $R_{x_q}=-\sum^{i}(EA_i-F_i)\left[\frac{u_p-u_q}{s_i}c_i+\frac{x_p-x_q}{2s_i}d_i\right]$

$$R_{y_q}=-\sum^{i}(EA_i-F_i)\left[\frac{v_p-v_q}{s_i}c_i+\frac{y_p-y_q}{2s_i}d_i\right]$$

$$R_{z_q}=-\sum^{i}(EA_i-F_i)\left[\frac{w_p-w_q}{s_i}c_i+\frac{z_p-z_q}{2s_i}d_i\right]$$

$$c_i=a_i+\tfrac{1}{2}b_i-\tfrac{3}{2}a_i^2$$

$$d_i=b_i-3a_i^2-3a_ib_i+5a_i^3$$

The expressions for R_{x_q}, R_{y_q}, and R_{z_q} are obtained by retaining terms of up to the third power of u, v, and w and neglecting those of higher order. It would be easy to recognize the R terms as the residual forces were a linear analysis of the problem carried out.

For a cable space structure three equations such as Eqs. 3-9 will be required per joint, and this will lead to a set of $3N$ simultaneous equations, where N is the total number of cable joints, the supporting structure being treated as rigid. The solution of these equations by a suitable iterative method (details discussed in Sec. 3-3) will give the values of u, v, and w. The substitution of these values into Eqs. 3-7 will give the change in forces.

It can be seen from the preceding analysis that the problem of analyzing cable structures reduces to the solution of nonlinear Eqs. 3-9 (or Eqs. 3-11 in Sec. 3-2.3 as the case may be). These equations when assembled for the whole or part of a structure can be expressed in the matrix form as[1]

$$\mathbf{K}\cdot\mathbf{U}=-\mathbf{P}+\mathbf{R} \quad (3\text{-}10)$$

where $\mathbf{K}$ = square stiffness matrix for structure; consists of coefficients of unknowns u, v, w, etc., in either Eqs. 3-9 or 3-11

[1] **Boldface letters indicate a matrix.**

U = vector of unknowns u, v, w
P = load vector
R = column vector containing residual terms R_{x_q}, R_{y_q}, R_{z_q}, etc.

The size of matrix **K** will be $2N \times 2N$ for a plane system and $3N \times 3N$ for a space system. Similarly the vectors **U**, **P**, and **R** will each have a size $2N \times 1$ or $3N \times 1$ for the above two cases.

There can be no direct solution of Eq. 3-10, since the right-hand side contains the term **R**, which is a function of the unknowns. An iterative numerical solution has therefore to be adopted. Some of the numerical procedures possible are discussed later in Sec. 3-3.

EXAMPLE 3-1

The simple two-link structure shown in Fig. 3-2a is used to illustrate the analytical procedure given in Sec. 3-2.1 in conjunction with the numerical schemes described later in Sec. 3-3. The data and the initial geometry for the problem are shown in Fig. 3-2a. The structure and the loading are symmetrical, and therefore only the vertical deflection at node 2 will have a nonzero value. The matrices **K**, **U**, **P**, and **R** all reduce to a single number each.

Numerical Scheme 1. We shall first compute the value of **K** from the left-hand side of Eq. 3-9c, which is the only equation applicable to the problem.[2]

$$F_1 = F_2 = 0$$

$$s_1 = s_2 = 25.495 \text{ ft } (7.77 \text{ m})$$

$$EA_1 = EA_2 = 2500 \text{ kips } (11{,}120 \text{ kN})$$

$$z_1 - z_2 = -5 \text{ ft } (-1.55 \text{ m})$$

$$z_3 - z_2 = -5 \text{ ft } (-1.55 \text{ m})$$

$$a_1 = \frac{1}{s_1^2}(z_1 - z_2)(-w_2^{(1)}) = \frac{1}{130} w_2^{(1)}$$

$$a_2 = \frac{1}{s_2^2}(z_3 - z_2)(-w_2^{(1)}) = \frac{1}{130} w_2^{(1)}$$

Substituting these in the left-hand side of Eq. 3-9c, we get $-7.543 w_2^{(1)}$.

[2] Superscript numbers in parentheses indicate the cycle number of iteration to which the value pertains.

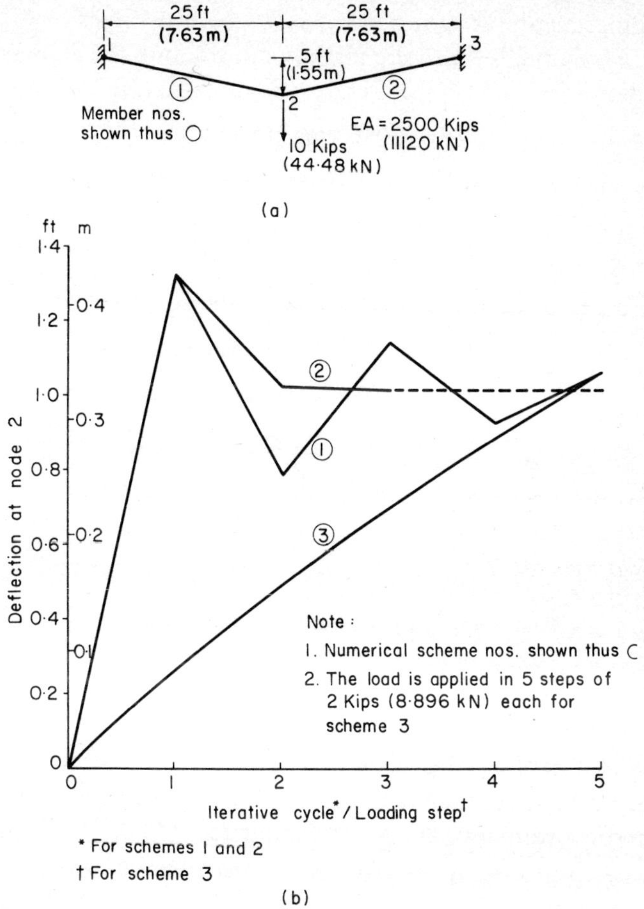

FIGURE 3-2 Comparison of numerical schemes (Example 3-1).

CYCLE 1 Assume $w_2^{(0)}$ and $R_{z_2}^{(0)} = 0$; hence, from Eq. 3-9*c*,

$$-7.543 w_2^{(1)} = -10.0$$

$$w_2^{(1)} = 1.326 \text{ ft } (0.404 \text{ m})$$

CYCLE 2 $R_{z_2}^{(1)}$ can now be computed on the basis of the value of $w_2^{(1)}$ computed in cycle 1.

$$a_1 = 0.0102 \quad a_2 = 0.0102$$

$$b_1 = 0.0027 \quad b_2 = 0.0027$$

Substituting the above values of a_1, a_2, b_1, b_2 and values of EA_1, EA_2, s_1, s_2, z_1, z_2, and $w_2^{(1)}$, which are already known, into the expression for R_{z_q}, we get

$$R_{z_2}^{(1)} = 4.086 \text{ kips } (18.15 \text{ kN})$$

Now we solve $-7.543w_2^{(2)} = -10.0 + 4.086$. The solution is

$$w_2^{(2)} = 0.784 \text{ ft } (0.239 \text{ m})$$

CYCLE 3 Repeating the procedure of cycle 2, we get

$$R_{z_2}^{(2)} = 1.394 \text{ kips } (6.2 \text{ kN})$$

$$-7.543w_2^{(3)} = -10.0 + 1.394$$

$$w_2^{(3)} = 1.142 \text{ ft } (0.348 \text{ m})$$

Similarly, the subsequent cycles of iteration are repeated.

Numerical Scheme 2.

CYCLE 1 $K^{(0)} = 7.543$ and $R_{z_2}^{(0)} = 0$. Therefore,

$$-7.543(w_2^{(1)}) = -10.0$$

$$w_2^{(1)} = 1.326 \text{ ft } (0.404 \text{ m})$$

CYCLE 2 To determine the value of $K^{(1)}$, the new values of length, force, and coordinates have to be computed.

$$z_1^{(1)} = z_3^{(1)} = 0 \qquad x_1^{(1)} = 0 \qquad x_2^{(1)} = 25 \text{ ft } (7.63 \text{ m})$$

$$x_3^{(1)} = 50 \text{ ft } (15.25 \text{ m})$$

$$z_2^{(1)} = z_2^{(0)} + w_2^{(1)} = 6.326 \text{ ft } (1.93 \text{ m})$$

$$s_1^{(1)} = s_2^{(1)} = 25.788 \text{ ft } (7.86 \text{ m})$$

$$F_1^{(1)} = F_2^{(1)} = 28.710 \text{ kips } (127.7 \text{ kN})$$

$$EA_1^{(1)} = EA_2^{(1)} = 2500 \text{ kips } (11{,}120 \text{ kN})$$

Now substituting the above values in Eq. 3-9c, we have[3]

$$2\left[-\frac{28.710}{25.788}\Delta w_2^{(1)} - (2500 - 28.710)\frac{(0-6.326)^2}{(25.788)^3}\Delta w_2^{(1)}\right]$$

$$= -13.759\,\Delta w_2^{(1)}$$

[3] The multiplication by 2 pertains to summation for the two members.

$K^{(1)}$ is thus computed to be -13.759. Similarly $R_{z_2}^{(1)}$ is computed[4] to be 4.085 kips (18.15 kN). Therefore,

$$-13.759\,\Delta w_2^{(1)} = 4.085$$

$$\Delta w_2^{(1)} = -0.297 \text{ ft } (0.091 \text{ m})$$

$$w_2^{(2)} = w_2^{(1)} + \Delta w_2^{(1)} = 1.029 \text{ ft } (0.314 \text{ m})$$

CYCLE 3 The procedure of the second cycle is repeated to obtain the value of $K^{(2)}$ as -12.282 and $R_{z_2}^{(2)}$ as 0.187. Therefore,

$$-12.282\,\Delta w_2^{(2)} = .0187$$

$$\Delta w_2^{(2)} = -0.015 \text{ ft } (0.0046 \text{ m})$$

$$w_2^{(3)} = w_2^{(2)} + \Delta w_2^{(2)} = 1.014 \text{ ft } (0.309 \text{ m})$$

The residual at the end of the third cycle is 0.008 kip.

Numerical Scheme 3. It has been shown in cycle 2 of the second scheme how the instantaneous stiffness of the system can be computed. The same procedure is used in this scheme. The results obtained by applying the load in five equal steps are summarized in Table 3-1.

It will be seen from Fig. 3-2*b* that the results in scheme 1 have failed to converge to a final answer in 5 cycles.[5] By using scheme 2 an accuracy of 0.008 percent is achieved within 3 cycles, and scheme

[4] All the properties of the structure substituted to compute $R_{z_2}^{(1)}$ are those which were obtained in the first cycle.

[5] The results converge to within 1.5 percent accuracy in 11 cycles.

TABLE 3-1 Results for Numerical Scheme 3 in Example 3-1

Step number	*K, kips/ft*	*− P, kips*	w_2*, ft*
1	−7.543	−2	0.265
2	−8.704	−2	0.230
3	−9.745	−2	0.205
4	−10.701	−2	0.187
5	−11.593	−2	0.173
		Total deflection	= 1.060 ft

1 kip = 4.448 kN

1 ft = 0.305 m

3 which actually represents 5 cycles of calculation (load is applied in five equal steps) overestimates the deflection by 4.55 percent. The results are shown plotted in Fig. 3-2*b*. This will establish the superiority of scheme 2 for this problem.

3-2.2 Determination of Initial Geometry

The formulation of equations in the preceding section is based on the assumption that the geometry of the system under the given set of prestressing forces is known. This is true in some cases, such as a dual-cable system, shown in Fig. 3-12*b* to *d*, or a hyperbolic paraboloid. Generally speaking, however, the initial geometry will need to be computed before the structure can be analyzed for applied loads. The problem can indeed be tackled by a suitable modification of the equations given in Sec. 3-2.1. This will, however, depend upon the manner in which the problem is specified, as explained in the following paragraphs.

Case I. The geometry of the supporting structure as well as the pretensioning forces in the cables are specified, and it is required to compute the network geometry. The nodal coordinates in the network then have to be assumed and suitably modified by a process of successive iteration, to be finally in equilibrium with the forces. To solve this problem, Eqs. 3-9 are modified by incorporating the following two conditions:

1. Equations 3-1 are not assumed to hold good, since it is not necessary that the assumed geometry satisfies static equilibrium.
2. Cable pretension is a given constant value, and so

$$F_i' = F_i$$

or

$$\Delta F_i = 0$$

Equations 3-9 thus become

$$\sum^{i} \frac{F_i}{s_i}\left[(u_p - u_q) - (x_p - x_q)a_i\right] = -P_{x_q} + R_{x_q} \qquad (3\text{-}11a)$$

$$\sum^{i} \frac{F_i}{s_i}\left[(v_p - v_q) - (y_p - y_q)a_i\right] = -P_{y_q} + R_{y_q} \qquad (3\text{-}11b)$$

$$\sum^{i} \frac{F_i}{s_i}\left[(w_p - w_q) - (z_p - z_q)a_i\right] = -P_{z_q} + R_{z_q} \qquad (3\text{-}11c)$$

where in this case P_{x_q}, P_{y_q}, P_{z_q} are the components of a part of the dead weight of the structure assumed concentrated at the node q, and

$$R_{x_q} = -\sum^{i} \frac{F_i}{s_i}\left[(x_p - x_q) - (u_p - u_q)a_i - \tfrac{1}{2}(x_p + u_p - x_q - u_q)e_i\right]$$

$$R_{y_q} = -\sum^{i} \frac{F_i}{s_i}\left[(y_p - y_q) - (v_p - v_q)a_i - \tfrac{1}{2}(y_p + v_p - y_q - v_q)e_i\right]$$

$$R_{z_q} = -\sum^{i} \frac{F_i}{s_i}\left[(z_p - z_q) - (w_p - w_q)a_i - \tfrac{1}{2}(z_p + w_p - z_q - w_q)e_i\right]$$

$$e_i = -b_i + 3a_ib_i - 5a_i^3$$

Case II. The manner of specifying the problem is only slightly different from case I. The forces in the cables are given as before, but it is specified that forces in some of the cable elements can change. The elements arc therefore placed into two categories:

$$\text{(a)}\quad F_i' = F_i \qquad \text{or} \qquad \Delta F_i = 0$$

$$\text{(b)}\quad F_i' \neq F_i \qquad \text{or} \qquad \Delta F_i \neq 0$$

A mixed condition will thus result at some of the nodes, the equations for which can be formulated as indicated in case I and in Sec. 3-2.1. Example 3-2 will illustrate clearly the two cases described.

EXAMPLE 3-2

The network shown in Fig. 3-3 is analyzed to illustrate the second case discussed above on the determination of initial geometry. Numerical scheme 2 (Sec. 3-3) is used for the iterative procedure. The data used for this case are as follows:

Pretension specified for members 1 to 16	50 kips (222.4 kN)
Pretension specified[6] for members 17 to 40	5 kips (22.24 kN)
Extensional stiffness *EA* for members 1 to 16	30,000 kips (133.44 MN)
Extensional stiffness *EA* for members 17 to 40	5000 kips (22.24 MN)

[6] This value is to remain constant.

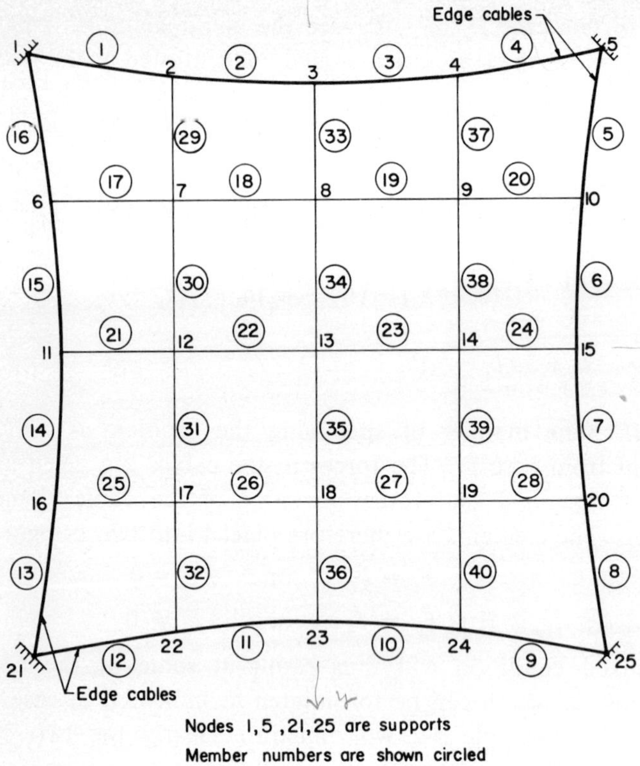

FIGURE 3-3 Cable net supported on edge cables (Example 3-2).

The coordinates of the edge cable are assumed by taking each of them to be parabolic in shape and considering each to lie in a separate single plane. The assumed coordinates and the changes Δx, Δy, Δz obtained in 3 cycles of computation are given in Table 3-2. Table 3-3 gives the change in edge-cable forces.

3-2.3 Approximations

It is possible, under certain conditions, to make approximations [45] in the rigorous form of analysis described in the preceding sections and thus reduce the computational effort. Although this introduces errors of varying degree in the results, the saving in time makes the approximations worth considering, at least, for preliminary analysis. Some of the possibilities are discussed below.

1. For shallow networks, with small dip/span ratios, the horizontal deflections u, v are much smaller than the vertical

TABLE 3-2 Assumed Coordinates and the Required Corrections for Example 3-2

Node No.	Assumed coordinates, ft: x	y	z	Δx, ft	Δy, ft	Δz, ft
1	– 20.0	– 20.0	– 4.0	0.0	0.0	0.0
2	– 10.0	– 18.5	– 2.0	– 0.0101	0.0017	0.0507
3	0.0	– 18.0	0.0	0.0	0.0052	0.0
4	10.0	– 18.5	2.0	0.0101	0.0017	– 0.0507
5	20.0	– 20.0	4.0	0.0	0.0	0.0
6	– 18.5	– 10.0	– 2.0	0.0017	– 0.0101	0.0507
7	– 10.0	– 10.0	– 1.0	– 0.0101	– 0.0101	– 0.0548
8	0.0	– 10.0	0.0	0.0	– 0.0101	0.0
9	10.0	– 10.0	1.0	0.0101	– 0.0101	0.0548
10	18.5	– 10.0	2.0	– 0.0017	– 0.0101	– 0.0507
11	– 18.0	0.0	0.0	0.0052	0.0	0.0
12	– 10.0	0.0	0.0	– 0.0101	0.0	0.0
13	0.0	0.0	0.0	0.0	0.0	0.0
14	10.0	0.0	0.0	0.0101	0.0	0.0
15	18.0	0.0	0.0	– 0.0052	0.0	0.0
16	– 18.5	10.0	2.0	0.0017	0.0101	– 0.0507
17	– 10.0	10.0	1.0	– 0.0101	0.0101	0.0548
18	0.0	10.0	0.0	0.0	0.0101	0.0
19	10.0	10.0	– 1.0	0.0101	0.0101	– 0.0548
20	18.5	10.0	– 2.0	– 0.0017	0.0101	0.0507
21	– 20.0	20.0	4.0	0.0	0.0	0.0
22	– 10.0	18.5	2.0	– 0.0101	– 0.0017	– 0.0507
23	0.0	18.0	0.0	0.0	– 0.0052	0.0
24	10.0	18.5	– 2.0	0.0101	– 0.0017	0.0507
25	20.0	20.0	– 4.0	0.0	0.0	0.0

1 ft = 0.305 m

TABLE 3-3 Changes in Edge-Cable Tension for Example 3-2

Member number	Change in force, kips (kN)	Member number	Change in force, kips (kN)
1	1.2920 (5.75)	5	1.2920 (5.75)
2	0.7037 (3.11)	6	0.7037 (3.11)
3	0.7037 (3.11)	7	0.7037 (3.11)
4	1.2920 (5.75)	8	1.2920 (5.75)

deflections w. If u, v are ignored, the problem is much simplified inasmuch as the number of equations reduces from $3N$ to N. It has been shown [50] that this approximation is justified within the limitations stipulated in many practical cases.

2. It can also be assumed that

   a. There is no change in the length of the cable elements, that is, $s' = s$.

   b. Residuals are negligible and $R = 0$.

   When the initial stiffness of a network is large and the loads applied are comparatively small, the resulting deflections will also be small. The change in the link lengths is as such small and justifies assumption *a*. The residual term R which is composed of higher-order terms in u, v, w is naturally small too. Neglecting R means linearizing Eqs. 3-9, and this is justified [50, 89] when the conditions described above hold good. Linearization of equations, where possible, can be of great practical advantage since it admits the use of the principle of superposition and therefore the use of influence lines or surfaces.

3. In order to facilitate the formulation of equations, it can be assumed for two-way networks [89] that the cables lie in vertical planes. This will, in general, have justification only when the dip/span ratio is small. The assumption will, however, always hold good for hyperbolic paraboloids.

4. The analysis of a cable network by the method suggested will require three equations per node to be set up in a general case. Since most practical networks will consist of a large number of cables, the number of governing equations will also be large. For example, in a two-way network, if there are 20 cables in each direction, the number of nodes and the number of corresponding equations will be 400 and 1200, respectively. This is a sizable problem, whereas the numbers mentioned are only moderate for a practical size of network.

   In most cases, therefore, the size of the problem will need to be reduced for the purposes of analysis. This can be done by grouping the cables and treating several cables as one. The number of nodes as well as equations will thus become smaller. It is obvious that errors will be caused on account

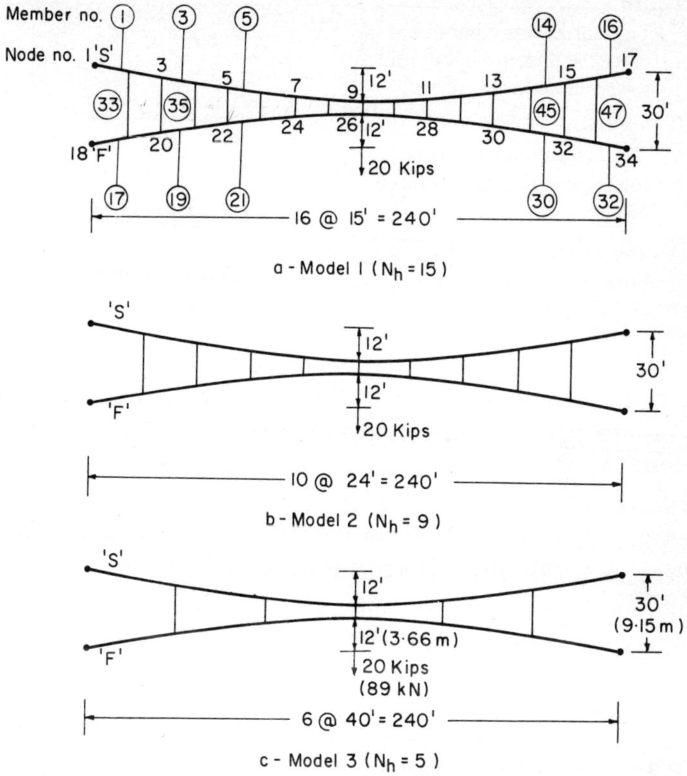

FIGURE 3-4 Analytical models of a plane cable truss to study the effect of varying number of hangers (Example 3-3). (Node and member numbers have relevance to Sec. 3-4.1.)

of this simplification, but experience with the analysis of cable systems has shown [42, 70] that these errors are within reasonable limits. Example 3-3 illustrates this point.

EXAMPLE 3-3

The effect of grouping members to reduce the number of nodes in the analytical model and thus to reduce the computation work is studied by analyzing the three systems shown in Fig. 3-4. The 15 hangers in model 1 are grouped into 9 hangers, in terms of both pretension and stiffness, for model 2. These are further grouped into 5 hangers for model 3. Some of the data for the three systems are as follows:

The horizontal component of pretension in the upper and lower cable for all three models	128 kips (570 kN)
The extensional stiffness *EA* for upper cable for all three models	66,000 kips (293.7 MN)
The extensional stiffness *EA* for lower cable for all three models	44,000 kips (195.8 MN)

The force of pretension is 3.2 kips (14.23 kN), 5.12 kips (22.8 kN), and 8.54 kips (38 kN), and the extensional stiffness is 4000 kips (17.79 MN), 6400 kips (28.47 MN), and 10,600 kips (47.15 MN), in each of the hangers for models 1, 2, and 3, respectively.

The main results for the three models are compared in Table 3-4. It will be seen that the differences are small, whereas the saving in computation time is considerable. It may be noted, too, that the number of equations of equilibrium necessary are 60, 36, and 20, respectively, for the three models.

TABLE 3-4 Results for Example 3-3

	Model 1	*Model 2*	*Model 3*
Maximum downward deflection, ft (m)	1.552 (0.474)	1.534 (0.468)	1.493 (0.456)
Maximum upward deflection, ft (m)	−0.118 (−0.036)	−0.109 (−0.033)	−0.104 (−0.032)
Tension change in segment marked “S”, kips (kN)	46.989 (209)	47.048 (209.4)	47.208 (210)
Tension change in segment marked “F”, kips (kN)	− 17.171 (−76.4)	− 17.457 (−77.6)	− 18.233 (−81.0)

3-2.4 Temperature Effects

The effects of temperature rise or fall in a cable system can be incorporated into the equilibrium equations quite easily. If the temperature rises above normal, it will tend to relieve the strain due to prestress, and vice versa. The force of tension F_i, in a typical element i, will therefore be modified as

$$F_i = F_i - EA_i \alpha t$$

or

$$F_i = F_i - F_i^t$$

where α = coefficient of linear expansion
t = change in temperature
F_i^t = change in force due to temperature in a typical element i

As a result, Eq. 3-9a will become

$$\sum^{i} \left[\frac{F_i - F_i^t}{s_i} (u_p - u_q) - \frac{F_i^t}{s_i} (x_p - x_q) + \frac{EA_i - F_i + F_i^t}{s_i} (x_p - x_q) a_i \right]$$
$$= -P_{x_q} + R_{x_q} \qquad \text{(3-12)}$$

where $R_{x_q} = -\sum^{i} \dfrac{(EA_i - F_i + F_i^t)}{s_i} \left[(u_p - u_q)(a_i + \frac{1}{2} b_i - \frac{3}{2} a_i^2) \right.$

$$\left. + (x_p - x_q)(\tfrac{1}{2} b_i - \tfrac{3}{2} a_i^2 - \tfrac{3}{2} a_i b_i + \tfrac{5}{2} a_i^3) \right]$$

Similar changes will occur in Eqs. 3-9b and 3-9c.

3-2.5 Effect of Slackness in Cable Elements

As discussed earlier, the cable elements cannot be expected to take any compressive forces. The structure will usually be pretensioned adequately so that the net force in any of the cables does not become negative as a result of normal loading on the structure. Despite this, if under some adverse loading condition one or more members become slack, the occurrence should be appropriately taken into account in the analysis. If it is noticed at the end of a particular cycle of iteration in the analysis that the net force in a member is compressive, that is, $\Delta F_i < -F_i$, then in subsequent cycles of computations the existence of this member can be ignored, by taking

$$EA_i = 0$$

and

$$\Delta F_i = -F_i$$

It will be obvious that numerical schemes 2 and 3 in Sec. 3-3 are better suited to take into account this physical change in the structure. Example 3-4 illustrates the order of errors that can occur if the above corrective measures are not used for slack members.

EXAMPLE 3-4

If a member or part of a cable structure becomes slack, this has to be considered in the analysis as indicated in the foregoing portion of the text. This example shows the errors that may occur if the correction is ignored. When the structure shown in Fig. 3-4*a* bears a load of 20 kips (89 kN) at the quarter point of the upper cable, the fourth hanger from the left will become slack. *The analysis of the problem is made here, for comparison in two ways,*

Case I: By incorporating the necessary correction

Case II: By ignoring the correction required

The main results obtained for the two cases are compared in Table 3-5. It will be seen that, if the procedure in case II is used, the force obtained in the fourth hanger is compressive, which is inadmissible, and that the upper-cable deflection is underestimated by 7.98 percent.

3-3 NUMERICAL PROCEDURES

It was established in Sec. 3-2 that the governing equations of equilibrium for a cable system are not amenable to a direct solution because of nonlinearity. Numerical iterative schemes therefore have to be used and some of the possible schemes are given herein.

TABLE 3-5 Results for Example 3-4

	Case I	*Case II*
Deflection of upper cable under load, ft (m)	1.766 (0.538)	1.625 (0.496)
Deflection of lower cable under load, ft (m)	1.261 (0.384)	1.595 (0.486)
Final force in hanger under load, kips (kN)	0.0	– 4.577 (–20.35)

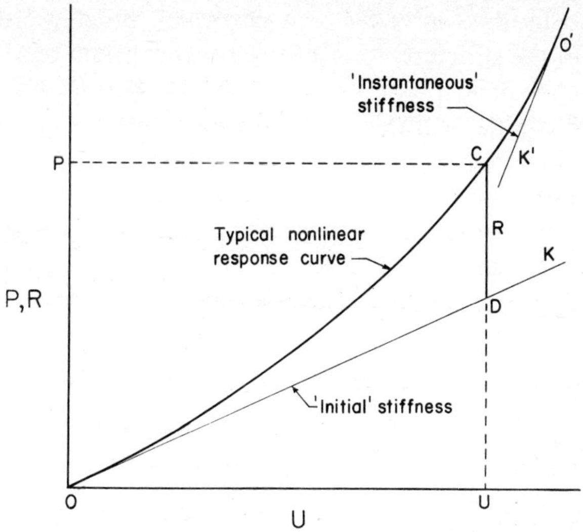

FIGURE 3-5 Typical nonlinear response curve of a cable.

The basic steps in any iterative scheme will be as follows:

1. Assume the value of **U**.[7]
2. Compute **R** on the basis of step 1.
3. In order to get a better estimate for **U**, solve

$$\mathbf{K} \cdot \mathbf{U} = -\mathbf{P} + \mathbf{R} \tag{3-13}$$

4. Obtain a more correct value of **R** on the basis of values obtained in step 3.
5. Repeat steps 3 and 4 until the desired accuracy is achieved, i.e., the values of **U** or **R** in two subsequent cycles agree with sufficient accuracy.

The variations possible are in the manner of setting up and solving Eq. 3-13 in each cycle of iteration. These possibilities can be discussed with the aid of simple graphical representations, without going into rigors of their mathematics. The load-deflection characteristics of the structure will be represented by nonlinear curves in multidirectional space. One such curve is shown in Fig. 3-5 (and also in Figs. 3-6 to 3-8, illustrating the numerical schemes). It is well known that

[7] **K**, **U**, **P**, and **R** are matrices, as defined earlier.

the tangent to this curve at any particular point represents the stiffness **K** of the structure, for the state of equilibrium represented by that particular point. For example, the tangent at the origin 0 represents the initial stiffness of the system, while the tangent at any other point **K**′ is representative of the *instantaneous* stiffness for the state of equilibrium at 0′. For any applied load **P** the correct value of deflection is **U** and the corresponding point that we seek on the curve is *C*. The intercept *CD* between the tangent at the origin and the curve represents the value of nonlinear terms contained in **R**.

3-3.1 Scheme 1: A Modified Newton-Raphson Method

A solution with this method will lead to a first-order convergence to the final results. A graphical representation of the steps in this method is given in Fig. 3-7. The steps involved are as follows:

1. Assume $\mathbf{U}^{(0)} = \mathbf{R}^{(0)} = 0$.
2. Solve $\mathbf{K} \cdot \mathbf{U}^{(1)} = -\mathbf{P}$ to get the value of $\mathbf{U}^{(1)}$.
3. Make appropriate substitution of $\mathbf{U}^{(1)}$ to compute $\mathbf{R}^{(1)}$.
4. Solve $\mathbf{K} \cdot \mathbf{U}^{(2)} = -\mathbf{P} + \mathbf{R}^{(1)}$ to obtain a better estimate of **U**, that is, $\mathbf{U}^{(2)}$.
5. Repeat steps 3 and 4 until the desired accuracy, which has to be specified in any particular problem, is achieved.

Computations in this method are carried out on the basis of the initial stiffness of the structure; i.e., the stiffness matrix for a particular structure will be computed once only and all subsequent calculations will be based on that. This is an advantage of the scheme over schemes 2 and 3, described later. There are, however, certain disadvantages too: any changes, elastic or physical, that may occur in the structure after loading, cannot be incorporated into the stiffness matrix (see Sec. 3-2.1).

It will be noted from Fig. 3-6 that the values of **U** and **R** are overestimated and underestimated in alternate cycles of iteration and, as such, keep oscillating about their correct values. The convergence is therefore slow, and the procedure cannot be considered suitable for large degrees of nonlinearity. It is also possible that for larger applied loads $\mathbf{R}^{(1)} > \mathbf{P}$; therefore, $\mathbf{P} - \mathbf{R}^{(1)}$ will be negative and will

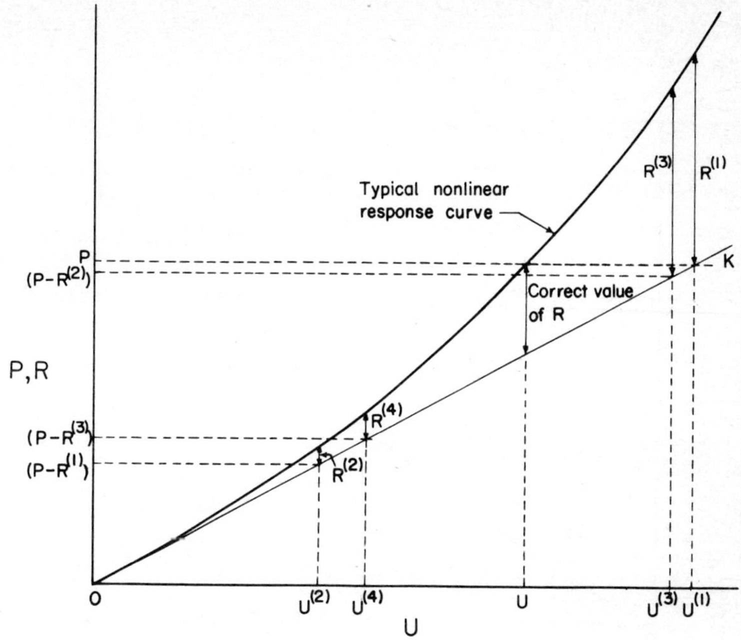

FIGURE 3-6 Diagrammatic representation of numerical scheme 1.

lead to a negative value of $\mathbf{R}^{(2)}$. The value of the right-hand side in Eq. 3-11 for the subsequent cycle will be greater than **P**, and $\mathbf{R}^{(3)} > \mathbf{R}^{(1)}$. It is easy to see that divergence will occur in such cases. A modification of scheme 1 has been suggested to overcome the problem of divergence or slow convergence [42, 44, 50, 70], and has been shown to work satisfactorily for many problems [41]. According to this modification, after the first two cycles of computation are completed, the value of **U** in a particular cycle is assumed as the mean of the values obtained in the previous two cycles.

3-3.2 Scheme 2: The Newton-Raphson Method

The procedure outlined here is the Newton-Raphson method, and it is fundamentally different from the previous scheme. The computations in this method are based on the *instantaneous* stiffness of the structure, derived anew in each iterative cycle, unlike the computations in scheme 1, which are concerned with the initial stiffness only. A much faster convergence rate can be obtained through the use of this method even for problems where scheme 1 may have failed

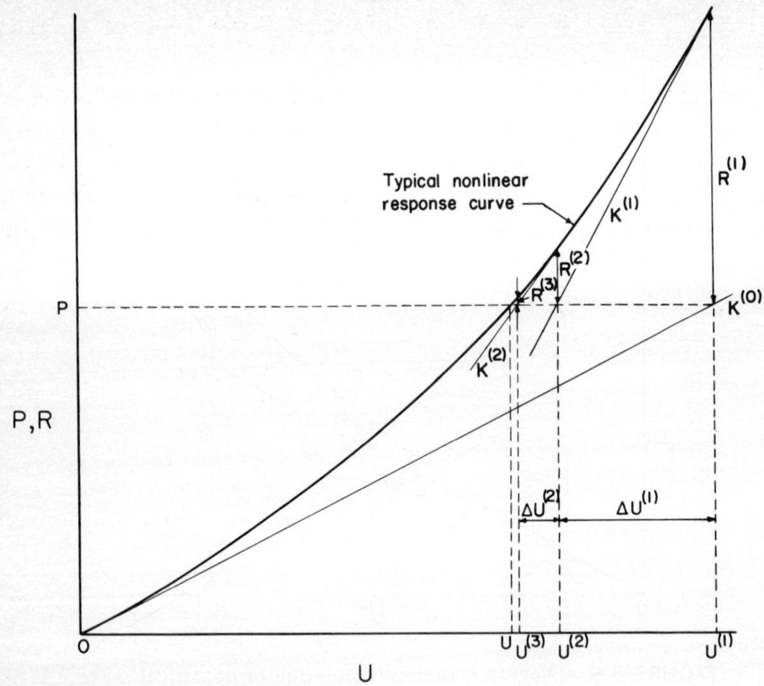

FIGURE 3-7 Diagrammatic representation of numerical scheme 2: Newton-Raphson method.

entirely. The method, however, entails greater computational effort because the stiffness matrix has to be computed in every cycle of iteration. On the other hand, any changes in the elastic or physical properties of the structure, in whatever stage of computation they may occur, can be easily accounted for in the stiffness matrix. The steps involved in this scheme are given below and are represented graphically in Fig. 3-7.

1. Assume $\mathbf{U}^{(0)} = \mathbf{R}^{(0)} = 0$.
2. Solve $\mathbf{K}^{(0)} \cdot \mathbf{U}^{(1)} = -\mathbf{P}$ to evaluate $\mathbf{U}^{(1)}$.
3. With the aid of the obtained value of $\mathbf{U}^{(1)}$ compute $\mathbf{R}^{(1)}$.
4. Solve $\mathbf{K}^{(1)} \cdot \Delta\mathbf{U}^{(1)} = \mathbf{R}^{(1)}$ to obtain $\Delta\mathbf{U}^{(1)}$, which is the correction to $\mathbf{U}^{(1)}$. Thus evaluate $\mathbf{U}^{(2)} = \mathbf{U}^{(1)} + \Delta\mathbf{U}^{(1)}$, and $\mathbf{R}^{(2)}$.
5. Solve $\mathbf{K}^{(2)} \cdot \Delta\mathbf{U}^{(2)} = \mathbf{R}^{(2)}$, and so on.

This procedure is carried on until the values of $\Delta\mathbf{U}$ or $\mathbf{R}$ become smaller than a specified limit.

Illustrated below is the manner in which the instantaneous stiffness matrix can be generated, say, for example, for the νth cycle of iteration. This is done with reference to Eqs. 3-9: the coefficients in the left-hand side of these equations constitute the stiffness matrix. The left-hand side of Eq. 3-9*a* will become, for the νth cycle of iteration,

$$\sum^{i}\left[\frac{F_i^{(\nu-1)}}{s_i^{(\nu-1)}}\left(u_p^{(\nu)}-u_q^{(\nu)}\right)+\left(EA_i^{(\nu-1)}-F_i^{(\nu-1)}\right)\frac{x_p^{(\nu-1)}-x_q^{(\nu-1)}}{s_i^{(\nu-1)}}\,a_i^{(\nu)}\right] \tag{3-14}$$

where

$$a_i^{(\nu)}=\frac{1}{s_i^{(\nu-1)^2}}\Big[\left(x_p^{(\nu-1)}-x_q^{(\nu-1)}\right)\left(u_p^{(\nu)}-u_q^{(\nu)}\right)$$
$$+\left(y_p^{(\nu-1)}-y_q^{(\nu-1)}\right)\left(v_p^{(\nu)}-v_q^{(\nu)}\right)$$
$$+\left(z_p^{(\nu-1)}-z_q^{(\nu-1)}\right)\left(w_p^{(\nu)}-w_q^{(\nu)}\right)\Big]$$

The exponent $(\nu-1)$ relates to values obtained at the end of the $(\nu-1)$th cycle and the exponent (ν) for the unknowns u, v, w relates to their values to be obtained from the νth cycle of iteration.

3-3.3 Scheme 3: Incremental Loading

Most practical problems for cable-roof systems expected to be met in practice should be satisfactorily handled by scheme 2. In case it does not suffice, a further improvement is possible by applying the load in several steps instead of a single step. A diagrammatic explanation of the procedure is given in Fig. 3-8*a*. Whereas $\mathbf{K}^{(0)}$ represents the initial stiffness matrix, $\mathbf{K}^{(1)}$, $\mathbf{K}^{(2)}$, etc., represent the instantaneous stiffness matrices computed after each step of loading. It will be seen from the figure that in each step of loading there is a certain residual error (e_1, e_2, e_3, etc., in Fig. 3-8*a*). If the step length is small, this error will be small but it will nevertheless exist. If the step length is large, the error will also be large. This can, however, be remedied by incorporating scheme 2 into each step of this method (see Fig. 3-8*b*). The method then has all the advantages of scheme 2

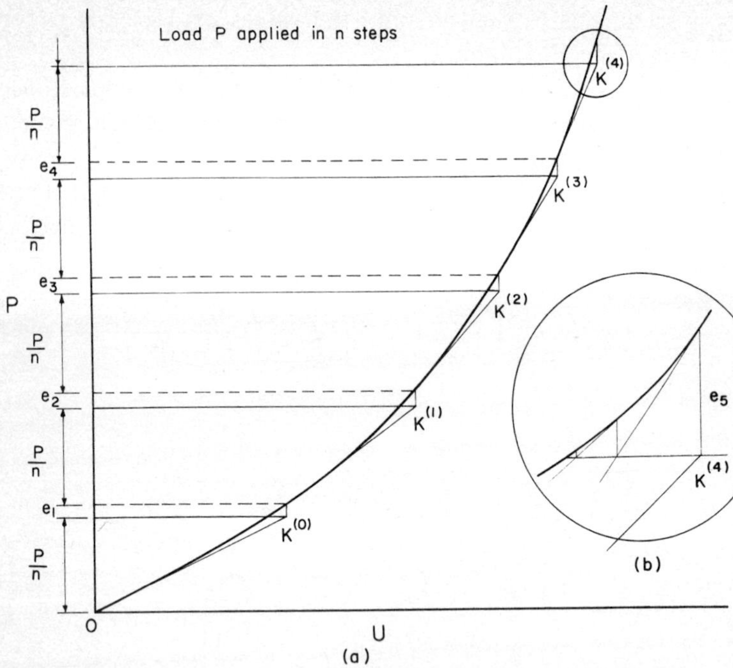

FIGURE 3-8 Diagrammatic representation of numerical scheme 3: incremental loading.

and has an additional advantage in that there is no limit to the degree of nonlinearity that it can handle. There is one problem: to decide the optimum load increment in order to achieve maximum accuracy with as few cycles of computation as possible. This can only be found out from experience gained by the application of this method to several types of problems.

The class of analysis problems in which creep is to be considered is handled more conveniently by the incremental procedure. Creep occurs in cable structures on account of sustained stress due to permanent loads and is classified as *stationary*. Occurrence of fire leads to high temperatures and provokes creep as well. In either instance this is a time-dependent phenomenon. The problem is best handled by giving small increments in time and treating the system as piecewise linear[8] for each step. It may be noted that in the case of fire the temperature rise can be expressed as a function of time; and

[8] See Ref. 103, p. 251.

associated with the temperature at any instant, strains and stresses in cable elements are determined, taking into account

1. Redistribution of stresses on account of temperature strains, which is essentially accompanied by high-temperature creep
2. The creep rate
3. The appropriate stress-strain law

It is quite possible that one or more of the cable elements may slacken because of high-temperature creep, and the discussion under Sec. 3-2.5 has relevance in the context of such problems.

Termination Procedure. In using any of the described numerical schemes, computation will be terminated when the desired degree of accuracy has been achieved. Suppose the check is being carried out in the νth cycle, then the computations can be stopped if the conditions given below are satisfied.

$$\left| \frac{u_q^{(\nu)} - u_q^{(\nu-1)}}{u_q^{(\nu)}} \right| < \nabla \tag{3-15}$$

or

$$\left| \frac{R_{x_q}^{(\nu)}}{P_{x_q}} \right| < \nabla \tag{3-16}$$

where $u_q^{(\nu)}$ = deflection in the νth cycle
$u_q^{(\nu-1)}$ = deflection in the $(\nu - 1)$th cycle
$R_{x_q}^{(\nu)}$ = residual force in the x direction in the νth cycle
P_{x_q} = applied load in the x direction, at a typical node q
∇ = specified accuracy

Usually there will be a large number of variables, both deflections or forces, on which the above checks can be made. It may not be necessary, though, to check each of them, and a check on the residuals (such as indicated in Eq. 3-16), corresponding to the larger deflection values, could be considered quite adequate.

3-4 RESPONSE VS. DESIGN PARAMETERS

In this section, three pretensioned cable systems, typically representative of a large majority of cable-roof structures, are studied, in order to gain a better insight into their structural response when loaded. One example each of three different types of structures is chosen for a detailed presentation of results and their analysis. Also given are curves representing the relationship between structural response and design parameters in a nondimensional form. The examples are

1. A dual-cable counterstressed system with vertical hangers
2. A cable truss
3. A hyperbolic paraboloid network

All three structures have a maximum cable span of 240 ft (73.2 m) and have been analyzed for two types of loading:

a. Unit vertical load of 1 kip (4.448 kN) applied consecutively at different node points to obtain a general response pattern

b. Different dispositions of a uniformly distributed vertical load

Combinations of uniformly distributed load, used in this analysis are based on the following values:

Dead load = 10 lb/ft^2 (480 N/m^2) of projected area

Snow load = 15 lb/ft^2 (720 N/m^2) of projected area

Wind load = 10 lb/ft^2 of projected area

and the loading is reduced to a system of equivalent concentrated loads applied at node points. Structures 1 and 2 above are assumed to be spaced at 20 ft (6.1 m). The analysis for all cases is done by using the stiffness method explained in Sec. 3-2 with an accuracy of 1 percent or better in obtained deflections.

3-4.1 The Dual-Cable System

The structure, as shown in Fig. 3-4*a* (Example 3-3, model 1), consists of two cables of reverse curvature, connected with uniformly spaced vertical hangers. The structure is analyzed for eight cases of loading. Cases I to IV consist of a single vertical load of 1 kip (4.448 kN) placed consecutively at nodes 3, 5, 7, and 9. The other four loading cases are tabulated in Table 3-6.

TABLE 3-6 Loading Cases V to VIII for Dual-Cable System

	Load, kips														
	Node Number*														
	2	3	4	5	6	7	8	9	10	11	12	13	14	15	16
Loading case V: Dead load only	3	3	3	3	3	3	3	3	3	3	3	3	3	3	3
Loading case VI: Dead load + snow load over full span	7.5	7.5	7.5	7.5	7.5	7.5	7.5	7.5	7.5	7.5	7.5	7.5	7.5	7.5	7.5
Loading case VII: Dead load + snow load over left half of span	7.5	7.5	7.5	7.5	7.5	7.5	7.5	5.25	3	3	3	3	3	3	3
Loading case VIII: Dead load + wind pressure on left half and suction over right half of span	6	6	6	6	6	6	6	6	0	0	0	0	0	0	0

* All other nodes have zero load.

1 kip = 4.448 kN

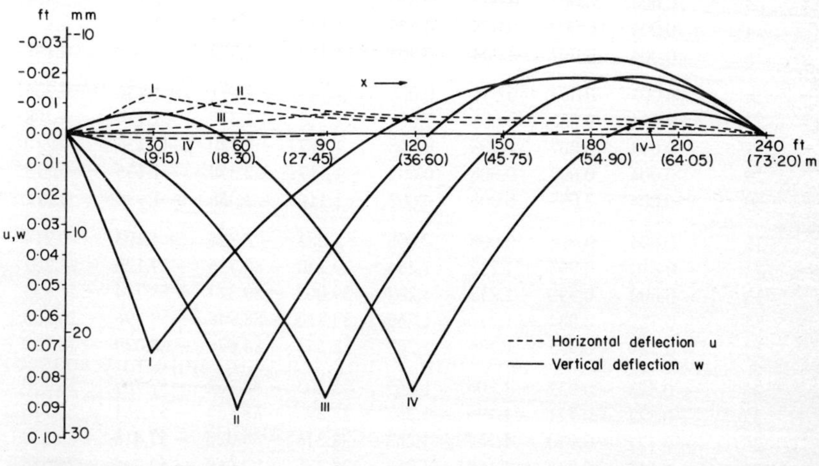

FIGURE 3-9 Deflections for the dual-cable system for loading cases I to IV (Sec. 3-4.1).

TABLE 3-7 Dual-Cable System

Member number	Change in member forces, kips							
	Loading case							
	I	II	III	IV	V	VI	VII	VIII
1	0.963	1.553	1.885	1.972	63.538	161.151	119.926	78.047
2	0.955	1.547	1.883	1.971	63.144	160.017	118.634	77.004
3	0.871	1.539	1.879	1.970	62.806	159.044	117.554	76.141
4	0.872	1.530	1.874	1.969	62.517	158.224	116.677	75.449
5	0.871	1.473	1.867	1.967	62.280	157.560	116.026	74.930
6	0.869	1.472	1.858	1.964	62.098	157.057	115.546	74.586
7	0.867	1.470	1.828	1.959	61.972	156.716	115.301	74.423
8	0.864	1.467	1.829	1.954	61.907	156.544	115.278	74.447
9	0.861	1.464	1.829	1.954	61.907	156.544	115.390	74.510
10	0.858	1.461	1.828	1.959	61.972	156.716	115.499	74.466
11	0.856	1.459	1.828	1.964	62.098	157.057	115.648	74.421
12	0.854	1.456	1.827	1.967	62.280	157.560	115.834	74.376
13	0.852	1.454	1.826	1.969	62.517	158.224	116.053	74.333
14	0.850	1.452	1.826	1.970	62.806	159.044	116.305	74.292
15	0.849	1.451	1.825	1.971	63.144	160.017	116.587	74.251
16	0.848	1.449	1.823	1.972	63.538	161.151	116.906	74.209
17	– 0.006	0.006	0.008	0.008	– 1.083	– 2.326	– 2.073	– 1.727
18	– 0.473	0.006	0.008	0.008	– 1.115	– 2.368	– 2.136	– 1.795
19	– 0.006	0.001	0.008	0.008	– 1.117	– 2.370	– 2.143	– 1.805
20	0.004	– 0.476	0.008	0.008	– 1.117	– 2.370	– 2.148	– 1.811
21	0.004	0.001	0.004	0.008	– 1.117	– 2.371	– 2.150	– 1.815
22	0.004	0.006	– 0.479	0.008	– 1.117	– 2.371	– 2.152	– 1.817
23	0.004	0.007	0.004	0.005	– 1.117	– 2.373	– 2.151	– 1.811
24	0.004	0.007	0.008	– 0.479	– 1.117	– 2.373	– 1.646	– 0.807
25	0.004	0.007	0.008	0.005	– 1.117	– 2.373	– 1.140	0.197
26	0.004	0.007	0.008	0.008	– 1.117	– 2.371	– 1.136	0.206
27	0.004	0.007	0.008	0.008	– 1.117	– 2.371	– 1.136	0.208
28	0.004	0.007	0.008	0.008	– 1.117	– 2.370	– 1.136	0.210
29	0.004	0.007	0.008	0.008	– 1.117	– 2.370	– 1.135	0.213
30	0.004	0.007	0.008	0.008	– 1.116	– 2.368	– 1.134	0.217
31	0.004	0.006	0.008	0.008	– 1.083	– 2.326	– 1.103	0.214
32	– 0.610	– 0.993	– 1.213	– 1.264	– 39.220	– 89.756	– 59.126	– 27.202
33	– 0.600	– 0.990	– 1.213	– 1.267	– 39.005	– 89.327	– 58.714	– 26.833
34	– 0.525	– 0.984	– 1.212	– 1.269	– 38.810	– 88.948	– 58.396	– 26.500
35	– 0.524	– 0.975	– 1.209	– 1.269	– 38.643	– 88.626	– 58.029	– 26.212
36	– 0.523	– 0.922	– 1.204	– 1.269	– 38.503	– 88.362	– 57.767	– 25.970
37	– 0.522	– 0.921	– 1.196	– 1.267	– 38.393	– 88.159	– 57.561	– 25.776
38	– 0.521	– 0.920	– 1.167	– 1.263	– 38.316	– 88.020	– 57.415	– 25.636
39	– 0.518	– 0.918	– 1.168	– 1.258	– 38.275	– 87.949	– 57.335	– 25.555
40	– 0.515	– 0.915	– 1.168	– 1.258	– 38.275	– 87.949	– 57.317	– 25.516

TABLE 3-7 (Continued)

Member number	Change in member forces, kips — Loading case I	II	III	IV	V	VI	VII	VIII
41	– 0.512	– 0.912	– 1.167	– 1.263	– 38.316	– 88.020	– 57.335	– 25.462
42	– 0.509	– 0.908	– 1.166	– 1.267	– 38.393	– 88.159	– 57.339	– 25.396
43	– 0.506	– 0.905	– 1.164	– 1.269	– 38.503	– 88.362	– 57.473	– 25.321
44	– 0.504	– 0.901	– 1.162	– 1.269	– 38.643	– 88.626	– 57.586	– 25.235
45	– 0.501	– 0.897	– 1.159	– 1.269	– 38.810	– 88.948	– 57.725	– 25.141
46	– 0.499	– 0.849	– 1.156	– 1.267	– 39.005	– 89.327	– 57.890	– 25.036
47	– 0.496	– 0.891	– 1.152	– 1.264	– 39.220	– 89.756	– 58.073	– 24.921

1 kip = 4.448 kN

The values of upper-cable deflection components along the *x* and *z* axes are plotted in Fig. 3-9 for cases I to IV, and in Fig. 3-10 for cases V to VIII. The values of change in member forces for all cases are given in Table 3-7. Whereas the structure has the appearance of a truss, it carries loads predominantly by cable action. The main features of the results are as follows:

1. It is seen that the tension change in the top (sagging) cable is much larger than that in the lower (hogging) cable, for both

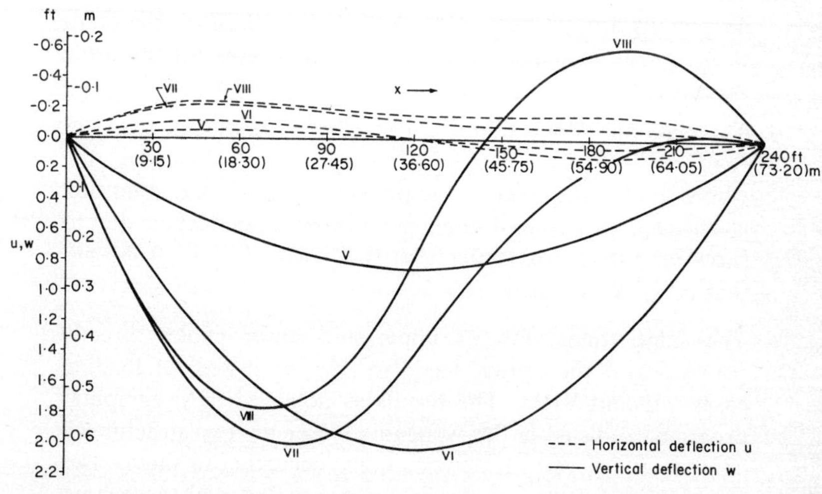

FIGURE 3-10 Deflections for the dual-cable system for loading cases V to VIII (Sec. 3-4.1).

cases of loading. The sagging cable is therefore rightly termed the *primary* cable, and the hogging cable is called the *secondary* cable. It is also seen that the tension change in a particular cable has little variation from segment to segment. Therefore, the design tension for the entire cable can be adopted as the same value without any appreciable loss in economy.

2. The suspender immediately above or below the load experiences a larger change in tension, while all the other suspenders undergo a much smaller and nearly equal change. Therefore, even if a large vertical load applied to the top cable causes the suspender below it to become slack, the two main cables will still continue to interact through the other suspenders. However, a different situation can occur when the system is loaded with a uniformly distributed load over the entire span. Under such a load, all suspenders may become slack and thus the lower cable may also become slack, in which case the structure will act as a single cable for any subsequent loading.
3. The horizontal deflections are much smaller in comparison to the vertical deflections, particularly for uniform loads. The maximum ratio of horizontal to vertical deflection occurs for case I and equals 0.17. This justifies the assumption often made in the analysis of cable systems, in which horizontal deflections are ignored. The deflection pattern of this dual-cable system is similar to that of a single cable for the same loading.
4. The maximum deflection of 2.15 ft (0.656 m), that is, 1/111 of span, occurs for the case of dead load + snow load over the whole span. However, if the prestress + dead load condition is assumed to be the datum, the additional deflection due to snow load (live-load deflection) is equal to 1.25 ft (0.381 m), that is, 1/193 of span.
5. The same truss, with its upper and lower cables directly connected at the center, has also been analyzed for loading cases VII and VIII. The results as obtained by a computer program are listed in the Appendix. For the two structures a comparison of vertical deformations occurring in these cases reveals that for the symmetric component of load there is no difference, while for the unsymmetric or antisymmetric component the deflections are reduced nearly to half for the truss

with connected cables. Horizontal deformations are also reduced considerably.

6. It has been shown [51] for this type of structure that the difference between lower- and upper-cable deflections at any vertical section for a particular case of loading is very small, whether the load is applied to the lower or the upper cable. This is also true for a cable truss [74] of the type treated in Sec. 3-4.2.

A study of the dual-cable system to determine its static response for variation in some important parameters now follows. The geometric, rigidity, or load parameters involved are shown in Fig. 3-11 and are defined below.

Upper-cable sag	$= f_u$
Lower-cable sag	$= f_l$
Extensional rigidity for upper cable	$= (EA)_u$
Extensional rigidity for lower cable	$= (EA)_l$
Extensional rigidity for verticals	$= (EA)_v$
Horizontal component of upper-cable pretension	$= H_u$
Dead load	$= q$ per unit length
Live load	$= p$ per unit length
Change in upper-cable tension	$= \Delta T_u$
Change in lower-cable tension	$= \Delta T_l$
Number of verticals	$= N_h$

Results of the study are given in a nondimensional form in Fig. 3-12*a* to *e* for the maximum negative and positive vertical deflection and force change in the cable, for loads applied to the upper cable.

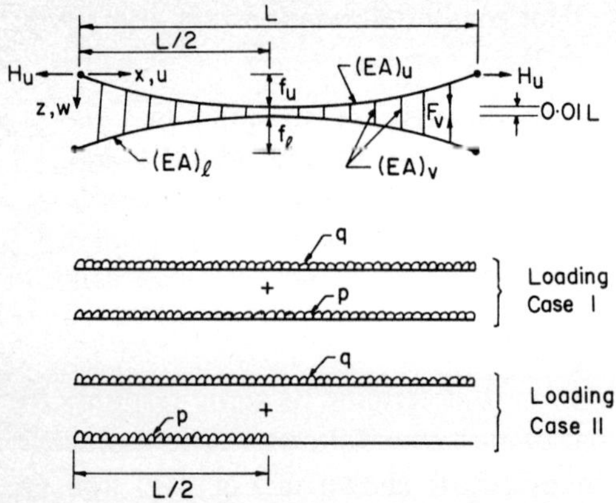

FIGURE 3-11 Analytical model and loading cases for the dual-cable system (see results in Fig. 3-12).

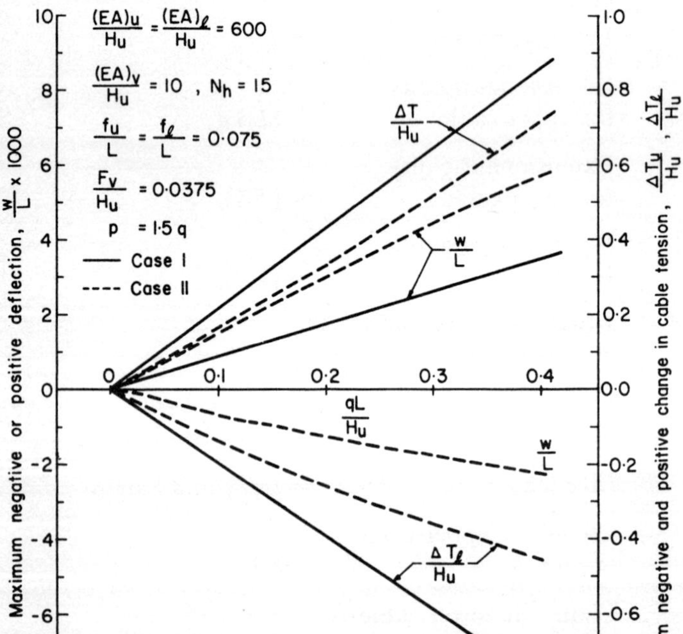

FIGURE 3-12*a* Response for dual-cable system (Fig. 3-11): effect of variation in load intensity.

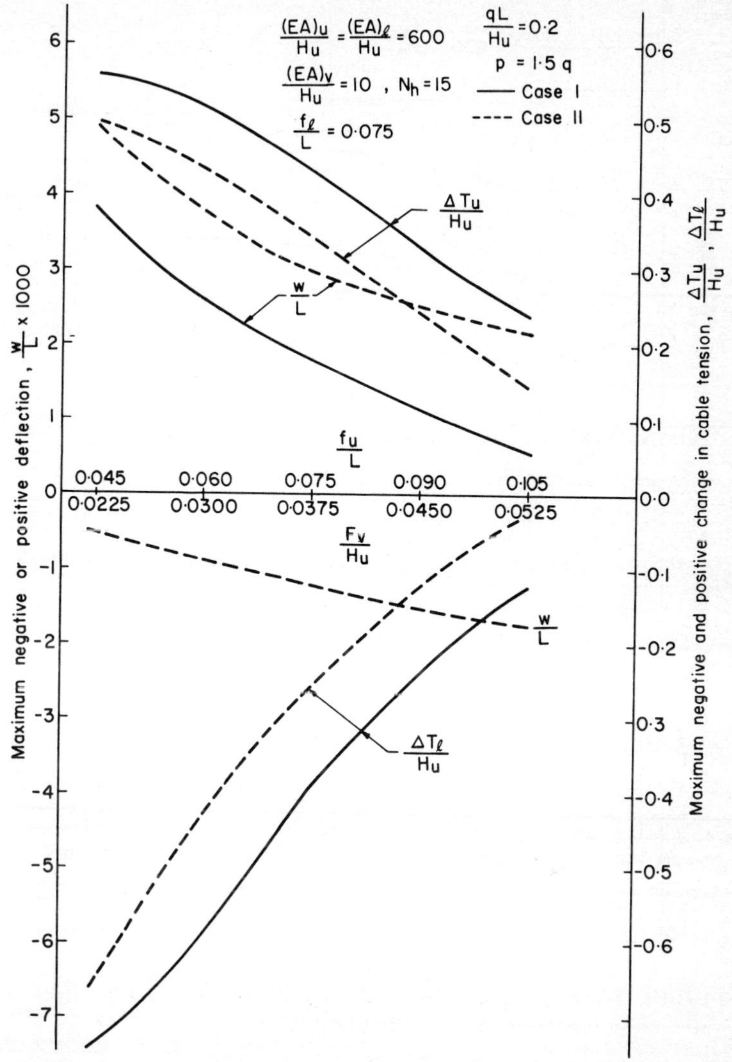

FIGURE 3-12*b* Response for dual-cable system (Fig. 3-11): effect of variation in upper-cable sag/span ratio.

3-4.2 Cable Truss

The cable truss analyzed is shown in Fig. 3-13. It differs from the dual-cable system in that the web members are inclined rather than vertical. The pretension in each web member is 8 kips (35.5 kN), and the horizontal component of top- and bottom-cable pretension is

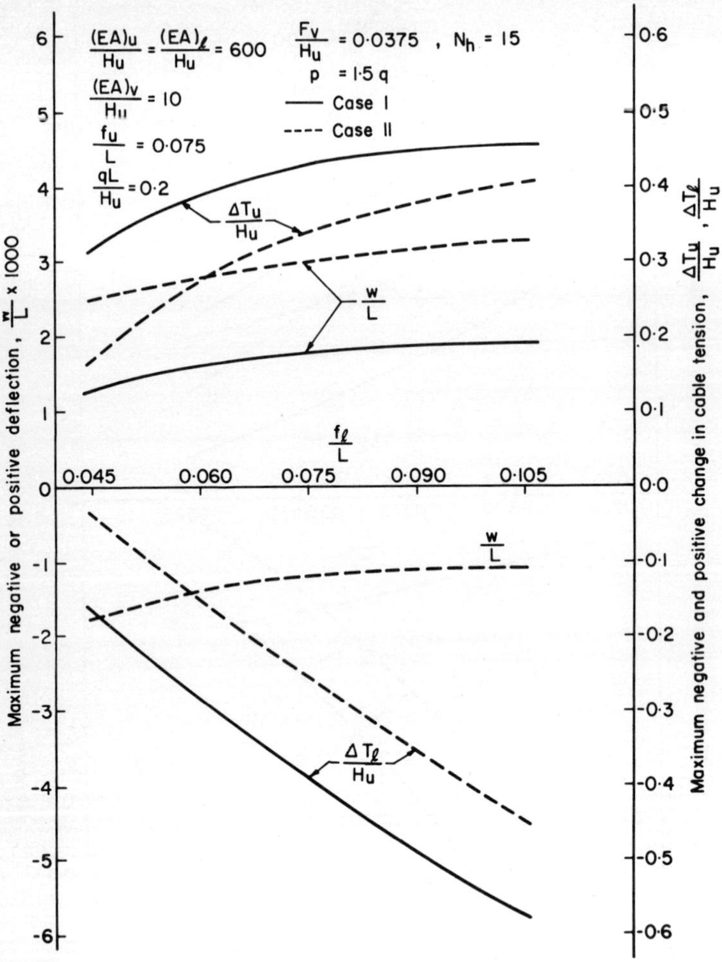

FIGURE 3-12c Response for dual-cable system (Fig. 3-11): effect of variation in the lower-cable sag/span ratio.

approximately 197 kips (876 kN). The cable pretension in this structure is therefore 54 percent greater than in the dual-cable system. The extensional stiffness of the top cable, bottom cable, and the web members is, respectively, 66,000 kips (293.6 MN), 44,000 kips (195.7 MN), and 4400 kips (19.6 MN). The structure is analyzed for eight cases of loading. Cases I to IV consist of a concentrated load of 1 kip (4.448 kN) each, placed at nodes 2 to 5 consecutively. The other four cases are tabulated in Table 3-8.

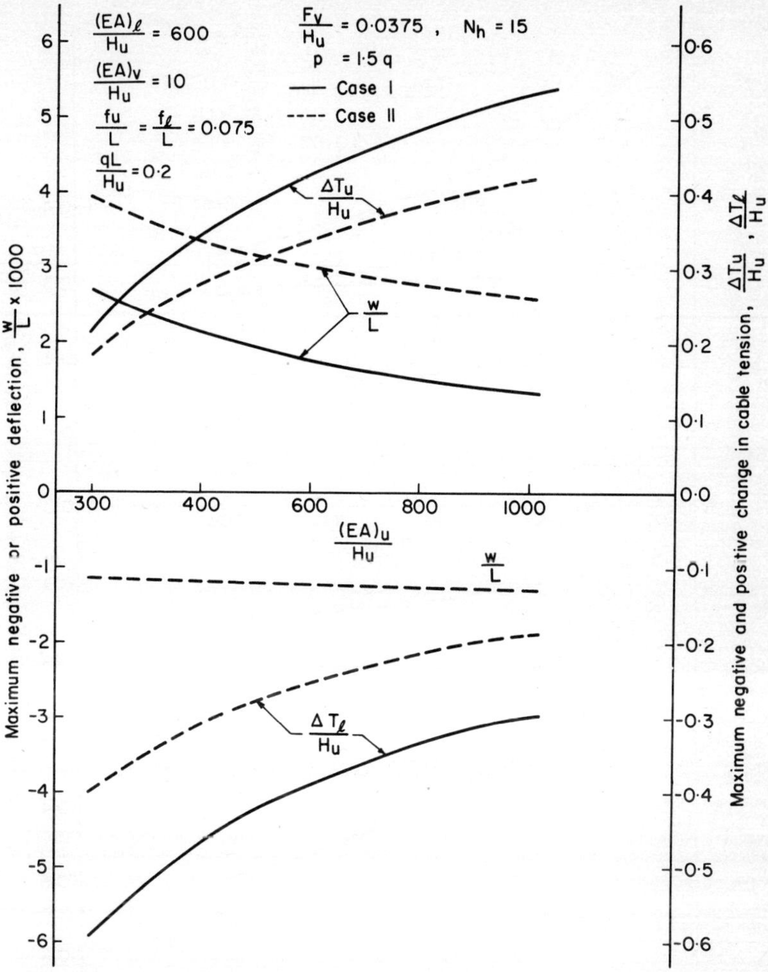

FIGURE 3-12*d* Response for dual-cable system (Fig. 3-11): effect of variation in the upper-cable rigidity.

The resulting deflections are plotted for cases I to IV in Fig. 3-14 and for cases V to VIII in Fig. 3-15. The change in member forces for all cases is tabulated in Table 3-9. The main features of the results are enumerated below.

1. It may be noted that the one web member immediately under each loaded node becomes slack, even though the degree of pretension in this structure has been purposely increased, over

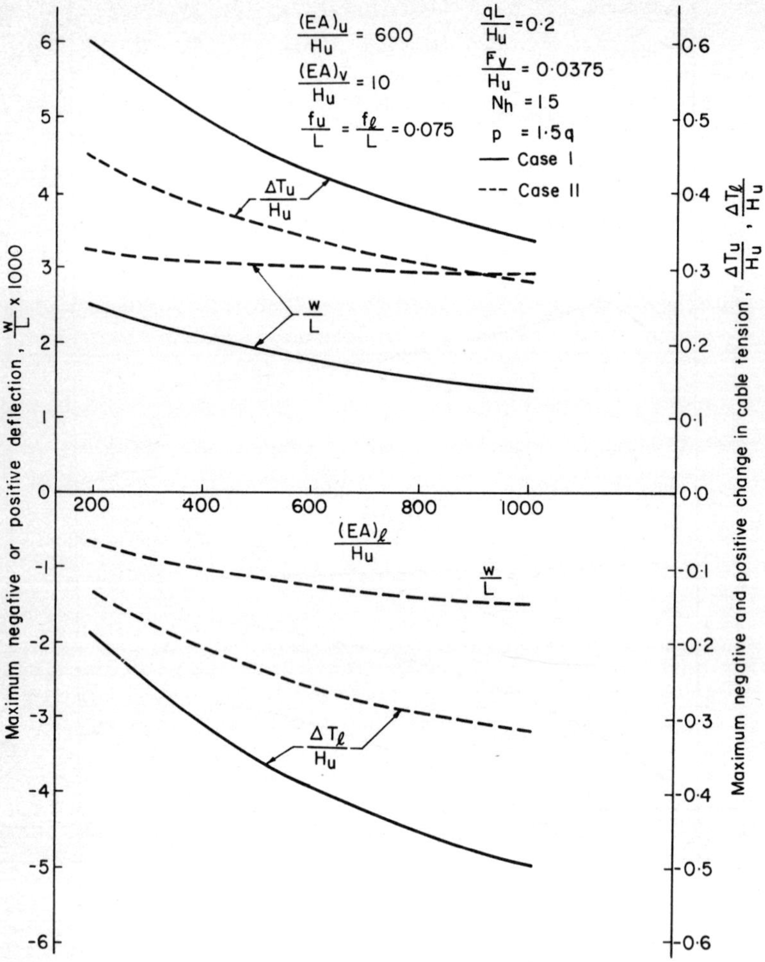

FIGURE 3-12*e* Response for dual-cable system (Fig. 3-11): effect of variation in the lower-cable rigidity.

that in the dual-cable system, to avoid slack members. Whereas this slackness cannot be said to have caused the failure of the truss, the occurrence of a large number of slack members is not desirable, as it leads to a decreased participation of the lower cable in sharing the load.

2. In the same manner as for the dual-cable system, the top cable can be said to be the *primary* cable in the truss.

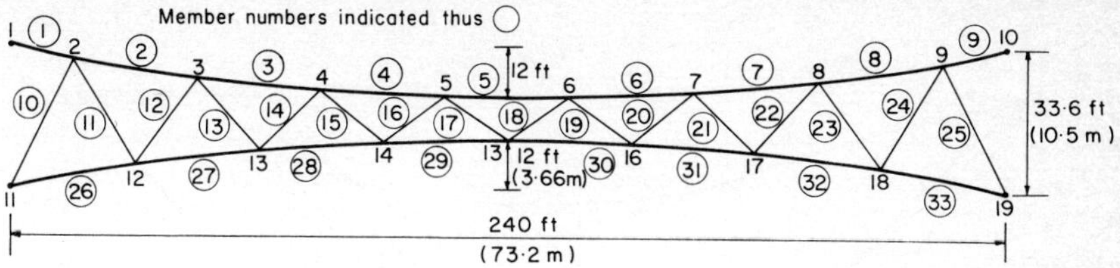

FIGURE 3-13 Cable truss studied in Sec. 3-4.2.

3. In the truss, generally, alternate diagonals show an increase or a decrease in tension. This is distinctly different from the pattern of tension changes in the hangers of the dual-cable system.

4. The maximum deflection of 1.78 ft (0.542 m), that is, span/135, occurs for the case of dead load + snow load. However, if the geometry at dead load is treated as datum, the additional deflection due to snow load only (see point 4, Sec. 3-4.1) is 1.29 ft (0.393 m), that is, span/186.

5. The maximum ratio of horizontal/vertical deflection occurs for loading case I and equals 0.21.

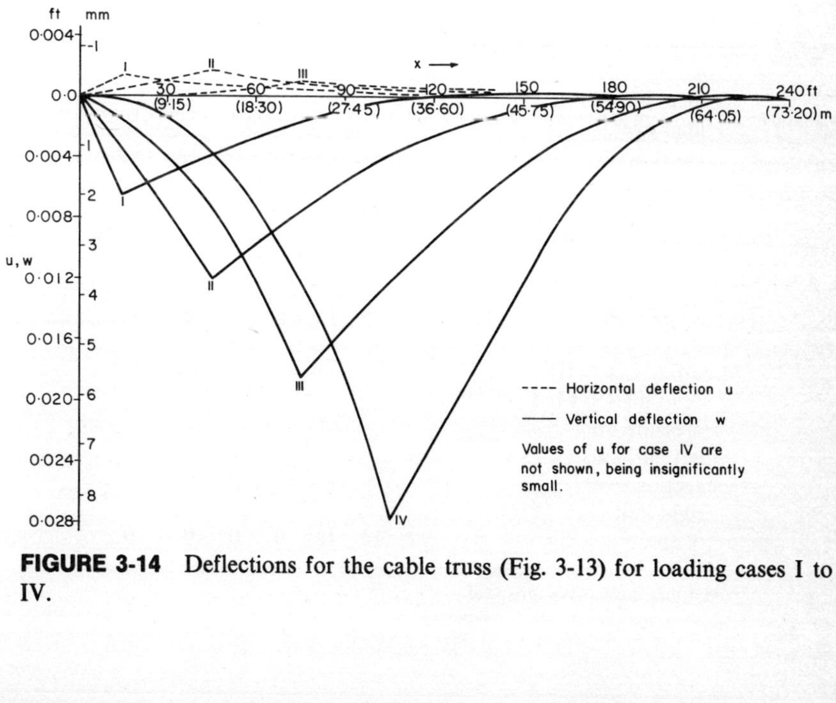

FIGURE 3-14 Deflections for the cable truss (Fig. 3-13) for loading cases I to IV.

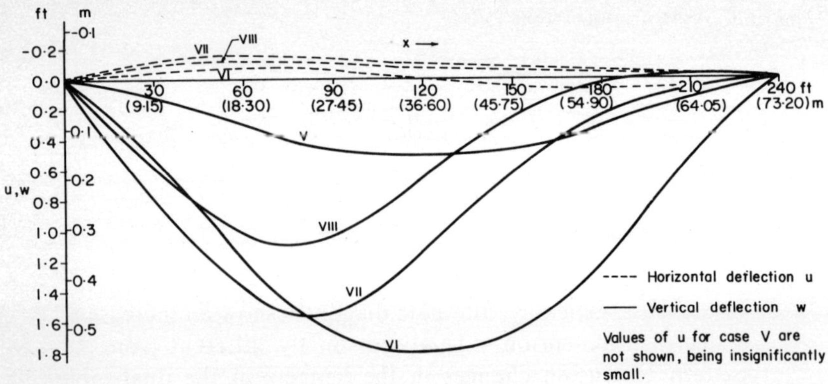

FIGURE 3-15 Deflections for the cable truss (Fig. 3-13) for loading cases V to VIII.

TABLE 3-8 Loading Cases V to VIII for Cable Truss

Loading cases	*Load, kips* — *Node number** 2	3	4	5	6	7	8	9
Loading case V: Dead load only	4.5	6	6	6	6	6	6	4.5
Loading case VI: dead load + snow load over entire span	11.25	15	15	15	15	15	15	11.25
Loading case VII: dead load + snow load over left half of span	11.25	15	15	15	6	6	6	4.5
Loading case VIII: dead load + wind pressure on left half and suction over right half of span	9	12	12	12	0	0	0	0

* All other nodes have zero load.

1 kip = 4.448 kN

TABLE 3-9 Cable Truss

Member number	Change in member forces, kips							
	Loading case							
	I	II	III	IV	V	VI	VII	VIII
1	0.642	1.344	1.757	1.673	48.209	147.627	104.952	73.616
2	0.247	0.819	1.495	1.599	42.386	140.018	96.965	66.142
3	0.253	0.336	0.708	1.214	33.899	131.565	88.901	58.263
4	0.238	0.352	0.083	0.234	23.987	122.409	80.874	50.307
5	0.196	0.352	0.305	0.110	18.089	119.200	79.407	49.182
6	0.142	0.333	0.516	0.625	23.987	122.409	86.284	54.921
7	0.100	0.303	0.631	1.091	33.899	131.565	96.287	56.459
8	0.076	0.274	0.650	1.241	42.386	140.018	102.287	54.401
9	0.064	0.249	0.622	1.236	48.209	147.627	104.007	50.960
10	− 0.827	− 0.462	− 0.187	− 0.006	− 6.181	− 8.000	− 8.000	− 8.000
11	− 0.071	0.628	0.335	0.117	5.004	4.386	4.684	4.859
12	0.011	− 0.911	− 0.494	− 0.212	− 8.000	− 8.000	− 8.000	− 8.000
13	− 0.002	− 0.069	0.741	0.380	5.354	3.242	2.717	3.215
14	− 0.002	− 0.016	− 1.050	− 0.506	− 8.000	− 8.000	− 8.000	− 8.000
15	0.017	− 0.006	− 0.178	0.740	4.689	2.257	1.620	1.894
16	− 0.020	0.008	0.133	− 0.964	− 7.573	− 8.000	− 8.000	− 8.000
17	0.030	0.008	− 0.133	− 0.559	− 0.514	− 4.482	− 6.231	− 6.605
18	− 0.030	− 0.005	0.140	0.458	− 0.514	− 4.482	1.054	4.414
19	0.035	0.019	− 0.110	− 0.417	− 7.573	− 8.000	− 8.000	− 2.392
20	− 0.028	− 0.015	0.098	0.363	4.689	2.257	5.302	2.237
21	0.027	0.025	− 0.046	− 0.225	− 8.000	− 8.000	− 8.000	0.584
22	− 0.019	− 0.020	0.030	0.161	5.354	3.242	4.286	− 0.766
23	0.019	0.030	0.011	− 0.061	− 8.000	− 8.000	4.368	3.287
24	− 0.012	− 0.024	− 0.018	0.017	5.004	4.386	1.486	− 2.937
25	0.015	0.035	0.052	0.056	− 6.181	− 8.000	0.231	5.447
26	− 0.114	− 0.887	− 1.320	− 1.217	− 26.147	− 52.375	− 36.882	− 30.113
27	− 0.141	− 0.076	− 0.880	− 1.037	− 19.152	− 45.442	− 29.785	− 23.018
28	− 0.141	− 0.128	0.390	− 0.410	− 9.600	− 37.000	− 21.635	− 14.628
29	− 0.111	− 0.140	0.134	1.011	− 0.698	− 27.766	− 13.318	− 6.143
30	− 0.061	− 0.130	− 0.094	0.159	− 0.698	− 27.766	− 19.402	− 15.374
31	− 0.009	− 0.101	− 0.268	− 0.491	− 9.600	− 37.000	− 29.677	− 19.231
32	0.024	− 0.070	− 0.322	− 0.766	− 19.152	− 45.442	− 38.018	− 18.312
33	0.041	− 0.042	− 0.309	− 0.813	− 26.147	− 52.375	− 41.451	− 15.168

1 kip = 4.448 kN

6. If deflections for the two systems are compared, it will be seen that for loading cases I to IV (in which very small loads are applied) the truss is distinctly stiffer than the dual-cable system. However, for the more practical cases (V to VIII) the difference between the two decreases. There are two things which may cause this change:

   a. Under the latter loading cases, many of the web members become slack, thus reducing the participation of the lower cable and causing a decrease in overall stiffness.

   b. Cable extensions become more important for these latter cases of loading.

   If the carrying cable is a flexural member, axial deformations would be comparatively small, and the difference in stiffness between the dual-cable system and the truss is likely to be greater [102].

3-4.3 Cable Networks

In the preceding sections, the response of plane cable systems was studied, and in this section, features of structural response of networks will be discussed. Cable networks as used in roofs are two- or multiway systems. Most of the networks are, however, two-way systems of negative gaussian curvature, and a typical example is a hyperbolic paraboloid. The mathematical model of a hyperbolic paraboloid studied in this context is shown in Fig. 3-16. The extensional stiffness for the sagging and hogging cables, respectively, is 66,000 kips (293.6 MN) and 44,000 kips (195.7 MN). The horizontal component of pretension in each cable is 180 kips (800 kN).

The structure is analyzed for 11 cases of loading. Cases I to VII deal with a 1-kip (4.448-kN) concentrated vertical load applied consecutively at nodes 1, 3, 7, 13, 2, 6, and 12. Cases VIII to XI are tabulated in Table 3-10.

The resulting deflections for cases I to VII are plotted in Fig. 3-17*a* to *c* and for cases VIII to XI in Fig. 3-18*a* to *c*. It may be noted that for clarity of the diagrams, very small deflection values are not shown. The change in member forces for all cases of loading is given in Table 3-11. The main features of the results are as follows:

1. The ratio of tension changes in the hogging and sagging cables is somewhat larger than for the two systems discussed

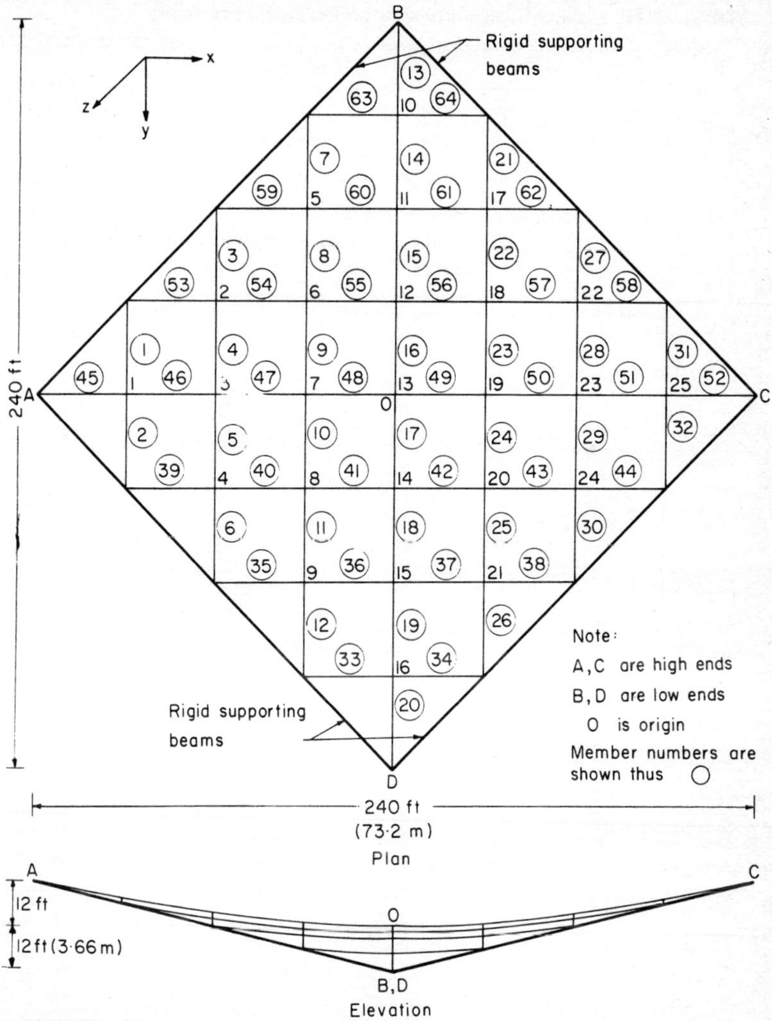

FIGURE 3-16 Hypar network studied in Sec. 3-4.3 (see results in Figs. 3-17 and 3-18).

already. This indicates that the sharing of load between the two sets of cables is better in the network.

2. Under the maximum loading (case IX) all cable segments have an adequate residual tension; no member becomes slack.
3. The maximum deflection of 2.75 ft (0.838 m), that is, span/88, occurs for loading case IX, which constitutes dead load plus

TABLE 3-10 Loads for the Hyperbolic Paraboloid Network

	Load, kips			
Node no.	Loading case VIII: Dead load only	Loading case IX: Dead load + snow load over whole surface	Loading case X: Dead load + snow load over ABD	Loading case XI: Dead load + wind pressure on ABD and wind suction over BCD
1	9	22.5	22.5	18
2	9	22.5	22.5	18
3	9	22.5	22.5	18
4	9	22.5	22.5	18
5	9	22.5	22.5	18
6	9	22.5	22.5	18
7	9	22.5	22.5	18
8	9	22.5	22.5	18
9	9	22.5	22.5	18
10	9	22.5	15.75	9
11	9	22.5	15.75	9
12	9	22.5	15.75	9
13	9	22.5	15.75	9
14	9	22.5	15.75	9
15	9	22.5	15.75	9
16	9	22.5	15.75	9
17	9	22.5	9	0
18	9	22.5	9	0
19	9	22.5	9	0
20	9	22.5	9	0
21	9	22.5	9	0
22	9	22.5	9	0
23	9	22.5	9	0
24	9	22.5	9	0
25	9	22.5	9	0

1 kip = 4.448 kN

snow load over the whole surface. For the dead-load geometry treated as datum, the additional deflection due to snow load is 1.60 ft (0.488 m), that is, span/150 (see point 4 in Sec. 3-4.1). The structure chosen would appear to be too flexible for the loading and may need to be stiffened for application in practice, depending upon the limitations specified on deformation.

4. The maximum ratio of horizontal/vertical deflection for cable *AC* occurs for case I. Its value is 0.168.

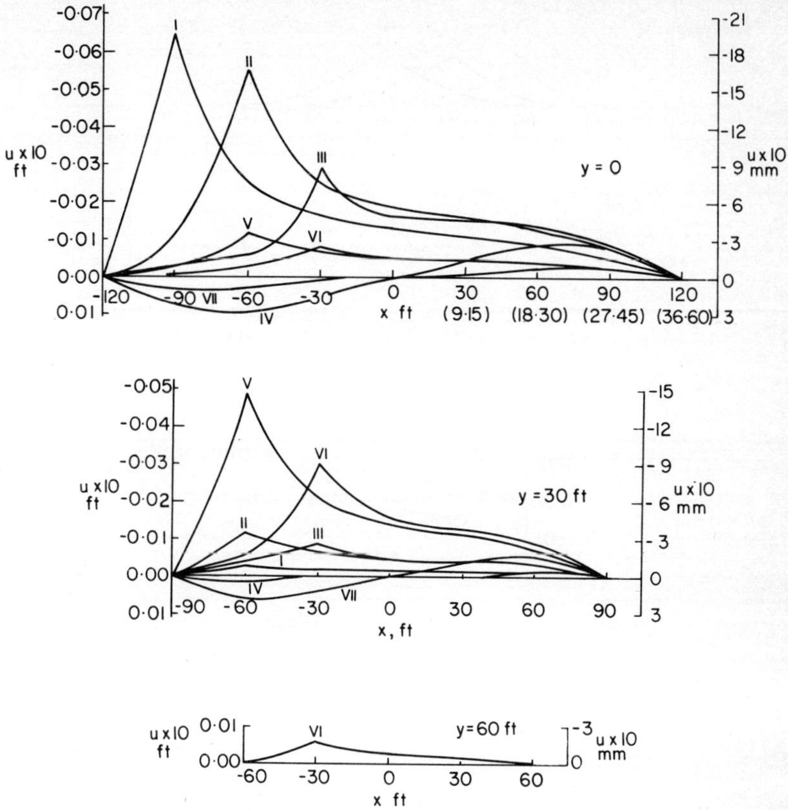

FIGURE 3-17*a* *u*-deflections for loading cases I to VII for the hypar network (Fig. 3-16).

In the same manner as described in Sec. 3-4.1 for a dual-cable system, a parametric study of static behavior of a hypar network has been carried out. The various parameters are shown in Fig. 3-19 and are defined below. The two loading cases are also shown in Fig. 3-19.

Diagonal span	$= L$
Central dip of both carrying and prestressing cables	$= f$
Extensional rigidity of all carrying cables	$= (EA)_s$
Extensional rigidity of all prestressing cables	$= (EA)_h$

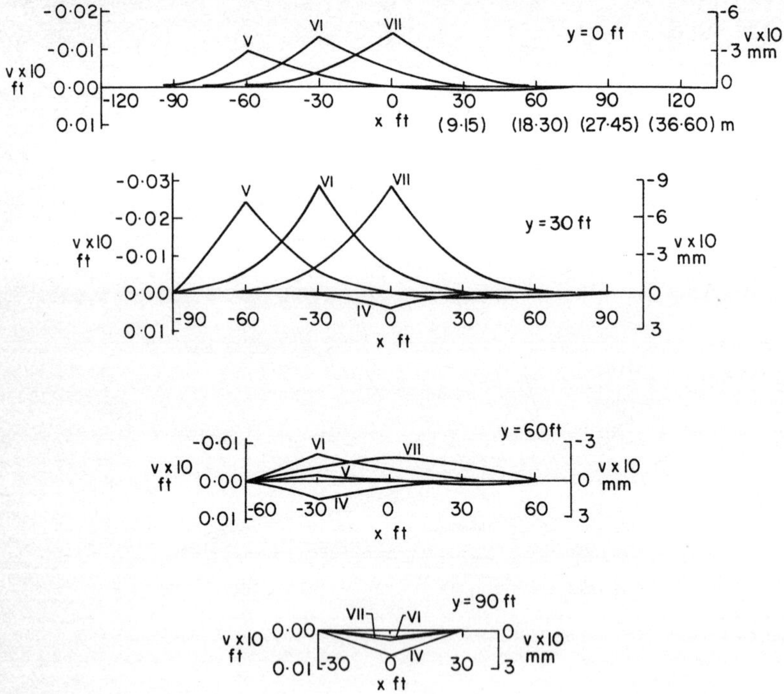

FIGURE 3-17*b* v-deflections for loading cases I to VII for the hypar network (Fig. 3-16).

Horizontal component of tension in each cable $= H$

Dead load $= q$ per unit projected area

Live load $= p$ per unit projected area

Change in tension $= \Delta T$

The nondimensional terms and the range of their variation may be seen in Fig. 3-20*a* to *d*, where the results for tension change and deflections are shown. The loading is reduced to nodal loads for purposes of analysis.

3-5 APPROXIMATE ANALYSIS

Modern developments in the rigorous analysis of cable systems are based on the discrete element formulation, leading either to a stiffness-matrix analysis or to the energy minimization approach, as discussed in Secs. 3-2 and 3-7. Continuum analysis has been used

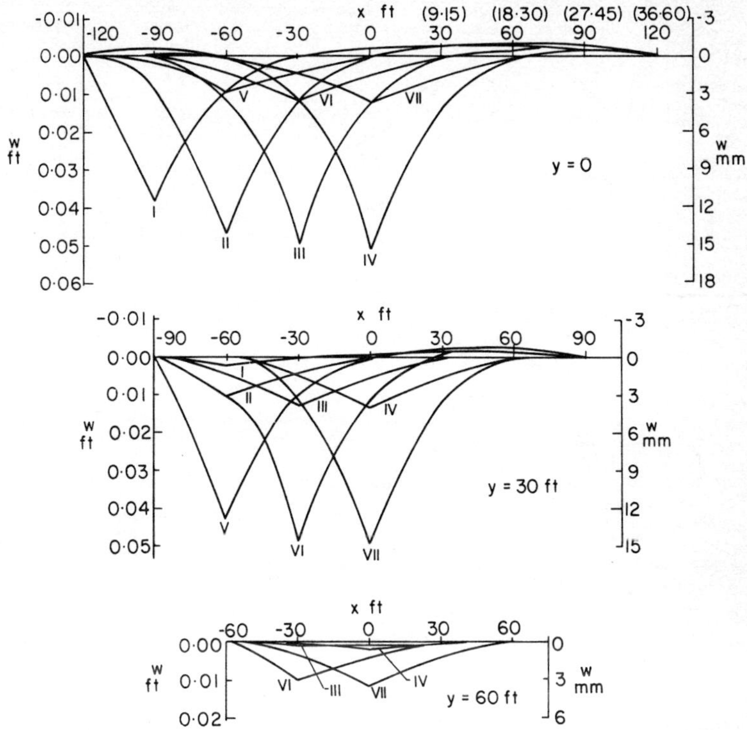

FIGURE 3-17*c* *w*-deflections for loading cases I to VII for the hypar network (Fig. 3-16).

for cable systems mainly for developing approximate methods. Whereas these approximate procedures admit several simplifying assumptions, their utility in producing closed-form formulas for speedy design office work must be recognized. Further, it will be seen that for the purpose of rigorous analysis of any form, values of design parameters are required. These can be estimated more easily with the help of the approximate analysis methods. Mollmann [66] and Schleyer [75] have given approximate solutions for certain cases of cable systems. By following similar procedures, formulas are derived in this section to cover a more comprehensive list of cases.

3-5.1 Plane Systems

For the purpose of this analysis, a plane system, such as those shown in Fig. 1-9, is considered as a single suspended cable of equivalent stiffness and the following assumptions are made [47]:

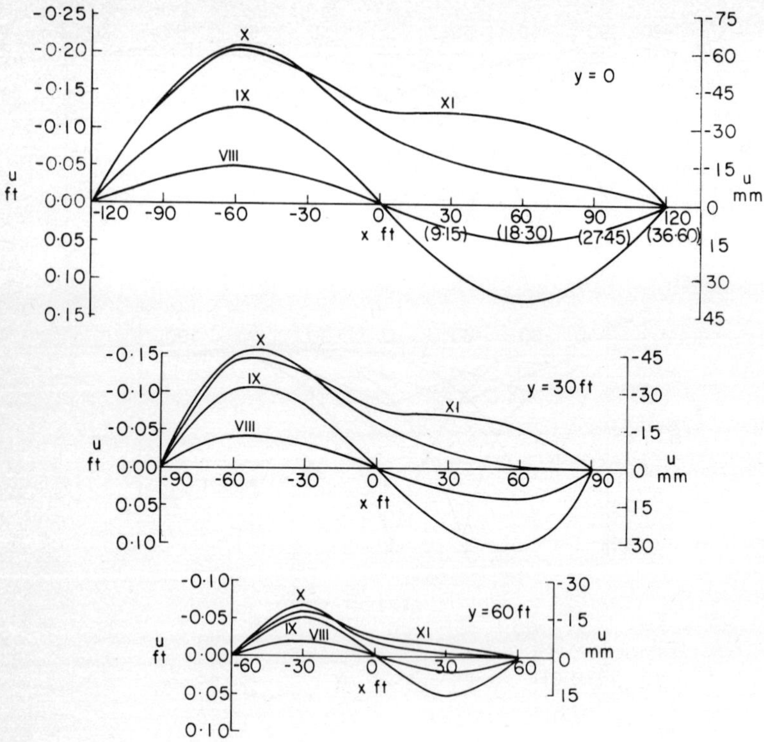

FIGURE 3-18*a* *u*-deflections for loading cases VIII to XI for the hypar network (Fig. 3-16).

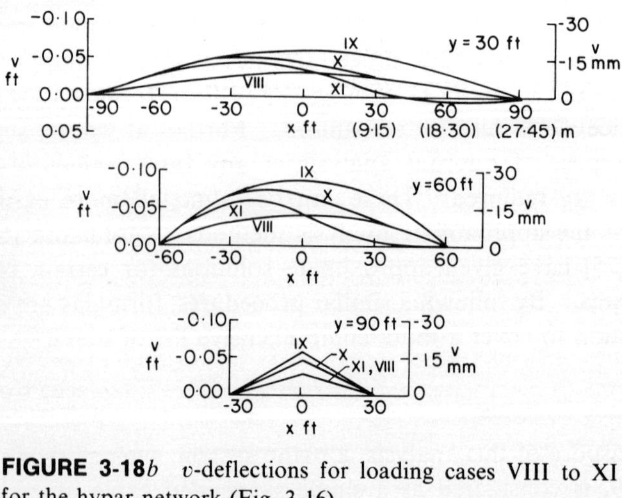

FIGURE 3-18*b* *v*-deflections for loading cases VIII to XI for the hypar network (Fig. 3-16).

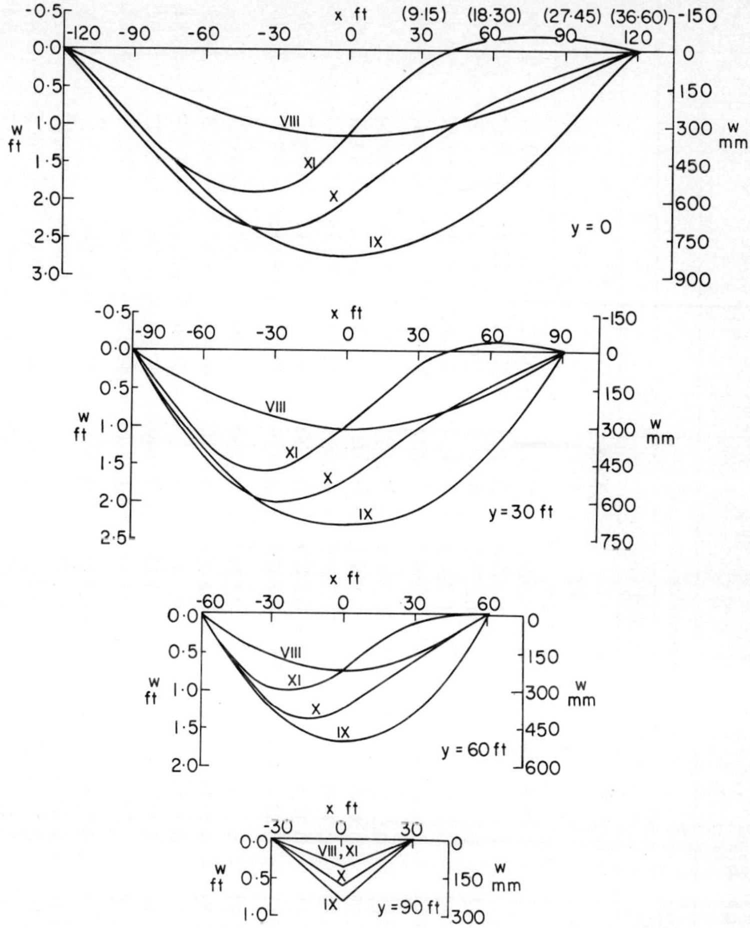

FIGURE 3-18*c* *w*-deflections for loading cases VIII to XI for the hypar network (Fig. 3-16).

1. Kinematic nonlinearity is neglected and the principle of superposition is admitted.
2. Horizontal deflections are neglected in comparison with vertical deflections.
3. The ties or struts connecting the main cables are replaced by a continuous membrane, and the prestress force is replaced by a continuous uniformly distributed load on the cable. This implies further that any change in the inclination of the web members on account of loading is neglected.

TABLE 3-11 Hyperbolic Paraboloid Network

Member number	Change in member force, kips										
	Loading case										
	I	II	III	IV	V	VI	VII	VIII	IX	X	XI
1	− 1.3895	− 0.3376	− 0.0347	0.0614	− 0.0707	− 0.0210	0.0173	− 10.62	− 12.24	− 10.11	− 12.08
2	− 1.3895	− 0.3376	− 0.0347	0.0614	− 0.0765	− 0.0224	0.0173	− 10.62	− 12.24	− 10.11	− 12.08
3	− 0.2421	− 1.2155	− 0.3912	− 0.0367	− 0.9956	− 0.2797	− 0.0158	− 30.77	− 57.96	− 60.14	− 55.34
4	− 0.2420	− 1.2140	− 0.3870	− 0.0365	− 0.9430	− 0.2843	− 0.0199	− 30.49	− 57.50	− 59.62	− 54.84
5	− 0.2420	− 1.2140	− 0.3870	− 0.0365	− 0.9393	− 0.2912	− 0.0222	− 30.49	− 57.50	− 59.62	− 54.84
6	− 0.2421	− 1.2155	− 0.3912	− 0.0367	− 0.9348	− 0.2966	− 0.0242	− 30.77	− 57.96	− 60.14	− 55.34
7	− 0.0274	− 0.2807	− 0.9320	− 0.3148	− 0.2067	− 0.8968	− 0.2580	− 45.49	− 93.97	− 90.51	− 74.69
8	− 0.0275	− 0.2819	− 0.9358	− 0.3108	− 0.2064	− 0.8948	− 0.2563	− 44.89	− 92.76	− 89.23	− 73.55
9	− 0.0271	− 0.2800	− 0.9371	− 0.3065	− 0.2096	− 0.8429	− 0.2600	− 44.60	− 92.16	− 88.57	− 72.96
10	− 0.0271	− 0.2800	− 0.9371	− 0.3065	− 0.2133	− 0.8373	− 0.2659	− 44.60	− 92.16	− 88.57	− 72.96
11	− 0.0275	− 0.2819	− 0.9358	− 0.3108	− 0.2150	− 0.8293	− 0.2709	− 44.89	− 92.76	− 89.23	− 73.55
12	− 0.0274	− 0.2807	− 0.9320	− 0.3148	− 0.2133	− 0.8244	− 0.2752	− 45.49	− 93.97	− 90.51	− 74.69
13	0.0234	− 0.0245	− 0.2339	− 0.7117	− 0.0109	− 0.1947	− 0.7159	− 48.66	− 103.93	− 77.77	− 46.81
14	0.0237	− 0.0247	− 0.2346	− 0.7170	− 0.0107	− 0.1956	− 0.7199	− 47.81	− 102.05	− 76.37	− 45.94
15	0.0239	− 0.0241	− 0.2339	− 0.7249	− 0.0107	− 0.1956	− 0.7218	− 47.31	− 100.97	− 75.56	− 45.44
16	0.0299	− 0.0177	− 0.2254	− 0.7224	− 0.0069	− 0.1940	− 0.6658	− 47.08	− 100.46	− 75.17	− 45.19
17	0.0299	− 0.0177	− 0.2254	− 0.7224	− 0.0079	− 0.1986	− 0.6599	− 47.08	− 100.46	− 75.17	− 45.19
18	0.0239	− 0.0241	− 0.2339	− 0.7249	− 0.0150	− 0.2073	− 0.6552	− 47.31	− 100.97	− 75.56	− 45.44
19	0.0237	− 0.0247	− 0.2346	− 0.7170	− 0.0158	− 0.2079	− 0.6467	− 47.81	− 102.05	− 76.37	− 45.94
20	0.0234	− 0.0245	− 0.2339	− 0.7117	− 0.0155	− 0.2060	− 0.6414	− 48.66	− 103.93	− 77.77	− 46.81

21	0.0530	0.0696	− 0.0234	− 0.3148	0.0598	− 0.0018	− 0.2580	− 45.49	− 93.97	− 50.50	− 8.69
22	0.0532	0.0700	− 0.0219	− 0.3108	0.0594	− 0.0031	− 0.2563	− 44.89	− 92.76	− 49.98	− 8.78
23	0.0531	0.0696	− 0.0213	− 0.3065	0.0571	− 0.0065	− 0.2600	− 44.60	− 92.16	− 49.75	− 8.86
24	0.0531	0.0696	− 0.0213	− 0.3065	0.0569	− 0.0086	− 0.2659	− 44.60	− 92.16	− 49.75	− 8.86
25	0.0532	0.0700	− 0.0219	− 0.3108	0.0562	− 0.0108	− 0.2709	− 44.89	− 92.76	− 49.98	− 8.78
26	0.0530	0.0696	− 0.0234	− 0.3148	0.0554	− 0.0129	− 0.2752	− 45.49	− 93.97	− 50.50	− 8.69
27	0.0665	0.1140	0.0958	− 0.0367	0.0809	0.0796	− 0.0158	− 30.77	− 57.96	− 26.33	7.90
28	0.0663	0.1133	0.0951	− 0.0365	0.0789	0.0763	− 0.0199	− 30.49	− 57.50	− 26.15	7.72
29	0.0663	0.1133	0.0951	− 0.0365	0.0790	0.0760	− 0.0222	− 30.49	− 57.50	− 26.15	7.72
30	0.0665	0.1140	0.0958	− 0.0367	0.0792	0.0757	− 0.0242	− 30.77	− 57.96	− 26.33	7.90
31	0.0542	0.0816	0.0872	0.0614	0.0251	0.0310	0.0173	− 10.62	− 12.24	− 7.89	7.35
32	0.0542	0.0816	0.0872	0.0614	0.0255	0.0313	0.0173	− 10.62	− 12.24	− 7.89	7.35
33	0.0014	− 0.0002	− 0.0096	− 0.0448	− 0.0018	− 0.0261	− 0.0776	26.36	70.99	48.36	26.20
34	0.0014	− 0.0002	− 0.0094	− 0.0448	− 0.0018	− 0.0263	− 0.0776	26.36	70.99	48.43	26.31
35	− 0.0042	0.0138	0.0671	0.1074	− 0.0063	− 0.0531	− 0.0616	58.66	150.21	106.74	63.62
36	− 0.0034	0.0171	0.0705	0.1067	− 0.0055	− 0.0508	− 0.0612	58.14	148.59	105.36	62.74
37	− 0.0028	0.0196	0.0730	0.1067	− 0.0051	− 0.0502	− 0.0612	58.14	148.59	105.63	63.07
38	− 0.0022	0.0215	0.0752	0.1074	− 0.0047	− 0.0498	− 0.0616	58.66	150.21	106.35	63.31
39	− 0.0101	0.2150	0.3920	0.4668	− 0.0132	0.0544	0.0849	79.06	201.81	145.23	89.33
40	0.0002	0.2234	0.3893	0.4625	0.0189	0.0549	0.0833	78.11	198.94	142.43	87.26
41	0.0088	0.2320	0.3923	0.4573	0.0218	0.0578	0.0825	77.66	197.59	141.40	86.63
42	0.0157	0.2390	0.3988	0.4573	0.0243	0.0600	0.0825	77.66	197.59	141.71	86.96
43	0.0206	0.2441	0.4040	0.4625	0.0262	0.0625	0.0833	78.11	198.94	142.33	87.08
44	0.0238	0.2478	0.4080	0.4668	0.0278	0.0644	0.0849	79.06	201.81	143.36	87.20
45	0.7324	0.9454	1.0273	1.0316	0.1778	0.2890	0.3360	81.70	210.60	152.36	94.56

TABLE 3-11 (Continued)

Member number	Change in member force, kips										
	Loading case										
	I	*II*	*III*	*IV*	*V*	*VI*	*VII*	*VIII*	*IX*	*X*	*XI*
46	0.5854	0.9434	1.0315	1.0376	0.1773	0.2900	0.3379	80.56	207.38	148.74	91.58
47	0.5737	0.8421	1.0328	1.0453	0.1831	0.2891	0.3370	79.80	205.10	146.30	89.93
48	0.5578	0.8318	0.9814	1.0494	0.1872	0.2924	0.3338	79.43	203.97	145.33	88.91
49	0.5438	0.8170	0.9744	1.0494	0.1929	0.2978	0.3338	79.43	203.97	145.47	88.90
50	0.5331	0.8044	0.9631	1.0453	0.1944	0.3000	0.3370	79.80	205.10	145.70	88.46
51	0.5263	0.7956	0.9536	1.0376	0.1944	0.3002	0.3379	80.56	207.38	146.16	87.92
52	0.5228	0.7903	0.9478	1.0316	0.1931	0.2985	0.3360	81.70	210.60	146.90	87.38
53	− 0.0101	0.2150	0.3920	0.4668	0.9810	1.2847	1.3484	79.06	201.81	145.23	89.34
54	0.0002	0.2234	0.3893	0.4625	0.8787	1.2825	1.3524	78.11	198.94	142.43	87.26
55	0.0088	0.2320	0.3923	0.4573	0.8698	1.2296	1.3540	77.66	197.59	141.40	86.63
56	0.0157	0.2390	0.3988	0.4573	0.8583	1.2232	1.3540	77.66	197.59	141.70	86.96
57	0.0206	0.2441	0.4040	0.4625	0.8494	1.2144	1.3524	78.11	198.94	142.33	87.08
58	0.0238	0.2478	0.4080	0.4668	0.8455	1.2093	1.2093	79.06	201.81	143.36	87.20
59	− 0.0042	0.0138	0.0671	0.1074	0.0383	0.4075	0.5689	58.66	150.21	106.74	63.63
60	− 0.0034	0.0171	0.0705	0.1067	0.0477	0.4115	0.5640	58.14	148.59	105.36	62.74
61	− 0.0028	0.0196	0.0730	0.1067	0.0558	0.4189	0.5640	58.14	148.59	105.63	63.08
62	− 0.0022	0.0215	0.0752	0.1074	0.0613	0.4248	0.5689	58.66	150.21	106.35	63.32
63	+ 0.0014	− 0.0002	− 0.0096	− 0.0448	0.0027	0.0351	0.0798	26.36	70.99	48.36	26.30
64	0.0014	− 0.0002	− 0.0094	− 0.0448	0.0032	0.0367	0.0367	26.36	70.99	48.43	26.31

1 kip = 4.448 kN

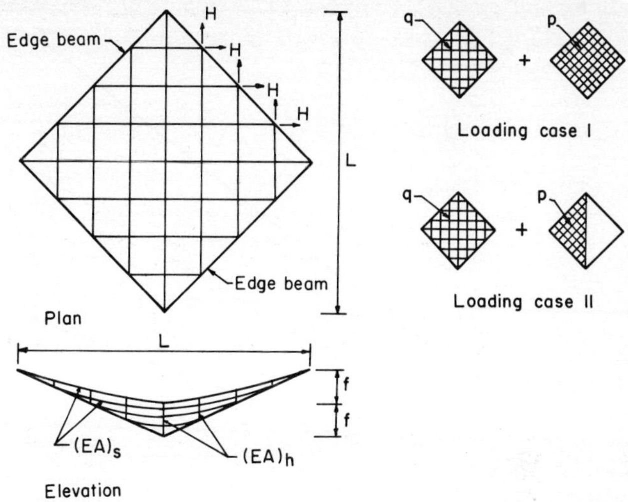

FIGURE 3-19 Analytical model and loading cases for the parametric study of hypar network (see results in Fig. 3-20).

4. Only vertical uniformly distributed loads are considered.

The basic expressions are first derived and are used later to obtain expressions for maximum deflection and change in cable tension in different cases.[9] For a single cable, the tension increment h, due to a vertical applied load p, is given as

$$h = -k \int \frac{d^2 z}{dx^2} w(x)\, dx \tag{3-17}$$

where $p = p_0\phi(x)$, $\phi(x)$ being the variation of load with x. In the above equation, z is the initial ordinate of the cable and w is the vertical deflection. It is now assumed that[10]

$$\frac{d^2 z}{dx^2} = -\Theta\phi(x) \tag{3-18a}$$

and

$$\frac{d^2 w}{dx^2} = -\overline{\Theta}\phi(x) \tag{3-18b}$$

[9] Table 3-12*b* is a collection of the resulting formulas [47] for two different types of plane trusses under various loading cases as shown in Table 3-12*a*. Although only trusses of the type shown in Fig. 1-9*b* are indicated, the given results are applicable to other plane systems as well (except the type shown in Fig. 1-9*d*).

[10] Relations assumed [75] in Eq. 3-18, though not rigorously justified, enable the given procedure to be used conveniently for deriving closed form expressions.

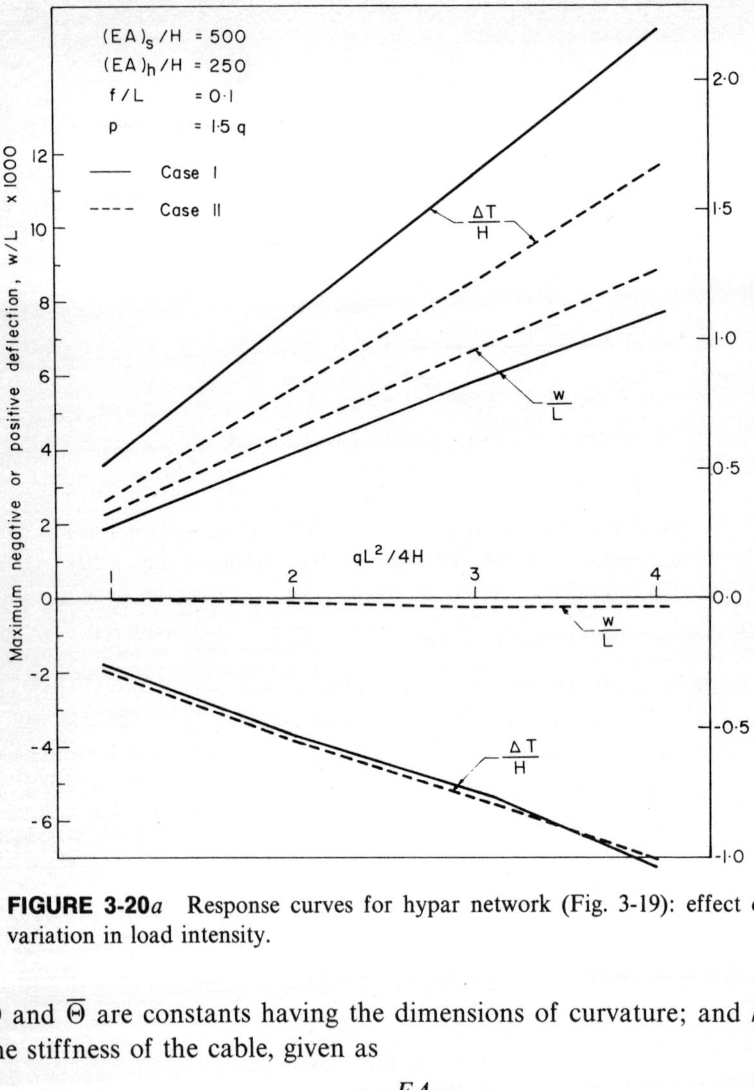

FIGURE 3-20*a* Response curves for hypar network (Fig. 3-19): effect of variation in load intensity.

Θ and $\overline{\Theta}$ are constants having the dimensions of curvature; and k is the stiffness of the cable, given as

$$k = \frac{EA}{L(1 + 8f^2/L^2)} \tag{3-19}$$

E, A, f, and L are Young's modulus of elasticity, the cross-sectional area, the sag, and the span of the cable, respectively. The horizontal component of tension increment given in Eq. 3-17 may be expressed for the upper cable as

$$h_u = \frac{\overline{\Theta}}{\Theta_u}\,\beta_u \tag{3-20a}$$

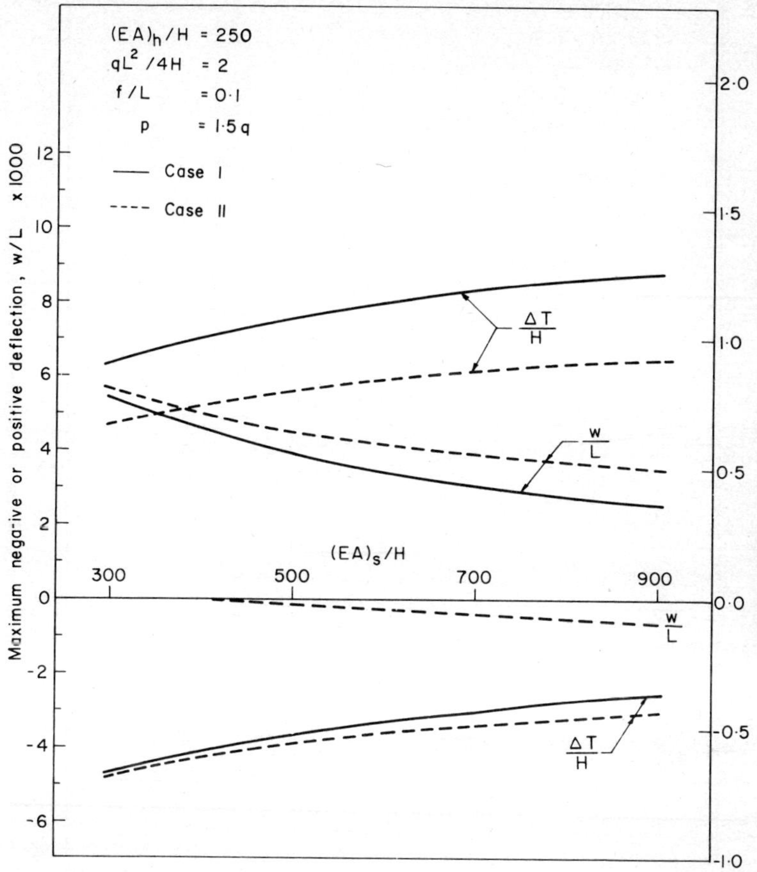

FIGURE 3-20*b* Response curves for hypar network (Fig. 3-19): effect of variation in rigidity of sagging cables.

$$\text{where } \beta_u = -k_u \frac{\Theta_u}{\overline{\Theta}} \int \left(\frac{d^2 z}{dx^2} \right)_u w(x)\, dx$$

Similarly for the lower cable,

$$h_l = -\alpha\nu \frac{\overline{\Theta}}{\Theta_u} \beta_u \tag{3-20b}$$

$$\text{where } \alpha = \frac{k_l}{k_u} \qquad \text{and} \qquad \nu = \frac{\Theta_l}{\Theta_u}$$

The subscripts u and l correspond to upper and lower cables, respectively, wherever they appear.

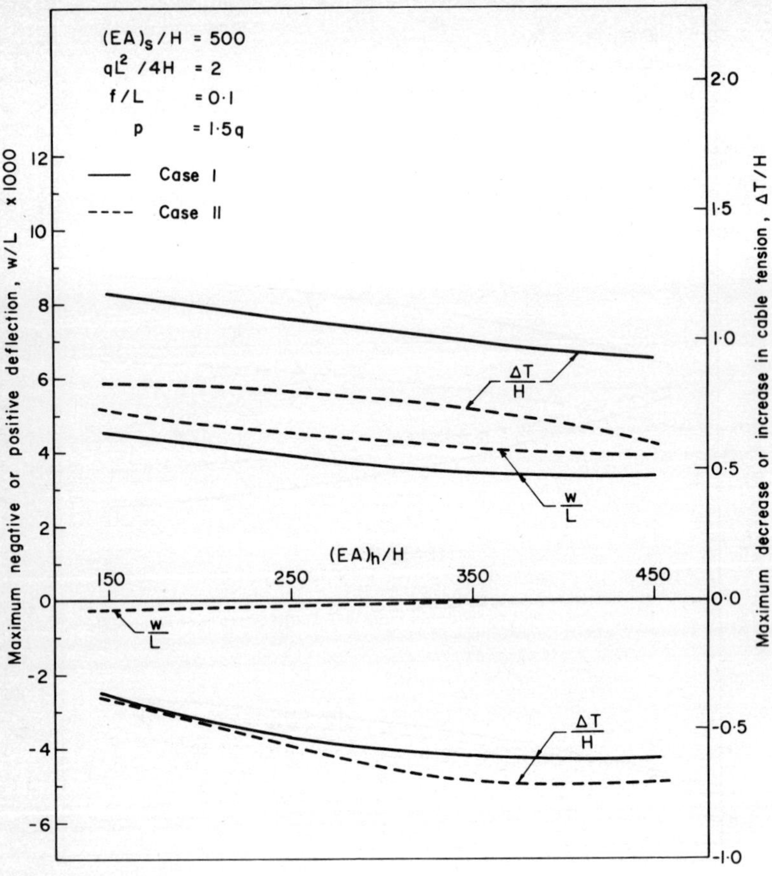

FIGURE 3-20*c* Response curves for hypar network (Fig. 3-19): effect of variation in rigidity of hogging cables.

The values of tension increments given in Eqs. 3-20*a* and 3-20*b* must satisfy the equation

$$\frac{\overline{\Theta}^2}{\Theta_u}\,\beta_u(1-\alpha\nu)+\overline{\Theta}\left[\,\beta_u(1+\alpha\nu^2)+H_u\left(1+\frac{1}{\nu}\right)+h_u'+h_l'\right]$$
$$-p_0+\Theta_u(h_u'-\nu h_l')=0 \qquad (3\text{-}21)$$

where h' is the horizontal component of tension increment due to temperature variations and H is the horizontal component of initial prestress in the cables. The value of $\overline{\Theta}$ works out to be very small, and hence the quadratic term in $\overline{\Theta}$ is neglected. For simplicity the

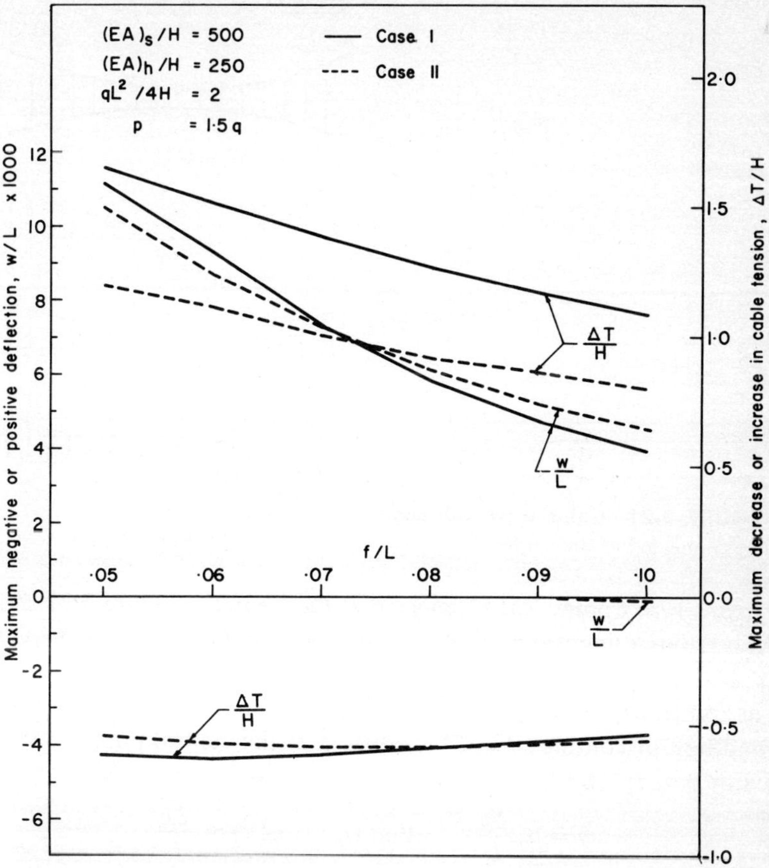

FIGURE 3-20*d* Response curves for hypar network (Fig. 3-19): effect of variation in the sag-span ratio.

effect of temperature variation may also be omitted at this stage in Eq. 3-21. The value of $\overline{\Theta}$ is thus obtained as

$$\overline{\Theta} = \frac{p_0}{\beta_u(1 + \alpha\nu^2) + H_u(1 + 1/\nu)} \tag{3-22}$$

and its substitution in Eqs. 3-20 will give the value of horizontal components of tension increment.

The force in web members (if vertical) is given as

force due to initial prestress + force due to external load

and equals $$\left[H_u(\Theta_u + \overline{\Theta}) - p_0\right]\phi(x) \tag{3-23}$$

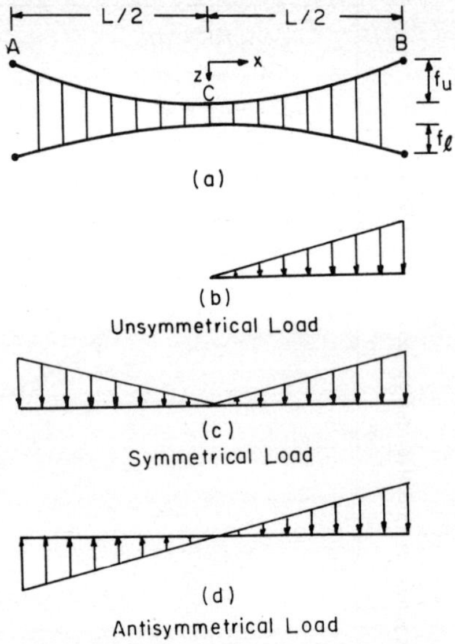

FIGURE 3-21 Cable truss with cables not connected at the center.

Trusses with Cables Not Connected at the Center. Figure 3-21 shows the various dimensions of the truss. A triangularly varying load is applied on one-half CB $(0 \leqslant x \leqslant L/2)$ of the truss (in Fig. 3-21b). The portions AC and CB will be considered separately, and the condition of continuity will be satisfied at the central point.

For $0 \leqslant x \leqslant L/2$,

$$p = p_0 \frac{2x}{L} \qquad \text{that is,} \qquad \phi(x) = \frac{2x}{L}$$

Now

$$\frac{d^2w}{dx^2} = -\overline{\Theta}\,\frac{2x}{L} \tag{3-24a}$$

Similarly for $0 \geqslant x \geqslant -L/2$, $\phi(x) = 0$, and

$$\frac{d^2w}{dx^2} = 0 \tag{3-24b}$$

Integrating Eqs. 3-24a and 3-24b twice and applying the boundary conditions, that is, $w = 0$ at $x = \pm L/2$, and the continuity condition at C, we get

$$w(x) = \frac{\overline{\Theta} L^2}{48}\left(1 + \frac{2x}{L} - \frac{16x^3}{L^3}\right) \qquad \text{for } 0 \leqslant x \leqslant \frac{L}{2} \tag{3-25a}$$

$$w(x) = \frac{\overline{\Theta} L^2}{48}\left(1 + \frac{2x}{L}\right) \qquad \text{for } 0 \geqslant x \geqslant -\frac{L}{2} \tag{3-25b}$$

Equations 3-25 give the deflections of the truss. The maximum deflection will be at $x = L/2\sqrt{6}$ and

$$w_{\max} = \frac{\overline{\Theta} L^2}{48}\left(1 + \frac{1}{3}\sqrt{\frac{2}{3}}\right)$$

The horizontal component of tension increment in the cables can be evaluated with the help of Eqs. 3-20, whence

$$\beta_u = -k_u \frac{\Theta_u}{\overline{\Theta}} \int \left(\frac{d^2z}{dx^2}\right)_u w(x)\, dx \qquad (3\text{-}26)$$

On integrating Eq. 3-26 the value of β_u is obtained as

$$\beta_u = k_u \Theta_u^2 \frac{13L^3}{2880}$$

and the horizontal components of tension increments are

$$h_u = \frac{13}{2880} \overline{\Theta} \Theta_u k_u L^3 \qquad (3\text{-}27a)$$

$$h_l = -\frac{13}{2880} \alpha \nu \overline{\Theta} \Theta_u k_u L^3 \qquad (3\text{-}27b)$$

SYMMETRIC LOAD In this case of loading (see Fig. 3-21c), the deflections given in Eqs. 3-25 are appropriately superimposed to give

$$w(x) = \frac{\overline{\Theta} L^2}{24}\left(1 - \frac{8x^3}{L^3}\right) \qquad 0 \leqslant x \leqslant \frac{L}{2} \qquad (3\text{-}28a)$$

$$w(x) = \frac{\overline{\Theta} L^2}{24}\left(1 + \frac{8x^3}{L^3}\right) \qquad 0 \geqslant x \geqslant -\frac{L}{2} \qquad (3\text{-}28b)$$

The deflected shape of the truss is a cubic parabola with maximum deflection at the central point, given as

$$w_{\max} = \frac{\overline{\Theta} L^2}{24}$$

also

$$\beta_u = \frac{1}{80} k_u \Theta_u^2 L^3 \qquad (3\text{-}29)$$

and the horizontal components of tension increments will be

$$h_u = \overline{\Theta} \Theta_u k_u \frac{L^3}{80} \qquad (3\text{-}30a)$$

$$h_l = -\alpha \nu \overline{\Theta} \Theta_u k_u \frac{L^3}{80} \qquad (3\text{-}30b)$$

where $\Theta_u = \dfrac{24 f_u}{L^2}$

ANTISYMMETRIC LOAD For the antisymmetric load (see Fig. 3-21d) the expressions for deflection, Eqs. 3-25, are superimposed with the appropriate values of x. The deflection for an antisymmetric load is thus given by

$$w(x) = \frac{\overline{\Theta} Lx}{12}\left(1 - 4\frac{x^2}{L^2}\right)$$

This expression gives the deflected shape of the truss with maximum deflection at

$$x = \pm \frac{L}{2\sqrt{3}} \qquad \text{and} \qquad w_{max} = \frac{\overline{\Theta} L^2}{36\sqrt{3}} \tag{3-31}$$

The horizontal component of tension increment will be zero, as the loads are acting in opposite directions in the two halves of the truss.

Trusses with Cables Connected at the Center. This type of truss (Fig. 3-22a) is commonly used to support the cladding in circular roofs. For symmetric loads the behavior of these trusses is similar to that of trusses with cables not connected at the center, and the expressions for deflection and horizontal component of tension increment already obtained hold good. However, this is not so for antisymmetric loads, and their case is treated herein.

To maintain equilibrium, the horizontal components of tension increment in both the cables, on the left and the right of the central point, must be the same. Let us suppose a horizontal contact force, $2\overline{T}$, is introduced (see Fig. 3-22b). Then the horizontal components

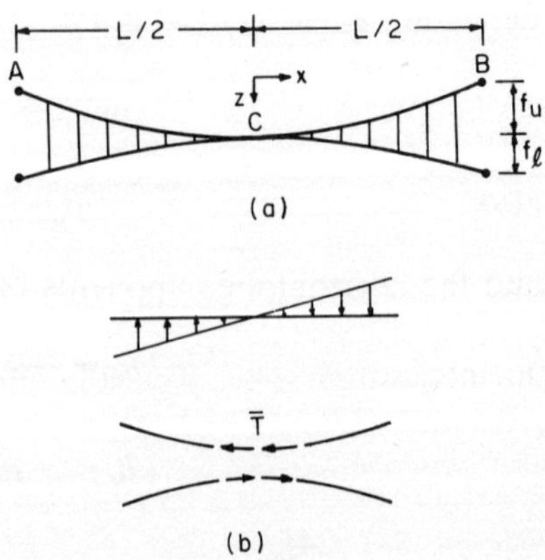

FIGURE 3-22 Cable truss with cables connected at the center.

of tension increment for the left and the right halves due to antisymmetric loads are

Left: $$h_u = -\overline{T} \qquad h_l = +\overline{T} \tag{3-32a}$$

Right: $$h_u = +\overline{T} \qquad h_l = -\overline{T} \tag{3-32b}$$

The values of horizontal displacement at the central points u_{uc} and u_{lc} are given by

$$u_{uc} = -\frac{\overline{T}}{2k_u} - \int_0^{L/2}\left(\frac{d^2z}{dx^2}\right)_u w\,dx \tag{3-33a}$$

$$u_{lc} = \frac{\overline{T}}{2\alpha k_u} + \nu\int_0^{L/2}\left(\frac{d^2z}{dx^2}\right)_u w\,dx \tag{3-33b}$$

Since $u_{uc} = u_{lc}$ at $x = 0$,

$$\overline{T} = -2\alpha k_u \frac{1+\nu}{1+\alpha}\int_0^{L/2}\left(\frac{d^2z}{dx^2}\right)_u w\,dx \tag{3-34}$$

For equilibrium, this value of $\overline{T}$ should satisfy the conditions

$$\overline{T}\left\{\frac{1}{2\alpha k_u}\,\frac{1+\alpha}{1+\nu}\left[H_u\left(1+\frac{1}{\nu}\right)+{}^s h_u + {}^s h_l\right]\right.$$
$$\left. -(1+\nu)\int_0^{L/2}\left(\frac{d^2z}{dx^2}\right)_u e_u\,dx\right\} + \int_0^{L/2}\left(\frac{d^2z}{dx^2}\right)_u a_M\,dx = 0 \tag{3-35}$$

where e = distance of cable below chord joining one end and central point
s = prefix pertaining to symmetric loads
a_M = bending moment on a simply supported beam of span $L/2$ due to external loads

For triangularly varying antisymmetric loads,

$$\phi(x) = \frac{2x}{L}$$

and

$$\frac{d^2w}{dx^2} = -\overline{\Theta}\,\frac{2x}{L} \tag{3-36}$$

On integrating twice and applying the boundary conditions, that is, $w = 0$ at $x = 0$ and at $x = L/2$,

$$w(x) = \frac{\overline{\Theta} L x}{12}\left(1 - \frac{4x^2}{L^2}\right) \tag{3-37}$$

Now from Eq. 3-34,

$$\bar{T} = -2\alpha k_u \frac{1+\nu}{1+\alpha} \int_0^{L/2} \left(\frac{d^2 z}{dx^2} \right)_u w(x)\, dx$$

which gives

$$\bar{T} = \frac{1}{180} \alpha \bar{\Theta} \Theta_u k_u \frac{1+\nu}{1+\alpha} L^3 \tag{3-38}$$

and the value of $\bar{\Theta}$ is given, from Eq. 3-35, as

$$\bar{\Theta} = \frac{p_0}{H_u(1 + 1/\nu) + 2\alpha k_u\left[(1+\nu)^2/(1+\alpha)\right](f_u^2/L)} \tag{3-39}$$

Effect of Flexibility of Supporting Structure and Temperature Variations. The effect of flexibility of the supporting structure, i.e., stays and columns, may be taken into account as suggested by Mollmann [66]. The stiffnesses of the upper and lower cables are to be modified as

$$k_u = \frac{(EA)_u \mu_u}{L(1 + 8f_u^2/L^2)} \tag{3-40a}$$

$$k_l = \frac{(EA)_l \mu_l}{L(1 + 8f_l^2/L^2)} \tag{3-40b}$$

and

$$\alpha = \frac{k_l}{k_u}$$

where

$$\frac{1}{\mu_u} = 1 + 2\left[\frac{(EA)_u}{(EA)_{bu}} \frac{L_{bu}}{L\cos^2\Theta_u} + \frac{(EA)_u}{(EA)_c} \frac{L_{cu}}{L} \tan^2\Theta_u \right] \tag{3-41a}$$

$$\frac{1}{\mu_l} = 1 + 2\left[\frac{(EA)_l}{(EA)_{bl}} \frac{L_{bl}}{\cos^2\Theta_l} + \frac{(EA)_l}{(EA)_c} \frac{L_{cl}}{L} \tan^2\Theta_l \right] \tag{3-41b}$$

Here $(EA)_c$, L_c are stiffness and length of the columns and $(EA)_b$ and L_b are those of backstays, as shown in Fig. 3-23. The modified values of k_u and k_l obtained from Eqs. 3-40 should be substituted as applicable in expressions given so far in Sec. 3-5.1.

The effect of temperature variation can be superimposed by directly adding the change in horizontal component of tension in the

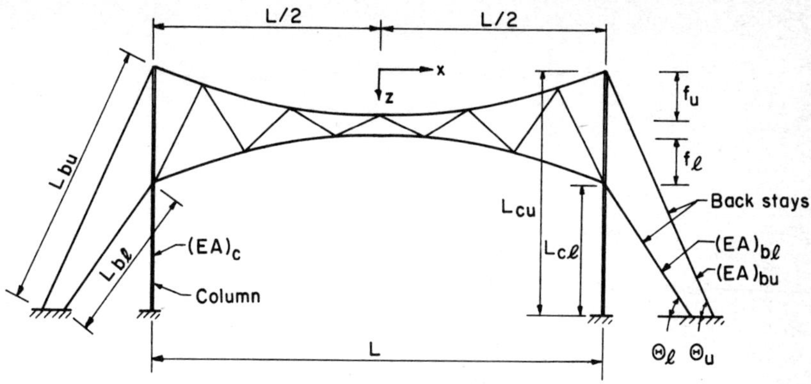

FIGURE 3-23 Parameters to be accounted for when estimating the effects of support flexibility.

cables, h', due to temperature variation. The final expressions for horizontal component of tension increments will be, from Eqs. 3-20,

$$h_u = \frac{\overline{\Theta}}{\Theta_u} \beta_u + h_u^t \tag{3-42a}$$

$$h_l = -\alpha\nu \frac{\overline{\Theta}}{\Theta_u} \beta_u + h_l^t \tag{3-42b}$$

In order to compare the approximate analysis given in the preceding part of this section with a more accurate nonlinear analysis, typical cases of a single cable and a cable truss are solved in Examples 3-5 and 3-6.

EXAMPLE 3-5

For the cable analyzed in Example 2-2, Fig. 2-15,

$$\text{Cable stiffness} \quad k_u = \frac{(EA)_u}{L(1 + 8f_u^2/L^2)} = \frac{20{,}000}{100(1 + 0.08)}$$

$$= 185 \text{ kips/ft } (2700 \text{ kN/m})$$

$$\Theta_u = \frac{8f_u}{L^2} = 0.008 \text{ ft}^{-1} \ (0.026 \text{ m}^{-1})$$

and as given in Table 3-12*b*,

$$\beta_u = \frac{1}{12}\,\Theta_u^2 L^3 k_u = 986.7 \text{ kips (4390 kN)}$$

Also,
$$h = \frac{\overline{\Theta}}{\Theta_u}\,\beta_u$$

and
$$\overline{\Theta} = \frac{p_0}{\beta_u + H_u}$$

For the initial uniformly distributed load of 0.1 kip/ft (1.46 kN/m), $H_u = 0$. Thus,

$$h_u = \frac{p_0}{\Theta_u} = \frac{0.10}{0.008} = 12.50 \text{ kips (55.5 kN)}$$

The unsymmetric load of 0.04 kip/ft (0.58 kN/m) in case II may be considered the combination of a symmetric load and an antisymmetric load, each of intensity 0.02 kip/ft (0.29 kN/m), and the results of the two superimposed.

For a symmetric load, $\overline{\Theta}$ is evaluated from case II in Table 3-12 by substituting $1/\nu$ and $\alpha = 0$, as

$$\overline{\Theta} = \frac{p_0}{\Theta_u^2(L^3/12)k_u + H_u} = \frac{0.02}{\left[(0.008)^2/12\right] \times 100^3 \times 185 + 12.5}$$

$$= 2 \times 10^{-5}$$

From Eq. 3-20*a*,

$$h_u = \frac{\overline{\Theta}}{\Theta_u}\,\beta_u$$

Substituting appropriate values,

$$h_u = \frac{2 \times 10^{-5} \times 986.7}{0.008} = 2.47 \text{ kips (11 kN)}$$

The deflections for this case are negligible.

For antisymmetric loads, the horizontal component of tension increment will be zero, and the maximum deflection, derived in the same manner as for a triangular load, occurs at $x = \pm L/4$ and is given by

$$w_{\max} = \frac{\overline{\Theta} L^2}{32}$$

where $\overline{\Theta} = p_0/H_u$. (Substitute $1/\nu = 0$ in expressions for case III of Table 3-12).

TABLE 3-12*a* **Structure Type and Load Cases for Approximate Formulas given in Table 3-12***b****

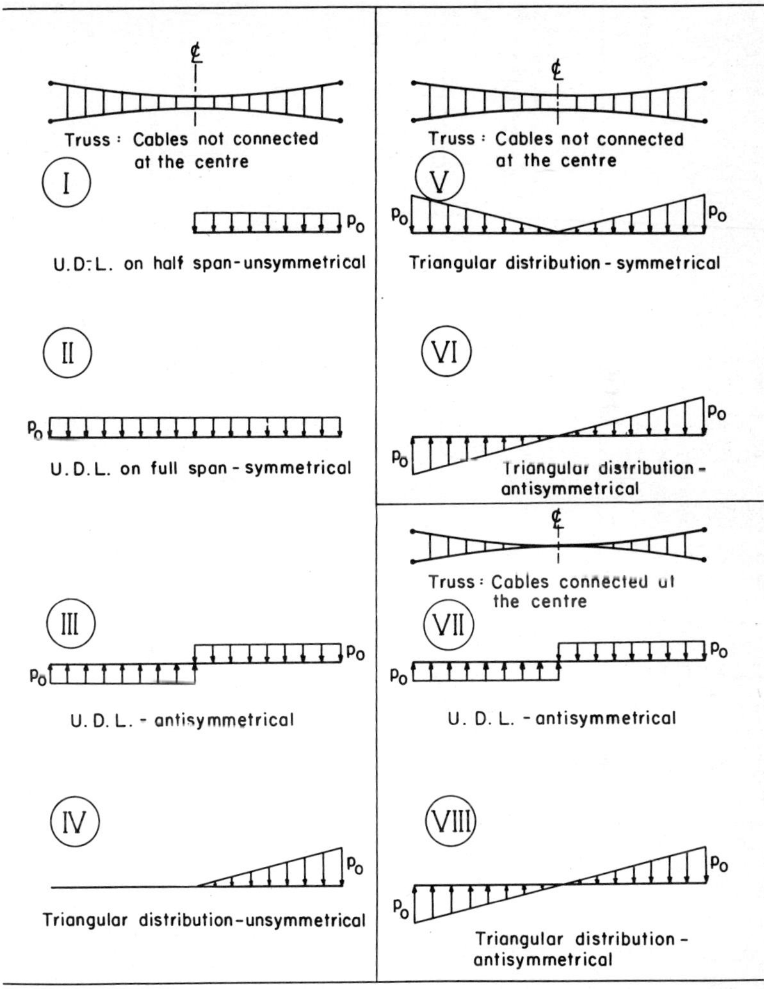

* Load is applied in each case to the upper cable.

TABLE 3-12b

Type of structure and loading case* (1)		Load function $\phi(x)$ (2)	Θ_u (3)	Cable deflection $w(x)$ (4)	Maximum deflection at $x =$ (5)	Maximum deflection $w_{\max}$ (6)	β_u (7)	Constant $\overline{\Theta}$ (8)
I	$0 > x > -\frac{L}{2}$	0	$\frac{8f_u}{L^2}$	$\frac{\overline{\Theta}L^2}{16}(1 + \frac{2x}{L})$	$\frac{L}{8}$	$\frac{9\overline{\Theta}L^2}{128}$	$\frac{5}{192}k_u\Theta_u^2L^3$	$\frac{p_0}{\left[\frac{5}{192}\Theta_u^2L^3k_u(1+\alpha\nu^2) + H_u\left(1+\frac{1}{\nu}\right)\right]}$
	$0 < x < \frac{L}{2}$	1		$\frac{\overline{\Theta}L^2}{16}(1 + \frac{2x}{L} - \frac{8x^2}{L^2})$				
II		1	$\frac{8f_u}{L^2}$	$\frac{\overline{\Theta}L^2}{8}(1 - \frac{4x^2}{L^2})$	0	$\frac{\overline{\Theta}L^2}{8}$	$\frac{1}{12}k_u\Theta_u^2L^3$	$\frac{p_0}{\left[\frac{\Theta_u^2}{12}L^3k_u(1+\alpha\nu^2) + H_u\left(1+\frac{1}{\nu}\right)\right]}$
III	$0 > x > -\frac{L}{2}$	-1	$\frac{8f_u}{L^2}$	$\frac{\overline{\Theta}Lx}{4}(1 + \frac{2x}{L})$	$\pm\frac{L}{4}$	$\pm\frac{\overline{\Theta}L^2}{32}$	0	$\frac{p_0}{H_u\left(1+\frac{1}{\nu}\right)}$
	$0 < x < \frac{L}{2}$	1		$\frac{\overline{\Theta}Lx}{4}(1 - \frac{2x}{L})$				
IV	$0 > x > -\frac{L}{2}$	0	$\frac{8f_u}{L^2}$	$\frac{\overline{\Theta}L^2}{48}(1 + \frac{2x}{L})$	$\frac{L}{2\sqrt{6}}$	$\frac{\overline{\Theta}L^2}{48}(1 + \frac{1}{3}\sqrt{\frac{2}{3}})$	$\frac{13}{2880}k_u\Theta_u^2L^3$	$\frac{p_0}{\left[\frac{13}{2880}\Theta_u^2L^3k_u(1+\alpha\nu^2) + H_u\left(1+\frac{1}{\nu}\right)\right]}$
	$0 < x < \frac{L}{2}$	$\frac{2x}{L}$	$\frac{24f_u}{L^2}$	$\frac{\overline{\Theta}L^2}{48}(1 + \frac{2x}{L} - \frac{16x^3}{L^3})$				

V	$0 > x > -\frac{L}{2}$	$-\frac{2x}{L}$	$24\frac{f_u}{L^2}$	$+\frac{\bar{\Theta}L^2}{24}(1 + 8\frac{x^3}{L^3})$	0	$\frac{\bar{\Theta}L^2}{24}$	$\frac{1}{80}k_u\Theta_u^2L^3$	$\frac{p_0}{\left[\frac{1}{80}\Theta_u^2L^3k_u(1+\alpha\nu^2) + H_u\left(1+\frac{1}{\nu}\right)\right]}$
	$0 < x < \frac{L}{2}$	$\frac{2x}{L}$		$\frac{\bar{\Theta}L^2}{24}(1 - 8\frac{x^3}{L^3})$				
VI		$\frac{2x}{L}$	$24\frac{f_u}{L^2}$	$\pm\frac{\bar{\Theta}Lx}{12}(1 - \frac{4x^2}{L^2})$	$\pm\frac{L}{2\sqrt{3}}$	$\pm\frac{\bar{\Theta}L^2}{36\sqrt{3}}$	0	$\frac{p_0}{H_u\left(1+\frac{1}{\nu}\right)}$
VII	$0 > x > -\frac{L}{2}$	-1	$\frac{8f_u}{L^2}$	$\frac{\bar{\Theta}Lx}{4}(1 + \frac{2x}{L})$	$\pm\frac{L}{4}$	$\pm\frac{\bar{\Theta}L^2}{32}$	$\bar{T} = \frac{\bar{\Theta}L^3}{48}k_u \times \alpha\frac{(1+\nu)}{(1+\alpha)}\Theta_u$	$\frac{p_0}{\left[H_u\left(1+\frac{1}{\nu}\right) + {}^sh_u + {}^sh_l + \frac{4}{3}\alpha k_u\frac{(1+\nu)^2}{(1+\alpha)}\frac{f_u^2}{L}\right]}$
	$0 < x < \frac{L}{2}$	1		$\frac{\bar{\Theta}Lx}{4}(1 - \frac{2x}{L})$				
VIII		$\frac{2x}{L}$	$\frac{24f_u}{L^2}$	$\frac{\bar{\Theta}Lx}{12}(1 - \frac{4x^2}{L^2})$	$\pm\frac{L}{2\sqrt{3}}$	$\pm\frac{\bar{\Theta}L^2}{36\sqrt{3}}$	$\bar{T} = \frac{1}{180}\bar{\Theta}L^3 \times k_u\alpha\frac{(1+\nu)}{(1+\alpha)}\Theta_u$	$\frac{p_0}{\left[H_u\left(1+\frac{1}{\nu}\right) + {}^sh_u + {}^sh_l + 2\alpha k_u\frac{(1+\nu)^2}{(1+\alpha)}\frac{f_u^2}{L}\right]}$

* See Table 3-12*a*.

TABLE 3-13 Results for Example 3-5

Load	*Parameters*	*Approximate method*	*More accurate theory*	*Pugsley's method* [80]
Symmetric UDL (0.1 kip/ft)	h, kips (kN)	12.500 (55.5)	12.500 (55.5)	12.500 (55.5)
Unsymmetric load (0.04 kip/ft on half span)	h, kips (kN)	2.470 (11.02)	2.500 (11.10)	2.550 (11.35)
	w_{max}, ft (cm)	0.417 (12.7)	0.400 (12.2)	0.400 (12.2)

So

$$w_{max} = \frac{p_0}{H_u}\frac{L^2}{32} = \frac{0.02}{14.97} \times \frac{100 \times 100}{32} = 0.417 \text{ ft (127 cm)}$$

These results are compared in Table 3-13 with those obtained in Example 2-2 by a more accurate nonlinear analysis and also by using the approximate formulas suggested by Pugsley [80].

EXAMPLE 3-6

The dimensions of the truss, used in a circular roof, are shown in Fig. 3-24*a*. The diameter of the hall is 340 ft (103.7 m), and angular spacing of these trusses is 6°. For a typical truss,

$$\Theta_u = \frac{24 f_u}{L^2} = 0.00457 \text{ ft}^{-1} \ (0.015 \text{ m}^{-1})$$

$$\Theta_l = \frac{24 f_l}{L^2} = 0.00353 \text{ ft}^{-1} \ (0.0115 \text{ m}^{-1})$$

$$k_u = \frac{(EA)_u}{L(1 + 8f_u^2/L^2)} = 190.5 \text{ kips/ft (2780 kN/m)}$$

$$k_l = \frac{(EA)_l}{L(1 + 8f_l^2/L^2)} = 100.8 \text{ kips/ft (1467 kN/m)}$$

$$\nu = \frac{f_l}{f_u} = 0.773$$

$$\alpha = \frac{k_l}{k_u} = 0.529$$

The following three cases of different load combinations are considered, and the results are given in Table 3-14.

Case I: Dead load + snow load on half the span

Case II: Dead load + snow load on full span

Case III: Dead load + wind pressure on left and suction on right

Dead load	$= 10\ \mathrm{lb/ft^2}\ (480\ \mathrm{N/m^2})$
Snow load	$= 15\ \mathrm{lb/ft^2}\ (720\ \mathrm{N/m^2})$
Wind load	$= 10\ \mathrm{lb/ft^2}$
Initial prestress in upper cable	$= 42.0$ kips (186.6 kN)
Spacing of truss at outer periphery	$= 170 \times 6\pi/180$
	$= 17.8$ ft (5.44 m)

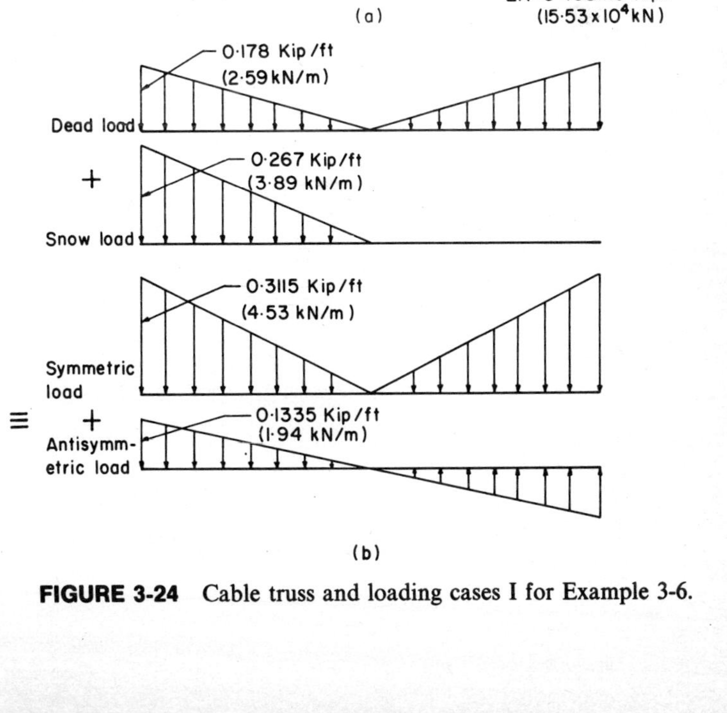

FIGURE 3-24 Cable truss and loading cases I for Example 3-6.

TABLE 3-14 Results for Example 3-6

	Loading case I		*Loading case II*		*Loading case III*	
Parameter	*Approximate method*	*Nonlinear method*	*Approximate method*	*Nonlinear method*	*Approximate method*	*Nonlinear method*
$(H_u + h_u)$ left, kips (kN)	115.5 (510)	112.6 (500)	115.8 (515)	120.0 (532)	100.6 (448)	95.5 (425)
$(H_u + h_u)$ right, kips (kN)	71.9 (320)	80.2 (356)	115.8 (515)	120.0 (532)	42.6 (190)	51.5 (229)
$(H_l + h_l)$ left, kips (kN)	11.4 (51)	21.4 (95)	24.3 (108)	30.2 (134)	13.2 (58)	22.8 (101)
$(H_l + h_l)$ right, kips (kN)	55 (247)	52.7 (235)	24.3 (108)	30.2 (134)	71.4 (317)	66.2 (295)
Maximum deflection, ft (m)	0.817 (0.249)	1.314 (0.405)	0.805 (0.245)	0.801 (0.245)	0.742 (0.236)	1.115 (0.352)

For illustration the calculations of case I are given in detail. They are as follows:

Each truss will carry a triangularly varying load of maximum intensity at the outer periphery as shown in Fig. 3-24*b*. For symmetric loads, the value of $\bar{\Theta}$ as given by case V in Table 3-12 is

$$\bar{\Theta} = \frac{p_0}{\frac{1}{80}\Theta_u^2 L^3 k_u(1 + \alpha\nu^2) + H_u(1 + 1/\nu)}$$

$$= 1.17 \times 10^{-4}\ \text{ft}^{-1}\ (3.84 \times 10^{-4}\ \text{m}^{-1})$$

Now,

$$h_u = \frac{1}{80}\bar{\Theta}\Theta_u k_u L^3 = 51.70 \text{ kips } (230 \text{ kN})$$

and

$$h_l = -\alpha\nu h_u = -21.20 \text{ kips } (-94.3 \text{ kN})$$

For antisymmetric loads, the horizontal component of tension increment is given from Eq. 3-35 as

$$\bar{T} = \frac{p_0 L^3}{360\{(1/2\alpha k_u\Theta_u)[(1+\alpha)/(1+\nu)] \times [H_u(1 + 1, \nu) + {}^s h_u + {}^s h_l] + (1+\nu)(f_u L/24)\}}$$

$$= \frac{0.1335 \times 340^3}{360(111.75 + 553)} = 21.8 \text{ kips } (96.9 \text{ kN})$$

Final values for the horizontal component of tension increment are

Top cable:	Left half = 51.7 + 21.8 =	73.5 kips (326.6 kN)
	Right half = 51.7 − 21.8 =	29.9 kips (133 kN)
Bottom cable:	Left half = −21.2 − 21.8 =	−43.0 kips (−191.2 kN)
	Right half = −21.2 + 21.8 =	0.6 kip (2.6 kN)

Values of h_u and h_l are similarly computed for the other two cases of loading. Table 3-14 gives the final tension for each case.

For the computation of deflections as well, the loading in case I is taken to consist of a symmetrical plus an antisymmetric component. In the case of the antisymmetric component (case VIII in Table 3-12), maximum deflection will occur at $x = -L/2\sqrt{3}$ and equals $\Theta L^2/36\sqrt{3}$. With the value of $\bar{T}$ already known, $\bar{\Theta}$ is obtained from

Eq. 3-38 and equals 1.94×10^{-4} ft^{-1}. The value of $\overline{\Theta}$ can also be obtained from Eq. 3-39 directly. Now

$$w = \frac{1.94 \times 10^{-4} \times 340^2}{36\sqrt{3}} = 0.360 \text{ ft } (11 \text{ cm})$$

For the symmetric load component (case V in Table 3-12),

$$w(x) = \frac{\overline{\Theta} L^2}{24}\left(1 + \frac{8x^3}{L^3}\right) \qquad \text{for} \quad x \leqslant 0$$

$$\text{At} \quad x = -\frac{L}{2\sqrt{3}}$$

$$w = \frac{0.8076 \overline{\Theta} L^2}{24}$$

As already computed, $\overline{\Theta} = 1.17 \times 10^{-4}$ ft^{-1} (3.84×10^{-4} m^{-1}). Therefore,

$$w = \frac{0.8076 \times 1.17 \times 10^{-4} \times 340^2}{24}$$

$$= 0.455 \text{ ft } (14 \text{ cm})$$

At $x = -L/2\sqrt{3}$,

$$\text{Total deflection} = 0.360 + 0.455 \text{ ft}$$
$$= 0.815 \text{ ft } (25 \text{ cm})$$

Maximum deflection for case II is similarly calculated at $x = 0$ and for case III at $x = L/2\sqrt{3}$. It may be noted from Table 3-14 that the results for the upper cable, which are more important from the design point of view, are in reasonable agreement, while the results for the lower cable are in somewhat greater error. The resulting maximum deflections are close for the symmetric load case only.

3-5.2 Networks

As in the case of plane structures, Mollmann [67] has done useful work in developing approximate analysis procedures for cable nets and has shown these to be reasonably accurate as compared to more rigorous nonlinear analyses. Described here is another, much simpler way of making a rough estimate of deflections and forces in cables under idealized static loading. The elements of the cable net are imagined to be equivalent to the main cables of a truss, not connected at the center, and the approximate method of analysis for cable trusses, discussed in the preceding part of this section, is thus applied to cable nets. The equivalence of cable trusses to the nets can be best explained through schematic sketches. Figure 3-25*a* shows the elements of a cable truss with the cables not connected at the center. The upper and lower cables of this truss are assumed to be equivalent to the two orthogonal cables, one sagging and the other hogging, shown in Fig. 3-25*b*, which are in turn equivalent to the unit membrane in Fig. 3-25*c*. The central cables of the net, shown thickened in Fig. 3-25*d*, are represented by this membrane.

The expressions for the truss derived earlier thus become applicable with the subscripts u and l replaced by x and y, respectively, to a net with a square plan.

Symmetric Uniformly Distributed Loads. The cable nets are mainly subjected to uniformly distributed loads caused by the self-weight of

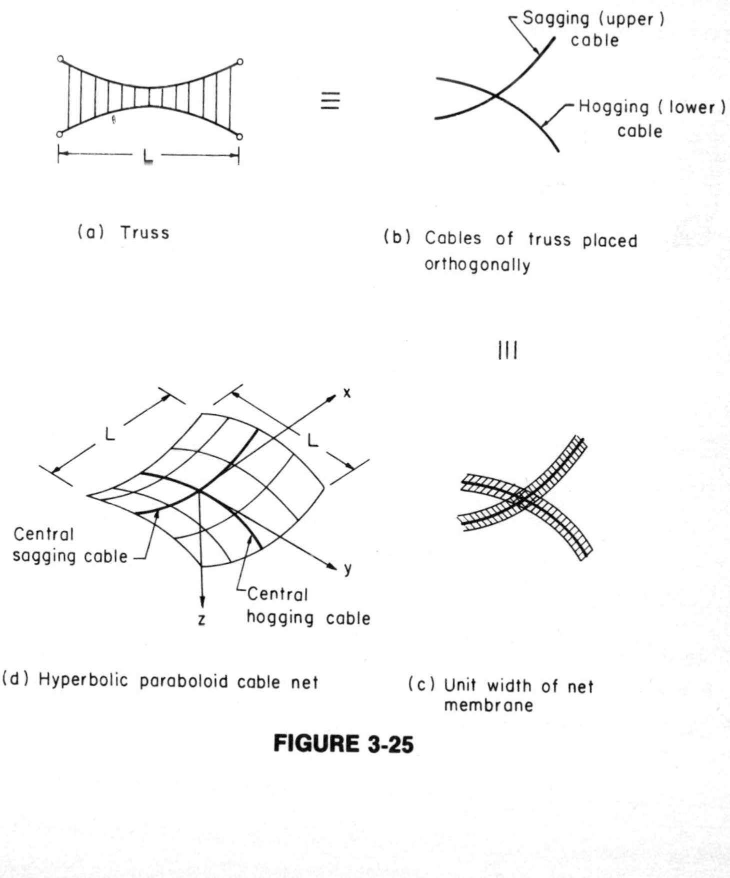

FIGURE 3-25

roof covering, cables, clamps, etc., and also by the accumulation of snow. For such loading,

$$h_x = \frac{1}{12}\,\overline{\Theta}\Theta_x k_x L^3 \tag{3-43a}$$

$$h_y = -\frac{1}{12}\,\overline{\Theta}\Theta_y k_y L^3$$

$$= -\alpha\nu h_x \tag{3-43b}$$

and

$$\beta_x = \frac{1}{12}\,\Theta_x^2 k_x L^3 \tag{3-44}$$

The deflection in the x and y directions is given by

$$w_x = \frac{1}{8}\,\overline{\Theta}L^2\left(1 - \frac{4x^2}{L^2}\right) \tag{3-45a}$$

$$w_y = \frac{1}{8}\,\overline{\Theta}L^2\left(1 - \frac{4y^2}{L^2}\right) \tag{3-45b}$$

The maximum deflection for symmetric loads ${}^s w_{max}$, for cable nets square in plan, is at the central point; that is, $x = 0$ and $y = 0$. This is given as

$${}^s w_{max} = \frac{\overline{\Theta}L^2}{8}$$

Schleyer [75] has indicated that in nets subjected to uniformly distributed loads the maximum deflection is between 1.10 to 1.25 of the average. An estimate of average value of deflections is thus possible from the maximum value.

Antisymmetric Uniformly Distributed Loads. The antisymmetric uniformly distributed loads are caused by uneven accumulation of snow or by wind loads. This case can be handled as described earlier, by using the expressions derived for plane trusses (with cables unconnected) in Sec. 3-5.1.

EXAMPLE 3-7

The hyperbolic paraboloid net shown in Fig. 3-16 and analyzed fully by a nonlinear method in Sec. 3-4.3 is solved by using the approximate procedure for these two loading cases:

1. Symmetrical load of 10 lb/ft^2 (480 N/m^2) over the entire surface
2. Load of 20 lb/ft^2 (960 N/m^2) over half the surface

The results for the central hogging and sagging cables are compared in Table 3-15.

TABLE 3-15 Results for Example 3-7

	Symmetric load		*Antisymmetric load*	
Parameter	*Approximate analysis*	*Nonlinear analysis*	*Approximate analysis*	*Nonlinear analysis*
h_x, kips (kN)	86.10 (383)	81.70 (363)	86.10 (383)	94.56 (421)
h_y, kips (kN)	− 57.40 (−255)	− 48.66 (−216)	− 57.40 (−255)	− 74.89 (−333)
w_{max}, ft (m)	1.20 (0.366)	1.16 (0.354)	1.45 (0.444)	1.88 (0.574)

3-6 FLEXIBILITY OF THE SUPPORTING STRUCTURE

Cable trusses are usually supported on columns, arches, or rings, whereas networks are supported on girders or arches. These supporting elements are much more rigid than cables, yet there is no doubt that the supporting structure interacts with the more flexible cable system and affects its load-bearing characteristics. The extent of this effect depends upon the relative rigidities. In cases where the supports are provided by a closed frame, an arch, or a ring, the cables and trusses forming the roof will also interact among themselves on account of deformation in the supporting structure. This is not so if cables or trusses are supported on separate columns. The effect of support flexibility assumes even greater importance during pretensioning operations. As the support is deformed by cable forces, cable tensions may relax or intensify.

It was pointed out in Secs. 1-2 and 3-2 that most analytical studies have assumed rigid supports for cables or their assemblies; only a few recent works [9, 68, 83] have been exceptions. A rigorous analysis is possible by treating the entire structure as a skeletal space frame with the attendant limitations of cable elements and nonlinearity duly

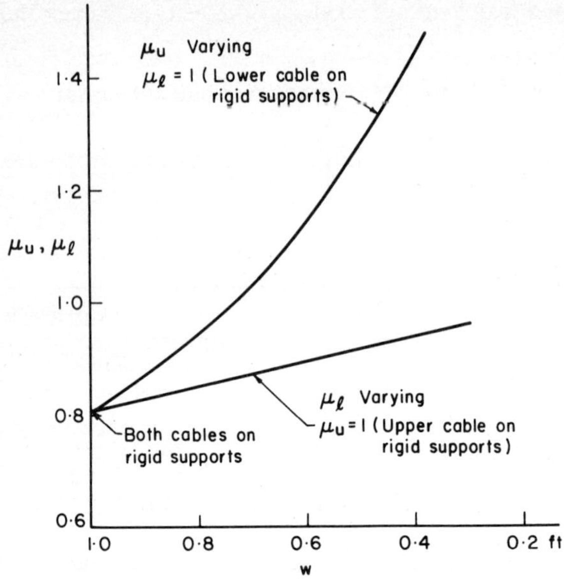

FIGURE 3-26

accounted for. This would indeed present numerical problems and would perhaps be expensive. The size of the problem can be curtailed to some extent by taking the behavior of the supporting structure as kinematically linear. This assumption will be borne out by a study of practical examples. An approximate procedure to account for support flexibility on cable trusses is described in Sec. 3-5.1. Using the formulas given in Eqs. 3-40 and 3-41, the truss in Example 3-6 is studied for the effect of support flexibility. The loading is the same as in case II of the example. Results are plotted in Fig. 3-26.

3-7 ENERGY METHOD

A procedure of analysis, alternative to the *tension-coefficient* approach described earlier, is the application of the principle of minimizing the total potential energy of the structure. The salient features of the method are described in this section. The total potential energy of any structure W_t can be expressed as

$$W_t = W_s + W_p$$

where W_s is the strain energy of the assembly and W_p is the potential energy of the applied loads. With respect to a particular state of equilibrium, say the unloaded geometry of the structure, W_t can be expressed as

$$W_t = \sum^{M} e_s - P^T U$$

where M = total number of cable elements
e_s = strain energy stored in a particular element
P^T = transpose of the vector of applied load
U = displacement vector corresponding to all the degrees of freedom in the structure, n

According to well-established energy principles, the equilibrium position of a structure corresponds to a minimum point in the n-dimensional total potential surface. This occurs when

$$\frac{\partial W_t}{\partial u_i} = G_i = 0 \qquad (i = 1, 2, 3, \ldots, n)$$

where u_i is the displacement corresponding to a particular degree of freedom i, and G_i is the corresponding gradient of the energy surface.

The minimization on such a surface may occur along any path directed towards the minimum and can be carried out with the help of standard gradient techniques used in optimization problems. These techniques include

1. The method of steepest descent
2. The method of relaxed steepest descent
3. The Newton-Raphson method
4. The method of conjugate gradients

The movement in a particular descent direction usually assumes the *instantaneous* matrix of the structure to remain a constant. This would be true for a structure with a linear load-deflection behavior, but not for nonlinear structures, such as cable systems. In these cases there is a need to control the step length, or, in other words, the movement along a certain descent direction. Experience with some of these methods in analyzing cable girders and nets has shown [7] that the Newton-Raphson method is the most efficient for girders, while the method of conjugate gradients is more efficient for the networks and grids. Step-length control improves the efficiency of the Newton-Raphson method.

The principle of minimization of total potential energy can, in a similar manner, be applied to cable systems supported on structures whose flexibility needs to be taken into account.

CHAPTER FOUR

Dynamic Analysis and Response*

4-1 GENERAL

In Chap. 1 it was stated that cable structures are prone to aerodynamic excitation or flutter since they are inherently flexible. Among this class of structures are suspension bridges, which have long and slender proportions; many failures of such bridges, because of aerodynamic forces, have occurred in the past [80]. No such failures have been recorded for roofs, but the flutter phenomenon has been observed in these also. Experience on such structures is too limited, however, to permit an attitude of complacency. Particularly if a structure is to be located in a region where gusty winds, seismic disturbances, or blasts are likely, the dynamic response of the systems needs careful evaluation, so that safe and serviceable structures can be designed. Appropriately, therefore, research on the dynamic properties of cable-suspended roofs has intensified over the last few years. This is evidenced in the fact that whereas in the first conference on the subject of cable roofs held in 1962 in Paris [25], only three papers were presented on dynamic aspects, one full session was devoted to this aspect in a Conference at Tokyo in 1971 [103]. Later in an international conference held in London in 1974, two full sessions were devoted to discussion of dynamic behavior and analysis of cable roof systems. Chapter 4 focuses on this subject and deals with procedures for determination of basic response characteristics of

* *Part of the work presented in this chapter has been carried out by V. K. Gupta, under the supervision of the author, for his doctoral research.*

typical cable systems. Methods of analysis are discussed with reference to pretensioned systems, but their application can be extended to simply suspended cables as well.

Vibrations may be caused in roofs by wind, blasts, or seismic disturbances. Wind or blasts will directly attack the roof and may cause flutter, whereas earthquake motion will first be transmitted to the supporting system and thence to the roof. In general, vertical, lateral, and torsional oscillations, or a combination thereof, are possible. However, in the case of cable roofs it would appear reasonable to expect only vertical or lateral symmetric or antisymmetric vibrations, to attain large amplitudes.

A complete dynamic study of cable roofs would consist of the following:

1. Determination of free-vibration characteristics of the cable system, with and without cladding
2. Assessment of damping for different practical situations
3. Determination of time-dependent loading, due to wind forces, blast, or seismic disturbance
4. Computation of dynamic response to applied time-dependent forces

Determination of free or forced dynamic response of structures can be obtained analytically; the methods of determination have been developed to a fair degree of fineness and are reported briefly in this chapter. The determination of wind or blast loads acting on roofs, or of earthquake excitation, and the assessment of damping must be based essentially on experimental data; there is as yet a wide gap between the requirement for and the availability of pertinent information.

4-2 ANALYSIS USING LUMPED-MASS MODELS

It was explained earlier (in Chaps. 2 and 3) that the behavior of cable systems is in general nonlinear owing to the possibility of large deformations. Any rigorous analysis for static or dynamic study, therefore, must take this into account. Linearization as a measure of approximation is of course possible. Further, as stated earlier, a cable system can be analyzed by treating it as a continuum or as a discrete system. Both these approaches have been followed in carrying out the dynamic analysis of cable systems. Membrane analysis is made, treating the system as a small-deflection problem, to arrive at approximate closed-form solutions, given in Sec. 4-3. Use of the discrete-system approach for analysis is more versatile and is developed in the following part of this section.

In general, the equation of dynamic equilibrium for a lumped-mass system can be written as

$$\mathbf{M}\ddot{\mathbf{U}} + \mathbf{C}\dot{\mathbf{U}} + \mathbf{K}\mathbf{U} = \mathbf{G} + \mathbf{P}(t) \tag{4-1}$$

where $\mathbf{M}$ = mass matrix
$\mathbf{C}$ = damping matrix
$\mathbf{K}$ = stiffness matrix
$\mathbf{G}$ = quadratic and cubic functions of joint displacements
$\mathbf{U}$ = displacement vector
$\mathbf{P}(t)$ = time-dependent load vector

Masses are assumed to be lumped at nodes in this analysis, but it is possible to obtain a consistent mass matrix. The evaluation of matrix **K** and the nonlinear terms **G** for a pretensioned system can be done exactly as described in Sec. 3-2. **K** can be computed corresponding to any particular equilibrium configuration. While generally the equilibrium position at dead load (point *A* in Fig. 4-1) is used as the datum, Tezcan et al. [93] suggest use of the equilibrium position at dead load + one-half live load (point *B* in Fig. 4-1). All the results given in this chapter are obtained by using modal analysis and are based on the assumption of small-amplitude vibrations about the configuration at prestress plus cable dead weight only.

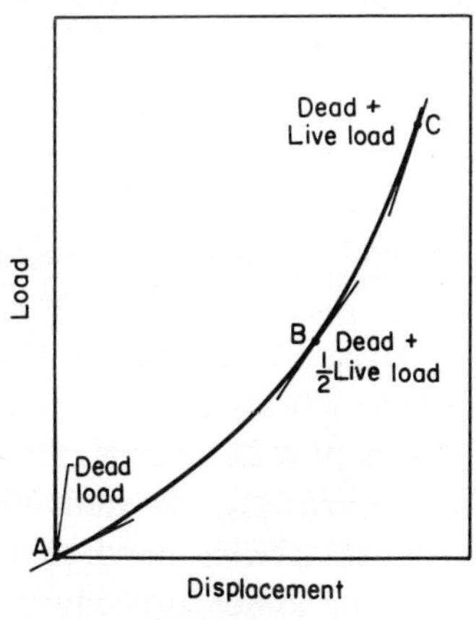

FIGURE 4-1

In case the problem is to be treated as linear, the terms **G** are dropped. Further, if damping is also neglected, we get

$$\mathbf{M\ddot{U}} + \mathbf{KU} = \mathbf{P}(t) \tag{4-2}$$

For the free-vibration problem the equation reduces to

$$\mathbf{M\ddot{U}} + \mathbf{KU} = 0 \tag{4-3}$$

The solution of Eq. 4-3 can be obtained by assuming the vibrations to be harmonic, with angular frequency ω. Equation 4-3 is reduced to the standard eigenvalue problem expressed as

$$\mathbf{AU} = \boldsymbol{\lambda}\mathbf{U} \tag{4-4}$$

where A is the modified stiffness matrix and $\boldsymbol{\lambda}$ is the characteristic eigenvalue scalar.

Methods of solution for such a problem can be broadly classified as follows:

1. Transformation methods, such as Jacobi's, Given's, and Householder's methods
2. Iteration methods, such as convergence iteration, Raleigh's quotient, inverse iteration, matrix deflation, Stadola's method, and Holzer's method

Transformation methods are preferred when all the eigenvalues and eigenvectors of a given system are required. These methods accomplish the solution by operating on the matrix, which needs to be stored and will usually require a large amount of computer space. The methods use a sequence of matrix transformation to find a similar matrix **B**, such that the evaluation of the determinant $\mathbf{B} - \omega^2\mathbf{M}$ is simpler than the calculation of $\mathbf{K} - \omega^2\mathbf{M}$. On the other hand, the iterative methods can avoid the necessity to store the entire matrix. These methods yield a sequence of scalars and vectors that converge to some particular eigenvalue and its corresponding eigenvector. The dynamic analysis of the idealized lumped-mass system has also been accomplished by using the principle of minimization of total potential energy and employing Jacobi's transformation method to solve the eigenvalue problem [8]. A study of the free-vibration characteristics of some typical pretensioned cable systems, obtained with the help of one or the other procedure of matrix analysis mentioned, now follows.

A study[1] of variation in the natural fundamental frequency η_1 of cable trusses (Fig. 4-2) with design parameters such as cable rigidity, pretension, aspect ratio, and mass intensity is shown in Figs. 4-3 to 4-5. In arriving at these results, the lumped-mass matrix has been used. It may be seen that the fundamental period of vibration in the unloaded state does not vary appreciably with the sag ratio for the two cables, but is affected by the cable rigidity and curvature and, to some extent, by pretension. A study of the mode shapes presented in the above mentioned paper [8] will further show that the degree of pretensioning can affect the modal shapes in some cases, while in others it has no significant effect. Further, contrary to expectation for conventional rigid structures, the fundamental mode is asymmetric in many cases.

Research work on similar lines has been carried out by the author

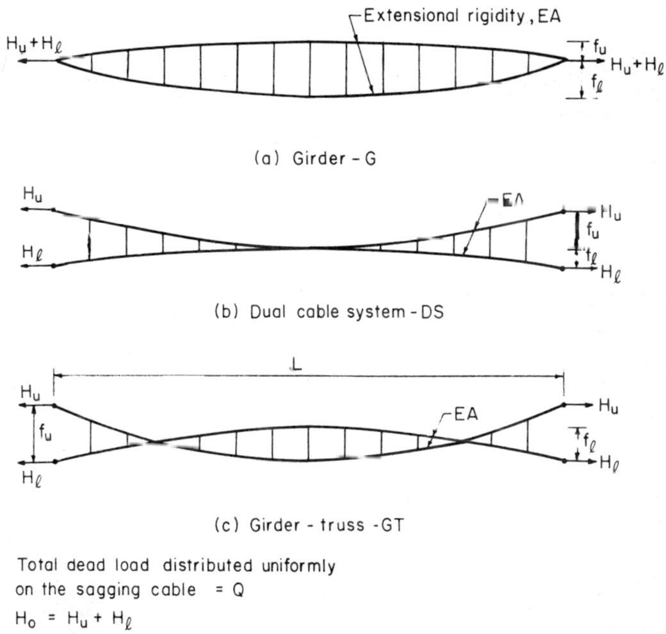

FIGURE 4-2 Truss forms examined for dynamic response (see Figs. 4-3 to 4-5).

[1] The results, which are based on Ref. 8, are reduced to a nondimensional form, on the basis of the original study carried out on 305-ft-span (100-m) trusses with $H_0 = 1600$ kN. Similarly, results given in Figs. 4-6 to 4-9 are for 100-m-span networks [49] with $H = 500$ kN.

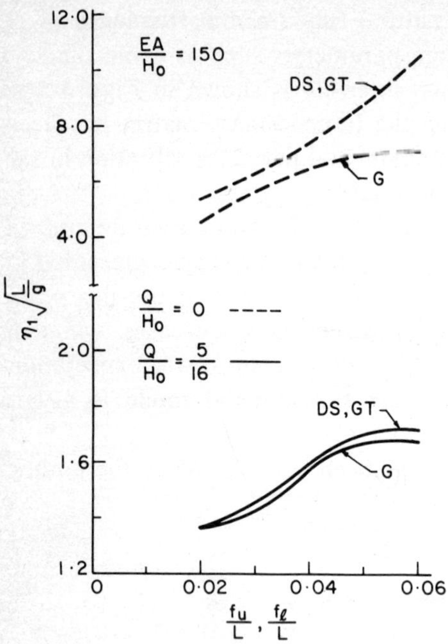

FIGURE 4-3 Variation in the fundamental frequency of vibration for trusses shown in Fig. 4-2, as a function of cable sags.

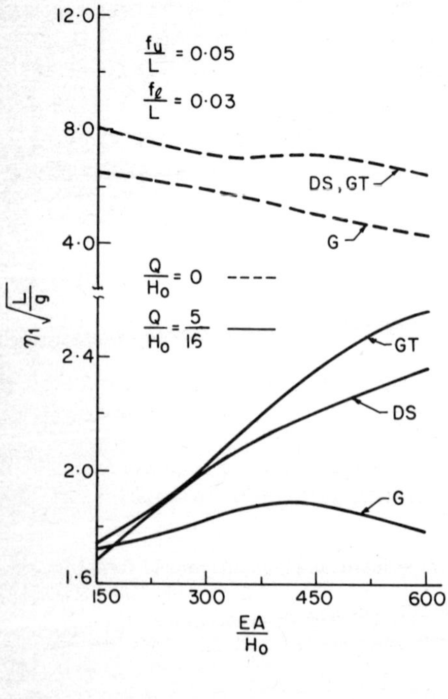

FIGURE 4-4 Variation in the fundamental frequency of vibration for trusses shown in Fig. 4-2, as a function of cable rigidity.

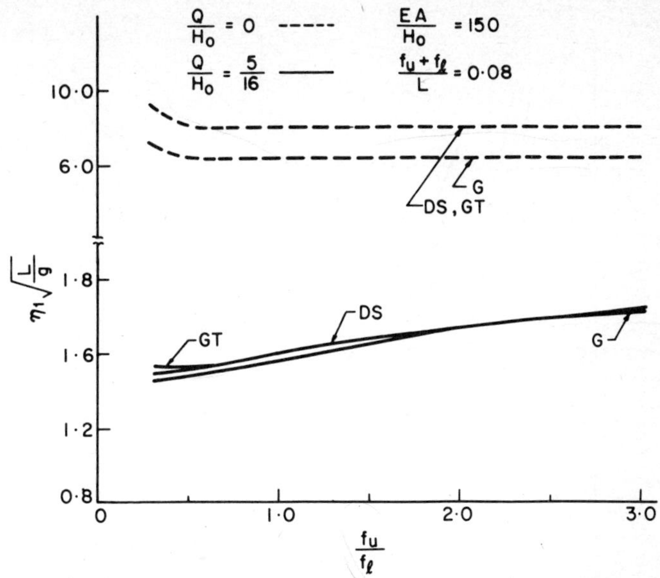

FIGURE 4-5 Variation in the fundamental frequency of vibration for trusses shown in Fig. 4-2, as a function of the ratio of cable sags.

on cable trusses [48] and hypar networks [49]. For the hypar network shown in Fig. 3-19 and the loading cases shown in Fig. 4-6, effects on frequencies of vibration η_1, η_2, η_3 of variation in design parameters (see footnote 1) are given in Fig. 4-6 to 4-9. Jensen [37–39] has also contributed significantly to the understanding of dynamic behavior of nets, through work carried out by him. Some significant features of the dynamic behavior of such networks are as follows:

1. The frequency of vibration decreases nonlinearly with increase in the intensity of applied load on the structure.
2. The frequency increases with increase in the dip/span ratio, providing additional proof that increased curvature affords greater stiffness to a network.
3. Dynamic characteristics are comparatively insensitive to changes of cable rigidity.
4. As borne out by dynamic studies of cable networks, the mode shapes for such structures are complex. For systems with simpler mode shapes it is possible to predict the number of

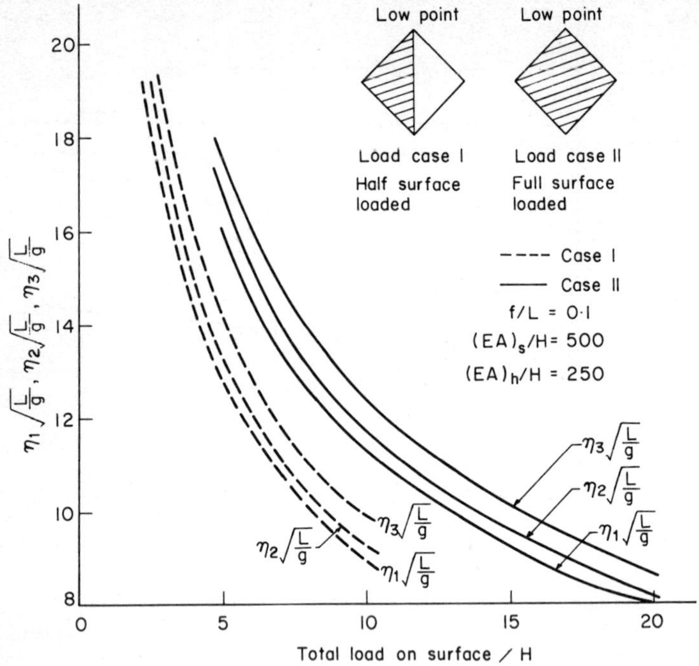

FIGURE 4-6 Vibration characteristics of cable nets (model for study same as in Fig. 3-19): effect of applied loads.

modes which would contribute significantly to the response of the structure. This may not be possible for networks because of the complexity of mode shapes. As a result, a large number of modes may need to be considered when computing dynamic response to applied time-dependent loads.

5. Jensen [40] has shown through experiments on cable networks that the frequency of vibration of a network plus cladding is 15 to 20 percent higher than that of the net alone.

A general conclusion that can be drawn from a study of results obtained for various types of cable systems is that the first mode of vibration tends to be antisymmetric and free of elastic stretching.

4-3 APPROXIMATE ANALYSIS

As shown in Sec. 4-2, modal analysis of cable systems can be accomplished by treating cable systems as lumped-mass systems and thus obtaining the eigenvalues and eigenvectors. While this information is valuable in understanding the dynamic characteristics of such

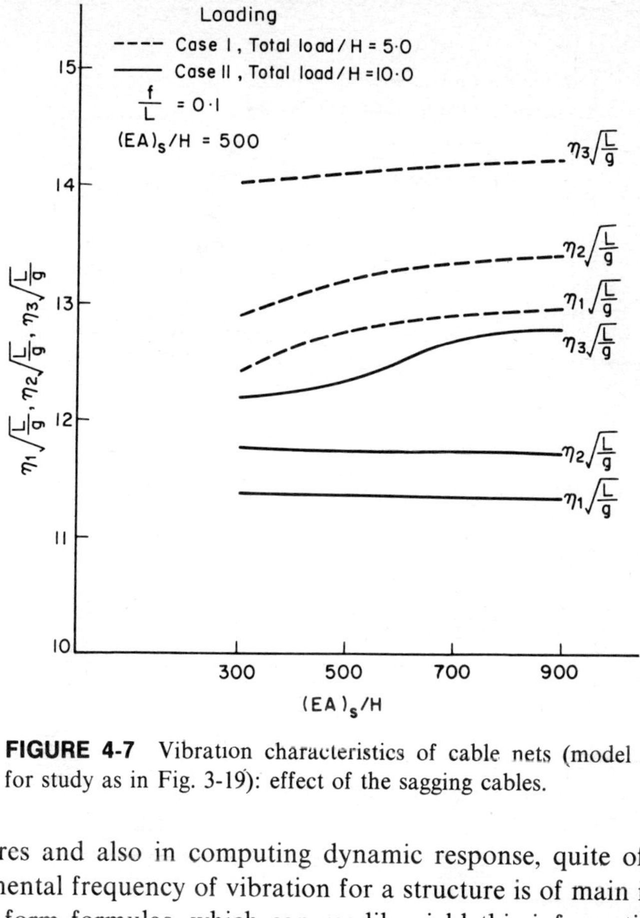

FIGURE 4-7 Vibration characteristics of cable nets (model for study as in Fig. 3-19): effect of the sagging cables.

structures and also in computing dynamic response, quite often the fundamental frequency of vibration for a structure is of main interest. Closed-form formulas, which can readily yield this information, are, therefore, of great value, even though the answer obtained may be an approximate one.

The fundamental frequency of vibration can be obtained in several ways, if linear behavior is assumed. A well-known procedure is to obtain the static deflections Δ_i for the lumped-mass system having N nodes, and then applying Rayleigh's method, which gives the circular frequency of vibration ω as

$$\omega^2 = g \frac{\sum_{i=1}^{N} P_i \Delta_i}{\sum_{i=1}^{N} P_i \Delta_i^2} \tag{4-5}$$

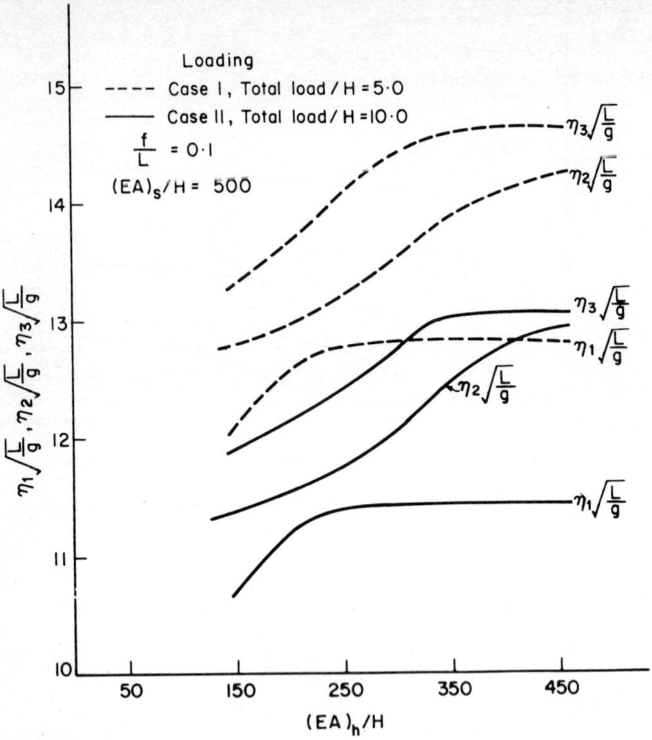

FIGURE 4-8 Vibration characteristics of cable nets (model for study same as in Fig. 3-19): effect of rigidity of the hogging cables.

where P_i = lumped mass at node i

Δ_i = deflection at node i

N = number of nodes in system

Another method is to apply the Rayleigh-Ritz [88] approach. With either method, the mode shape must be assumed. However, while the mode shape corresponding to the fundamental mode is easy to assume for conventional structures, it is not so for cable structures. This is a serious handicap in using the above procedures. Pugsley [79] has suggested the use of a semiempirical approach. The frequencies in the first three modes for a single cable, thus obtained, are as follows:

$$\eta_1 = \frac{1}{2\sqrt{2}} \sqrt{\frac{g}{f}\left(1 - \frac{3f^2}{L^2}\right)} \qquad \text{Hz}$$

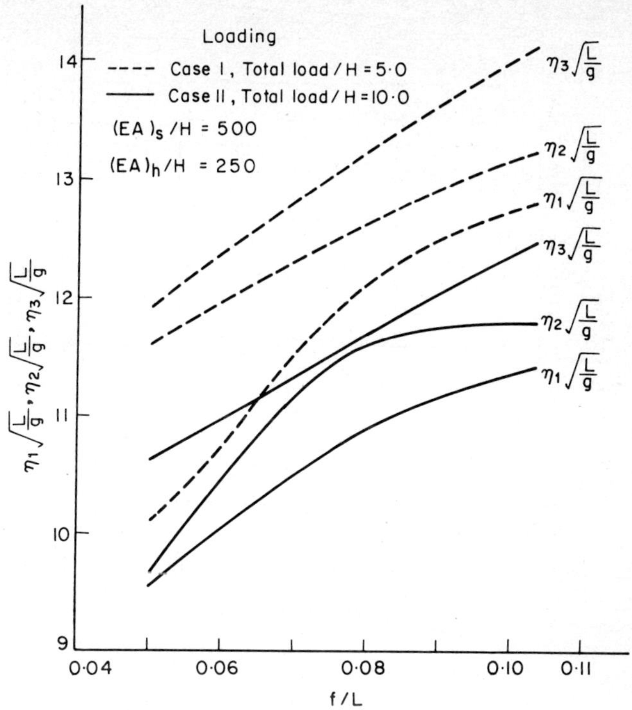

FIGURE 4-9 Vibration characteristics of cable nets (model for study same as in Fig. 3-19): effect of sag/span ratio.

$$\eta_2 = \frac{1}{2}\sqrt{\frac{g}{f}\left(1 - \frac{1.5f^2}{L^2}\right)} \qquad \text{Hz}$$

$$\eta_3 = \frac{1}{\sqrt{2}}\sqrt{\frac{g}{f}\left(1 - \frac{0.7f^2}{L^2}\right)} \qquad \text{Hz}$$

In what follows, the Rayleigh-Ritz approach proposed by Siev [88] is adopted to derive expressions for the frequencies of cable nets and trusses. In both these cases the structure is treated as a continuum.

4-3.1 Cable Net

The cable net considered for illustrating the method is a translational surface of two parabolic curves and is shown in Fig. 4-10. The analysis neglects edge displacements and horizontal loads as well as

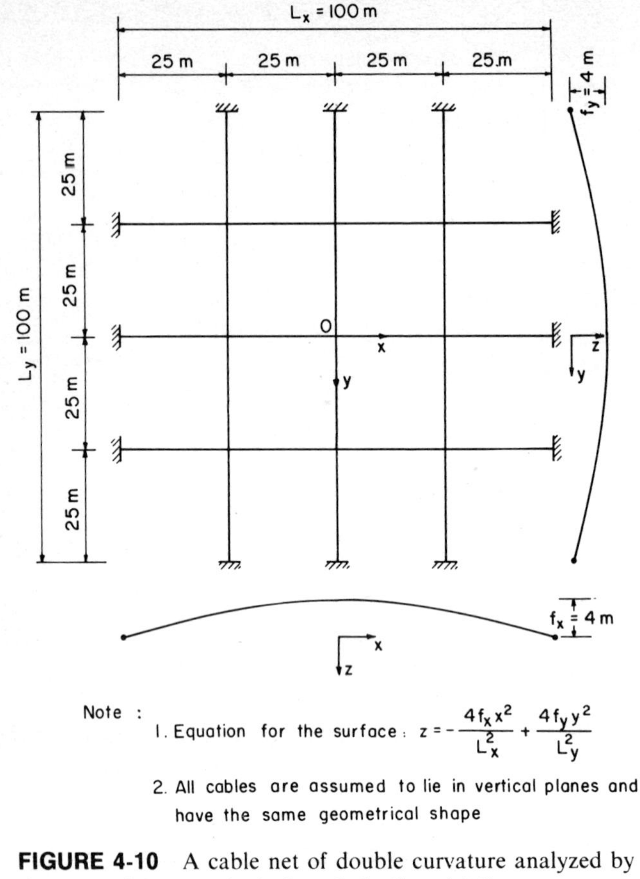

Note :

1. Equation for the surface : $z = -\dfrac{4f_x x^2}{L_x^2} + \dfrac{4f_y y^2}{L_y^2}$

2. All cables are assumed to lie in vertical planes and have the same geometrical shape

FIGURE 4-10 A cable net of double curvature analyzed by the approximate method of analysis (Sec. 4-3.1).

horizontal displacements. Only vertical loads and vertical displacements are considered.

For a surface with mass m per unit area and with horizontal components of tensile force in the cables of $H_x + h_x$ and $H_y + h_y$ in x and y directions, respectively, the equilibrium equation at any point can be written as

$$\bar{h}_x \frac{d^2z}{dx^2} + \bar{h}_y \frac{d^2z}{dy^2} + \overline{H}_x \frac{d^2w}{dx^2} + \overline{H}_y \frac{d^2w}{dy^2} = -m \frac{d^2w}{dt^2} \qquad (4\text{-}6)$$

where $\overline{H}_x$, $\overline{H}_y$ = horizontal component of tension in x and y directions per unit width of strip, that is, H_x/a, H_y/a, respectively

$\bar{h}_x$, $\bar{h}_y$ = change in horizontal component of tension in x and y directions per unit width of strip = h_x/a, h_y/a

a = spacing of cables in x and y directions

m = mass per unit area

w = vertical displacement of a point at any instant of time

Multiplying both sides of Eq. 4-6 by $0.5w$ and integrating the left-hand side, over the entire surface, we get

Strain energy of net W_1

$$= \tfrac{1}{2}\int_0^{L_x}\int_0^{L_y}\left(\bar{h}_x\,\frac{\partial^2 z}{\partial x^2}+\bar{h}_y\,\frac{\partial^2 z}{\partial y^2}+\overline{H}_x\,\frac{\partial^2 w}{\partial x^2}+\overline{H}_y\,\frac{\partial^2 w}{\partial y^2}\right)w\,dx\,dy \tag{4-7}$$

Kinetic energy W_2 can be worked out as $\frac{1}{2}$ mass × velocity^2, and for equilibrium it should be equal to W_1. It is known that the change in length of the cable

$$\Delta l_x \approx -\int_0^{L_x} w\,\frac{\partial^2 z}{\partial x^2}\,dx \tag{4-8}$$

After accounting for the change in length of the cable and assuming the curve of the cable to be parabolic, $\bar{h}_x$ and $\bar{h}_y$ can be expressed as

$$\bar{h}_x = -\,\frac{E\overline{A}_x}{L_x}\int_0^{L_x} w\,\frac{\partial^2 z}{\partial x^2}\,dx \tag{4-9a}$$

$$\bar{h}_y = -\,\frac{E\overline{A}_y}{L_y}\int_0^{L_y} w\,\frac{\partial^2 z}{\partial y^2}\,dy \tag{4-9b}$$

where $\overline{A}_x$, $\overline{A}_y$ = cross-sectional areas of cables along x and y directions, respectively, per unit width of strip

E = Young's modulus of elasticity

L_x, L_y = span of cable in x and y directions, respectively

Assuming the origin of the hypar to be at the center of the roof, O (Fig. 4-10), the equation of surface is given by

$$z = \frac{4f_x}{L_x^2}\,x^2 + \frac{4f_y}{L_y^2}\,y^2 \tag{4-10}$$

Curvatures in x and y directions are

$$\frac{\partial^2 z}{\partial x^2} = \Theta_x = -\frac{8f_x}{L_x^2} \tag{4-11}$$

$$\frac{\partial^2 z}{\partial y^2} = \Theta_y = \frac{8f_y}{L_y^2}$$

where f_x and f_y are the sags of cable in the x and y directions, respectively.

The origin is now assumed to be at one corner for convenience in calculations [88]. For a uniformly distributed load, the displacement is given by

$$w = \sum_i \sum_j C_{ij} \sin \frac{i\pi x}{L_x} \sin \frac{j\pi y}{L_y} \tag{4-12}$$

where i and j are odd integers.

Equation 4-10 remains unaltered because of the invariance of the second derivative. By combining Eqs. 4-11 and 4-12 with Eq. 4-9, one obtains the values of $\bar{h}_x$ and $\bar{h}_y$. Substitution in Eq. 4-7 now yields the value of strain energy W_1 as

$$W_1 = \frac{E\bar{A}_x L_x L_y \Theta_x^2}{\pi^2} \sum\sum \frac{C_{ij}^2}{i^2} + \frac{E\bar{A}_y L_x L_y \Theta_y^2}{\pi^2} \sum\sum \frac{C_{ij}^2}{j^2}$$

$$+ \frac{\bar{H}_x \pi^2 L_y}{8L_x} \sum\sum i^2 C_{ij}^2 + \frac{\bar{H}_y \pi^2 L_x}{8L_y} \sum\sum j^2 C_{ij}^2 \tag{4-13}$$

The maximum kinetic energy for the vibrating structure is given by

$$W_2 = \frac{1}{2} \text{ mass} \times (\text{velocity})^2$$

$$= \frac{m}{2} \omega^2 \int_0^{L_x} \int_0^{L_y} w_{\max}^2 \, dx \, dy \tag{4-14}$$

Assuming that the maximum amplitude of vibration is given by Eq. 4-12 and that $w = w_{\max} \sin \omega t$, the kinetic energy W_2 is obtained as

$$W_2 = \frac{m}{2} \omega^2 \frac{L_x L_y}{4} \sum_i \sum_j C_{ij}^2 \tag{4-15}$$

According to the Rayleigh-Ritz method, which proceeds by minimizing $W_2 - W_1$,

$$\frac{\partial(W_2 - W_1)}{\partial C_{ij}} = 0 \tag{4-16}$$

Therefore, by carrying out the differentiation in Eq. 4-16, and as an approximation, considering only the first term in the series in Eq. 4-12, we get

$$\omega^2 = \frac{\pi^2}{m}\left[\frac{\bar{H}_x}{L_x^2} + \frac{\bar{H}_y}{L_y^2} + \frac{8}{\pi^2}\left(\frac{\Theta_x^2 E\bar{A}_x}{\pi^2} + \frac{\Theta_y^2 E\bar{A}_y}{\pi^2}\right)\right] \tag{4-17}$$

The terms within parentheses arise from $\bar{h}_x$ and $\bar{h}_y$, respectively, and if i and j are even integers in Eq. 4-12, $\bar{h}_x$ and $\bar{h}_y$ are equal to zero and these terms vanish. Taking $i = j = 2$ (even integers), for C_{22} we get

$$\omega^2 = \frac{4\pi^2}{m}\left(\frac{\bar{H}_x}{L_x^2} + \frac{\bar{H}_y}{L_y^2}\right) \tag{4-18}$$

Equation 4-18 gives the frequency of vibration for an antisymmetric mode with asymmetry about both the x and y axes.

If i is odd and j is even in Eq. 4-12, $\bar{h}_y$ is equal to zero, and taking $i = 1$ and $j = 2$, for C_{12} we get

$$\omega^2 = \frac{\pi^2}{m}\left(\frac{\bar{H}_x}{L_x^2} + \frac{4\bar{H}_y}{L_y^2} + \frac{8\Theta_x^2 E\bar{A}_x}{\pi^4}\right) \tag{4-19}$$

Equation 4-19 gives the frequency of vibration for a symmetric-

TABLE 4-1 Comparison of Modal Analysis with Approiimate Analysis for a Cable Net

Serial no.	*Nature of mode*	*Frequency of vibration, Hz* — *Aprox. formulas*	*Modal analysis*
1	Symmetric	1.315	1.26
2	Antisymmetric	1.080	0.98
3	Symmetric about x axis and antisymmetric about y axis	1.15	1.095
4	Symmetric about y axis and antisymmetric about x axis	1.259	1.17

antisymmetric mode, with symmetry about the x axis and asymmetry about the y axis. Similarly, taking $i = 2$ and $j = 1$, for C_{21} we get

$$\omega^2 = \frac{\pi^2}{m}\left[\frac{4\overline{H}_x}{L_x^2} + \frac{\overline{H}_y}{L_y^2} + \frac{8\Theta_y^2 E\overline{A}_y}{\pi^4}\right] \tag{4-20}$$

Equation 4-20 gives the frequency of vibration for a symmetric-antisymmetric mode, with symmetry about the y axis and asymmetry about the x axis.

Results for a cable net of the type shown in Fig. 4-10 are obtained by modal analysis. Next, the net is reduced to an equivalent membrane [88] and analyzed by using Eqs. 4-17 to 4-20. The results are compared in Table 4-1. The data for the net are

EA = 262,500.0 tonnes (2575 MN) per cable

H_x = 580.0 tonnes (5685 kN) per cable

H_y = 326.0 tonnes (3195 kN) per cable

Load at each node = 18.75 tonnes (183 kN)

$L_x = L_y = 100$ m $f_x = f_y = 4$ m

The results for a flat net (Fig. 4-11) have also been computed using modal analysis as well as from Eq. 4-17. The resulting values of fundamental frequency are 57.6 Hz and 59 Hz, respectively.

The data for the problem are

EA = 15,000 kips (66.6 MN) for each cable

$H_x = H_y$ = 75 kips (333 kN) for each cable

Load at each node = 8.7 lb (38.6 N)

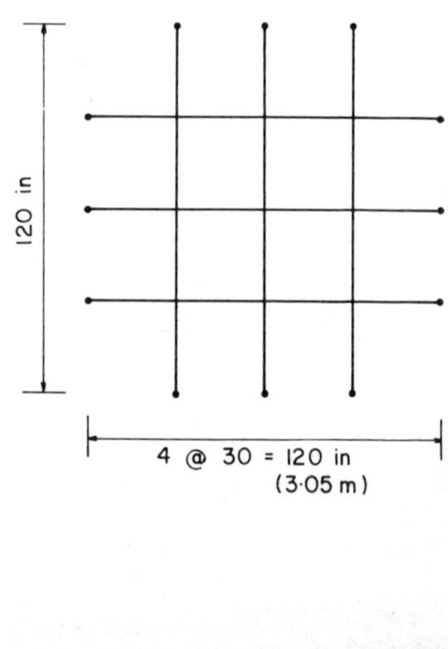

FIGURE 4-11 Cable grid.

4-3.2 Cable Truss

Using the approach adopted in Sec. 4-3.1, the equilibrium equation at any point for a truss can also be written as follows:

$$h_u\left(\frac{d^2z}{dx^2}\right)_u + h_l\left(\frac{d^2z}{dx^2}\right)_l + (H_u + H_l)\frac{d^2w}{dx^2} = -m\frac{d^2w}{dt^2} \tag{4-21}$$

where m is the mass per unit length of span. On simplification, Eq. 4-21 reduces to

$$\left(\frac{d^2z}{dx^2}\right)_u \frac{(EA)_u}{L}(1 + \kappa v^2)\int\left(\frac{d^2z}{dx^2}\right)_u w\,dx - (H_u + H_l)\frac{d^2w}{dx^2} = m\frac{d^2w}{dt^2} \tag{4-22}$$

TABLE 4-2 Comparison of Modal Analysis with Approximate Analysis for Cable Trusses

		Natural frequency, Hz			
		Double-wave antisymmetric vibration		Minimum frequency of symmetric‡ vibration. Modal analysis	Single-wave symmetric vibration. Approximate analysis (Eq. 4-23)
Serial No.	Data*	Modal analysis	Approximate analysis (Eq. 4-24)		
1	$f_u = f_l = 0.0625L$ $(EA)_u = (EA)_l = 300H_u$ $(EA)_v = 30H_u$ $F_v = 0.08H_u$ $qL^† = 1.2H_u$ $N_h = 5$	0.386	0.404	0.478	0.541
2	$f_u = 0.045L$ $f_l = 0.075L$ $(EA)_u = (EA)_l = 400H_u$ $(EA)_v = 11.2H_u$ $F_v = 0.0225H_u$ $qL^† = 0.5H_u$ $N_h = 15$	0.567	0.577	0.743	0.959
3	$f_u = 0.105L$ $f_l = 0.075L$ $(EA)_u = (EA)_l = 400H_u$ $(EA)_v = 11.2H_u$ $F_v = 0.0525H_u$ $qL^† = 0.5H_u$ $N_h = 15$	0.687	0.688	0.924	1.355

* For notation see Sec. 3-4.1.

† Load acting on the upper cable.

‡ Mode shape corresponding to this value is of the type

where L = span
$\nu = \Theta_l/\Theta_u$
$\kappa = (EA)_l/(EA)_u$
$(EA)_l$, $(EA)_u$ = cable rigidity for lower and upper cable, respectively
Θ_l, Θ_u = curvature for lower and upper cable, respectively

Solution of Eq. 4-22 is carried out by the same procedure adopted in Sec. 4-3.1 and an approximate formula for natural frequency of a dual-cable system is derived for a single-wave symmetric mode as

$$\omega^2 = \frac{\pi^2}{mL^2}\left[(H_u + H_l) + \left(\frac{8}{\pi^2}\right)\frac{\Theta_u^2(EA)_u L^2}{\pi^2}(1 + \nu^2\kappa)\right] \quad (4\text{-}23)$$

The closed-form expression for frequency of vibration with a double-wave antisymmetric mode is given by

$$\omega^2 = \frac{4\pi^2}{mL^2}(H_u + H_l) \quad (4\text{-}24)$$

Results of a few cases of dual-cable systems of the type analyzed in Sec. 3-4.1 have been computed by using Eqs. 4-23 and 4-24, and they have been compared in Table 4-2 with the results obtained from modal analysis. The correlation is reasonably good.

4-4 DAMPING

The damping capacity of tension-roof structures is an important parameter in the study of the dynamic stability of these structures. A structure derives its damping capacity through various highly complex energy-absorption mechanisms. In tension roofs, damping may be affected by such structural details as the joints and nodes, by prestressing and the cladding material chosen, and by the surrounding air. The damping can thus be classified as structural, material, or aerodynamic damping.

A commonly used measure of damping in a system is the logarithmic decrement δ, which is defined as

$$\delta = \log_e \frac{a_n}{a_{n+1}} \quad (4\text{-}25)$$

where a_n and a_{n+1} are the amplitudes of vibration in the nth and $(n + 1)$th cycles of vibration. For aerodynamic damping, which need be taken into account only for light structures, the logarithmic decrement can be expressed as

$$\delta = \tfrac{1}{2} C_d \frac{m'}{A'} \frac{A}{m} \tag{4-26}$$

where m'/A' = mean virtual mass per unit area of the vibrating air
m/A = mass per unit area of the structure
C_d = a coefficient varying between 0.03 and 0.07

Information on the value of δ to be expected in cable systems is extremely limited. However, some guiding values can be had from experiments reported by Jensen [40]. These are summarized below:

Cable trusses and networks (without cladding)	$\frac{1}{2}$ to 3 percent
Tent-type structures	17 to 20 percent

Jensen has further observed that the presence of cladding considerably enhances the value of damping. Damping does not appear to be affected by change in the mass of the system; however, change in the superimposed loading may affect these values considerably. For the Olympic Stadium roof at the 1972 Munich Olympics, a value of 10 percent, including air damping, was considered in the design [55]. (See also Sec. 5-6.)

4-5 TIME-DEPENDENT LOADS

This section includes a general discussion of the nature of loads caused by wind, blast, and earthquakes.

4-5.1 Wind Loading

Computations for the static or dynamic effects of wind loads can be only as accurate as the corresponding estimate of the load, the accuracy of which generally is quite elusive. The general nature of wind is turbulent, but for convenience in analysis, it may be considered to have a steady (or *time-averaged*) velocity superimposed with a fluctuating component. The mean wind leads to a steady (time-averaged) pressure and a static response in the structure. The fluctuating component, which is caused by the turbulence or gustiness of the wind, leads to a fluctuating pressure and hence a dynamic response.

The wind pressure is given by

$$p_w = \tfrac{1}{2}\rho C_p V^2 \tag{4-27}$$

where p_w = pressure
ρ = mass density of air
C_p = wind-pressure coefficient
V = velocity

The value of the pressure coefficient C_p is dependent primarily on the geometric location of the point of measurement in the structure and the Reynolds number for bodies placed within the boundary layer. For sharp-edged bodies, where the separation points are fixed, the Reynolds number is known to have little effect on pressure distribution over the body. As a result, the values of C_p obtained from a model can be applied to a prototype. The same advantage can be had for cable roofs, which are likely to have sharp-edged parapets or beams for architectural or structural reasons and are also placed well within the boundary layer.

Further, in cases where the roof is likely to vibrate with large amplitudes, the surrounding air may influence its behavior. This can be taken into account by the use of the expression

$$p_w = \tfrac{1}{2}\rho C_p V^2 + C_m \rho \frac{A_0}{D} \frac{dV}{dt} \tag{4-28}$$

where C_m = coefficient of an additional mass
A_0 = reference area for a virtual mass
D = typical dimension for the object

Jensen [40] has assigned a range of 2.5 to 7.5 for the coefficient C_m on the basis of experimental work. The mass of vibrating air is, understandably, a function of the geometric characteristics of the structure and the mode of vibration, and its effect is likely to be more significant for lightweight roofs. The mean value of the wind velocity V may be evaluated from measurements at site [21] or obtained from the relevant code of practice.

A representation of the fluctuating wind pressure is not easy to obtain. Although wind records as shown in Fig. 4-12 are available, there exists little information about the distribution of the wind

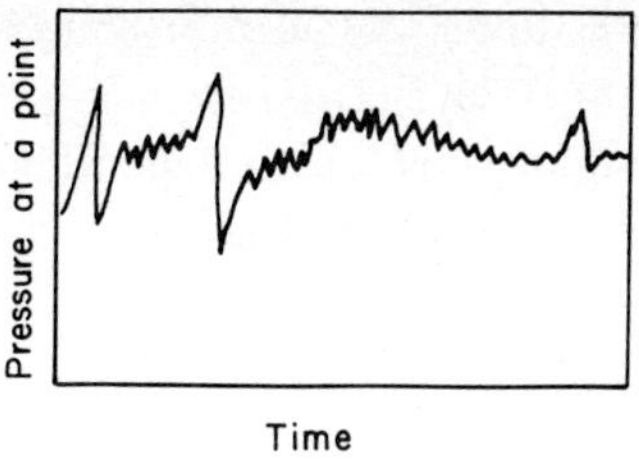

FIGURE 4-12 Typical wind pressure vs. time record.

velocity over the surface of the structure as a function of time. For large structures, this point can be of importance, because the elementary approach, which assumes that the same velocity acts on all points on a structure simultaneously, is perhaps oversimplified.

The spatial distribution of wind and its variation with time can be obtained by measuring the fluctuating pressures on a model placed in a *steady* turbulent flow. Although, ideally an aeroelastic model should be used in order to admit interaction between the dynamic response and the pressures, rigid models can be expected to give fair estimates. The turbulence of the approaching wind is affected by the physical characteristics of the region, such as the approach roughness. An idea of the nature of the wind is obtainable by field measurements. A great deal of work has already been done in this regard [20] and may be used for modeling the turbulence.

Once the spatial distribution of the wind pressures is known, it is best to obtain the dynamic wind loads with a statistical approach [16], by computing the variance, standard deviation, autocovariance function, spatial correlation function, and the power spectral density.

Whereas cognizance ought to be given to the wind-tunnel tests on aeroelastic models and meteorological measurements at site, this is time-consuming effort. Reliable estimates can be obtained by taking advantage of the major pioneering work carried out by Davenport [18–21] on the development of statistical characteristics of wind and estimation of gust factors. His results are analytically derived on the basis of field data and laboratory measurements. According to Davenport [21], the gust factor G is given as,

$$G = 1 + \Psi r \sqrt{\Omega + \gamma} \tag{4-29}$$

where Ψ = peak factor, which depends upon the fundamental natural frequency of the structure η_1 and the averaging period t. The period t should be between 5 min and 1 h. The values of Ψ are given in Fig. 4-13.

r = roughness factor, which depends upon the terrain roughness and the height of the structure (see Fig. 4-14)

Ω = excitation by background turbulence, which is dependent on the height of the structure and is given in Fig. 4-15.

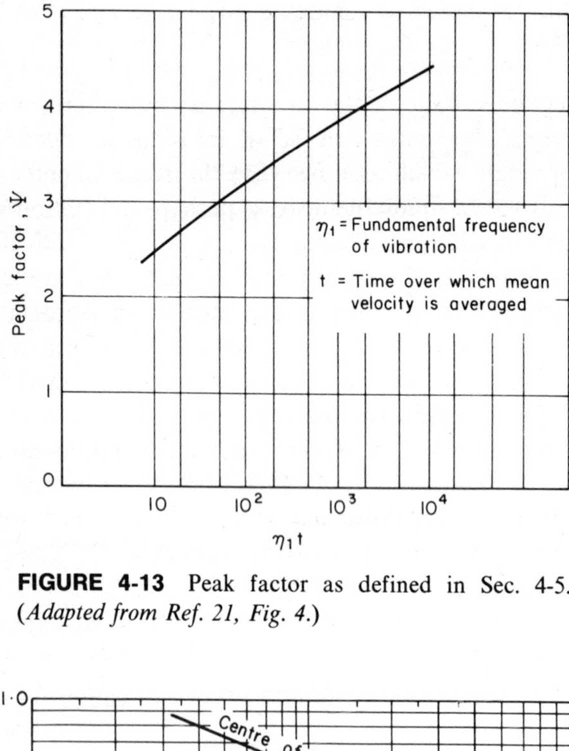

FIGURE 4-13 Peak factor as defined in Sec. 4-5.1. (*Adapted from Ref. 21, Fig. 4.*)

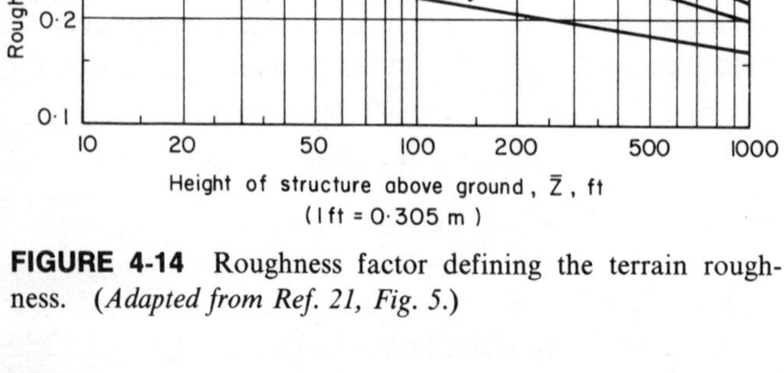

FIGURE 4-14 Roughness factor defining the terrain roughness. (*Adapted from Ref. 21, Fig. 5.*)

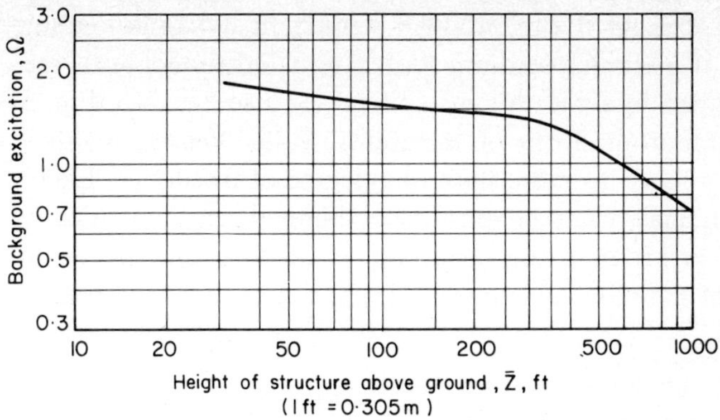

FIGURE 4-15 Background excitation factor as defined in Sec. 4-5.1. (*Adapted from Ref. 21, Fig. 6.*)

γ = excitation by turbulence resonant with structure and can be expressed as

$$\gamma = \frac{\vartheta W_g}{\zeta}$$

where ϑ = size reduction factor and depends upon the reduced frequency $\eta_1 Z / V_m$ and the width to height ratio D/Z (Fig. 4-16). V_m is the mean wind velocity at the top of the structure of height Z and width D.

W_g = gust energy ratio, dependent upon η_1 / V_m (Fig. 4-17).

ζ = critical damping ratio, covering both aerodynamic as well as mechanical damping

Once the gust factor is evaluated, the maximum pressure is obtained as G times the average pressure, and the structural response can be evaluated in the usual manner. The result thus obtained would be the total aerostatic equivalent of the effect of wind.

Many structures generate significant self-induced dynamic loading. Vortices form behind most bluff bodies within the practical range of velocities. In doing so, they cause a dynamic load across the direction of the flow, which produces an oscillatory effect on the complete structure. While this is unlikely to be important in suspended roofs, another form of instability that can be caused by vortices shed from the edges may be of concern. Often for structural or aesthetic reasons, some form of parapet is required at the roof edge. Vortices forming behind it can produce significant dynamic loading. In the case of the Munich stadium roof, after wind-tunnel tests, the edge rim was shaped to avoid this instability (see also Sec. 5-6, Fig. 5-10).

An important form of instability, called *flutter*, involves coupled motion related to more than one degree of freedom. Each degree of freedom may itself be positively damped aerodynamically, but the energy to promote the instability is drawn from the airstream as a result of coupling of modes. The phenomenon has been studied in

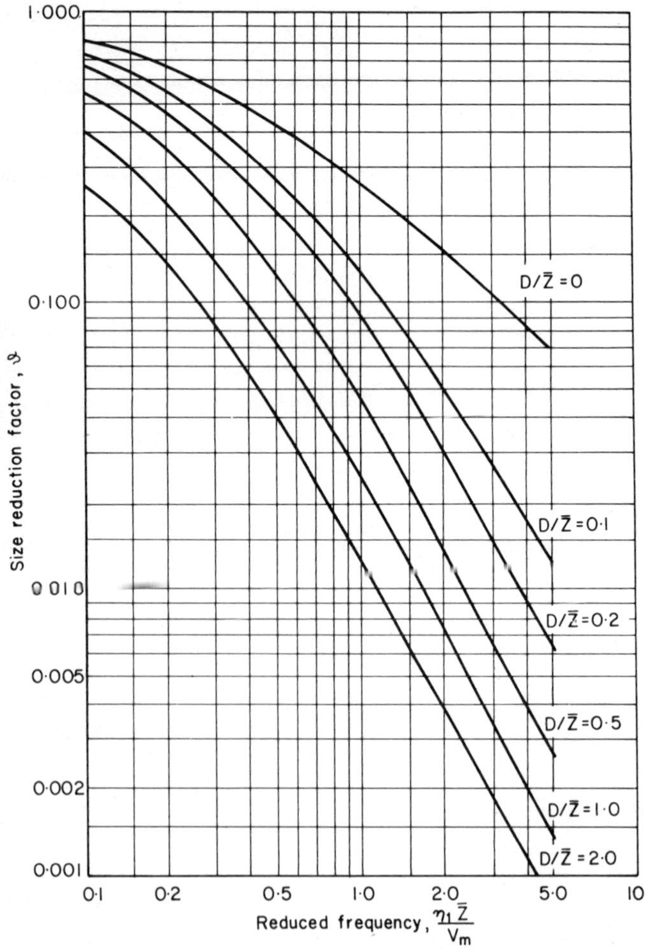

FIGURE 4-16 Size reduction factor vs. reduced frequency. (*Adapted from Ref. 21, Fig. 7.*)

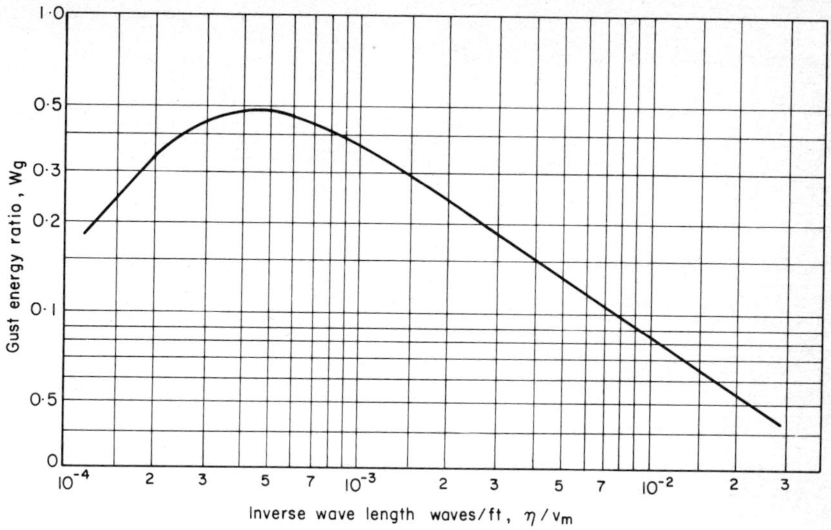

FIGURE 4-17 Gust energy ratio. (*Adapted from Ref. 21, Fig. 8.*)

detail for aircraft wings and suspension bridges. Also, theoretical predictions are possible with respect to *platelike* structures. They may, however, be difficult to make for suspended roofs on the basis of our present knowledge. Another form of aerodynamic instability that can occur locally in panels is known as *panel flutter*. This is caused by the intense turbulence that occurs locally in the boundary layer. Its prediction by analytical means is also difficult.

It is worth noting further that cable roofs, being low-frequency structures, are liable to respond to low-frequency gusts, which contain a larger proportion of the turbulent kinetic energy. The peak response will be obtained at the natural frequency of the structure. However, because of the random nature of the dynamic pressure, only quasi-resonance conditions may be expected unless the roof lies in the wake of another structure and buffetting occurs.

4-5.2 Blast Loading

A blast or an explosion near the surface causes the maximum damage; it loads the structure dynamically. The loading can be divided into three parts: (1) diffraction loading, (2) the effect of overpressure, and (3) drag loading. Diffraction loading results from the direct and reflected pressures from the object and lasts until the wave

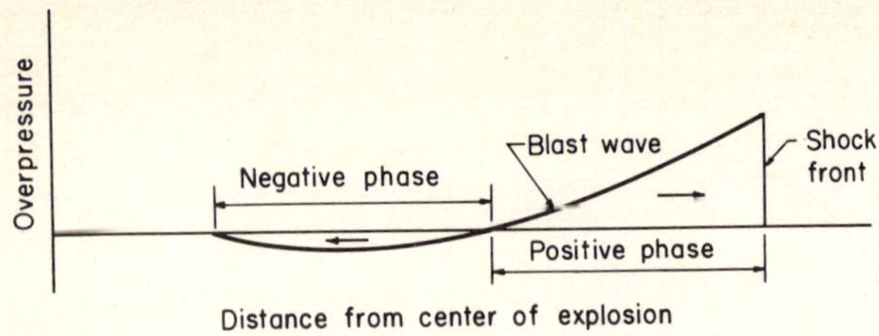

FIGURE 4-18 Overpressures due to a blast, as function of distance from the point of explosion.

reaches the rear part of the structure. The time involved is therefore a function of the building dimensions. Overpressure is created by the explosion around the blasting site and propagates in all directions. At a site the variation of overpressure is as shown in Fig. 4-18. Drag force is exerted by the air particles moving with high velocity V and is given by $\frac{1}{2} C_D \rho V^2$, where C_D is the drag coefficient. The total effect of blast can thus be seen from the pressure-time diagram given in Fig. 4-19. A numerical estimate of the blast pressures and the related time scales can be obtained by using results given by Hansen [72].

It should thus be possible to determine the velocity and pressure distribution of a wave front at the time it meets a structure. The loading caused on the structure is very much a function of the physical features of the structure itself. The problem is similar to that of determining loads due to wind except that the characteristics of the wave front are perhaps more easily predictable than those of the wind. Studies have been carried out to determine blast loads on box-type structures [72, 99], but these are unlikely to be of much help in the determination of blast loads on cable roofs.

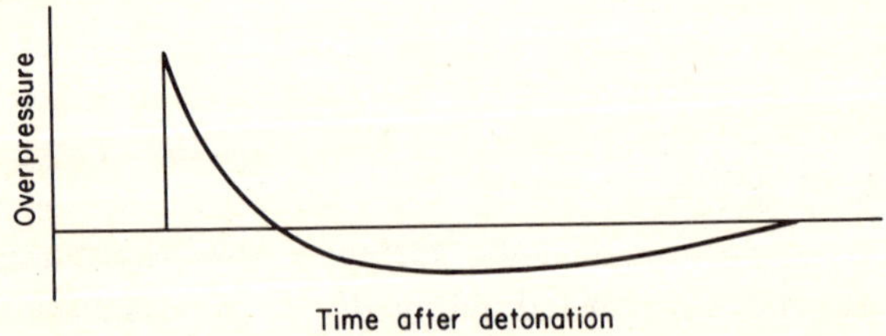

FIGURE 4-19 Overpressures due to a blast, as a function of time.

4-5.3 Earthquake Loading

An earthquake is a ground motion which causes the entire cable-roof system to vibrate along with the supporting structure. The response of the roof is therefore a function of the dynamic characteristics of the supporting structure as well as of the nature of the earthquake ground motion. For a dynamic analysis, a possible model of the structural system could be one in which ground motion is applied at the base of the supporting structure. The equations of motion could be written in the following fashion:

$$\mathbf{M}\ddot{\mathbf{U}} + \mathbf{C}\dot{\mathbf{U}} + \mathbf{K}\mathbf{U} = -\mathbf{M}\mathbf{a} \tag{4-30}$$

where $\mathbf{M}$ = diagonal mass matrix
$\mathbf{C}$ = damping matrix
$\mathbf{K}$ = stiffness matrix
$\mathbf{U}$ = relative displacement vector of lumped masses with respect to ground
$\mathbf{a}$ = ground acceleration

Equations of the above form are to be solved using any of the numerical integration techniques available. Alternately, response spectrum theory could be used to compute modal responses which may be combined to give the total dynamic response. In Eq. 4-30, $\mathbf{a}$ is to be chosen specifically for the site in question. This is often a difficult choice and depends on the seismic history and activity of the region and on the distance of the site from possible future epicenters. Recorded accelerograms with modifications and also artificial accelerograms are commonly employed in practice for use as input in Eq. 4-30.

It is seen from dynamic analysis of cable roofs that such structures usually have a long period of vibration (say, more than 0.25 s). Also, the supporting structure is relatively a much stiffer system. Therefore, high-frequency contents of earthquake ground motion (acceleration pulses usually have a range of frequency varying from about 3 to 10 Hz) will be amplified by it and low-frequency components will attenuate considerably by the time they reach the cable system and would thus cause no serious problem. However, if low-frequency contents dominate the ground motion, the supporting structure will transmit the same with little attenuation and analysis must be carried out as outlined in the preceding paragraphs. Seismic forces no doubt may be small on account of the flexibility of the cable system, but the dynamic displacements could still be large and may cause damage to the cladding or the details. Necessary precautions should be taken to safeguard against this.

4-6 DYNAMIC RESPONSE

The dynamic response of structures can be determined with reasonable accuracy employing lumped-mass models of the type discussed in Sec. 4-2. The dynamic loads on the structure are considered to be concentrated at the nodes too, and may be designated as $P_1(t)$, $P_2(t), \ldots, P_i(t)$.

After determining the natural frequencies (ω_j) and the mode shapes (ϕ_{ij}), the response in a mode is computed using the concept of *dynamic load factor* (DLF). The value of DLF depends upon the shape of the force-time curve, time t, and the frequency of the system; it can be found in any textbook on structural dynamics. The displacement response of any node i in the jth mode is easily shown to be

$$u_{ij} = \frac{\phi u_{ij}(\mathrm{DLF})_j \sum_{i=1}^{N} P_{i1}\phi u_{ij}}{\omega_j^2 \sum_{i=1}^{N} m_i\left(\phi u_{ij}^2 + \phi v_{ij}^2 + \phi w_{ij}^2\right)} \tag{4-31}$$

where P_{i1} = maximum value of force at node i
ϕu_{ij} = mode shape coefficient in x direction for ith mass in jth mode
ϕv_{ij} = mode shape coefficient in y direction for ith mass in jth mode
ϕw_{ij} = mode shape coefficient in z direction for ith mass in jth mode
$(\mathrm{DLF})_j$ = dynamic loading factor corresponding to jth mode
m_i = lumped mass at the ith node
N = number of nodes
ω_j = frequency in jth mode

The maximum dynamic displacement $u_{ij,\max}$ is obtained by using the maximum values of $(\mathrm{DLF})_j$ in Eq. 4-31, and modal member forces F_j are obtained by using Eq. 3-7. The final value of force is obtained by combining the modal forces according to the quadratic law and

TABLE 4-3 Change in Forces in Truss Members, kips × 10^3

		No. of modes superimposed							
		*Horizontal applied force**				*Vertical applied force**			
	Member no.	*1*	*3*	*5*	*10*	*1*	*3*	*5*	*10*
	1	1.06	1.06	1.26	1.33	Insignificant	155.42	155.54	155.56
	2	1.04	1.04	1.18	1.21		155.29	155.41	155.43
	3	0.96	0.96	1.04	1.04		155.25	155.40	155.42
	4	0.84	0.84	0.86	0.87		155.29	155.46	155.48
	5	0.69	0.69	0.70	0.71		155.45	155.62	155.64
	6	0.50	0.50	0.51	0.52		155.51	155.68	155.69
	7	0.29	0.29	0.30	0.31		155.50	155.66	155.66
	8	0.09	0.09	0.09	0.10		155.45	155.45	155.60
	17	0.98	0.98	1.00	1.01		104.22	104.34	104.34
	18	0.84	0.84	0.85	0.86		104.09	104.18	104.18
	19	0.64	0.64	0.65	0.66		103.69	103.76	103.76
	20	0.45	0.45	0.47	0.48		103.15	103.23	103.23
	21	0.29	0.29	0.30	0.31		102.59	102.69	102.69
	22	0.16	0.16	0.16	0.17		102.18	102.27	102.27
	23	0.08	0.08	0.08	0.08		101.94	102.04	102.04
	24	0.04	0.04	0.04	0.04		101.91	101.98	102.00
Hangers	33	0.41	0.41	0.78	1.37		2.88	3.81	4.58
	34	0.73	0.73	1.19	1.38		5.19	5.91	6.01
	35	0.95	0.95	1.16	1.39		6.09	6.13	6.49
	36	1.03	1.03	1.03	1.41		5.35	5.68	5.97
	37	0.94	0.94	1.15	1.39		3.96	4.82	5.03
	38	0.72	0.72	1.17	1.36		4.13	4.25	4.88
	39	0.39	0.39	0.76	1.35		5.81	6.06	6.07
	40	0.01	0.01	0.01	0.01		6.55	7.22	7.65

* Horizontal force gets maximum response from mode 1, which is asymmetric, and vertical force gets maximum response from mode 2, which is symmetric.

1 kip = 4.448 kN

equals $(\sum F_j^2)^{1/2}$. The total displacement response is obtained likewise.

As an example, the cable-truss structure in Fig. 3-4*a* is subjected at each of its upper-cable nodes (nodes 2 to 16) to a force of 1 kip diminishing linearly to zero in a period of 0.1 s. The load is deemed to represent, in an idealized form, a blast near the roof. The computed response for the left half of the truss is tabulated in Tables 4-3 and 4-4. Results of the first mode are compared with those obtained by superimposing the first 3, 5, and 10 modes, respectively. It may

TABLE 4-4 Deflection Response, $f_t \times 10^3$

Node no.	*No. of modes superimposed*															
	Horizontal applied force								*Vertical applied force*							
	1		*3*		*5*		*10*		*1*		*3*		*5*		*10*	
	u	*w*	*u*	*w*	*u*	*w*	*u*	*w*	*u*	*w*	*u*	*w*	*u*	*w*	*u*	*w*
2	0.0550	0.2925	0.0550	0.2925	0.0603	0.3207	0.0625	0.3324	Insignificant	Insignificant	0.1699	0.9312	0.1804	0.9882	0.1833	1.0134
3	0.0957	0.5410	0.0957	0.5410	0.1014	0.5717	0.1020	0.5745			0.2680	1.6331	0.2761	1.6758	0.2764	1.6771
4	0.1188	0.7076	0.1188	0.7076	0.1216	0.7192	0.1216	0.7211			0.2795	1.9353	0.2807	1.9353	0.2811	1.9407
5	0.1256	0.7666	0.1256	0.7666	0.1260	0.7666	0.1261	0.7710			0.2242	1.9210	0.2246	1.9376	0.2249	1.9413
6	0.1207	0.7088	0.1207	0.7088	0.1207	0.7214	0.1207	0.7232			0.1439	2.0197	0.1461	2.0517	0.1462	2.0541
7	0.1105	0.5428	0.1105	0.5428	0.1105	0.5744	0.1105	0.5774			0.0778	2.5346	0.0788	2.5379	0.0793	2.5430
8	0.1012	0.2939	0.1012	0.2939	0.1012	0.3223	0.1012	0.3337			0.0352	3.1309	0.0352	3.1395	0.0352	3.1396
9	0.0975	0.0000	0.0975	0.0000	0.0975	0.0000	0.0975	0.0000			0.0000	3.3819	0.0000	3.4060	0.0000	3.4103
19	0.0540	0.2900	0.0540	0.2900	0.0590	0.3168	0.0608	0.3269			0.1657	0.9137	0.1753	0.9667	0.1777	0.9798
20	0.0939	0.5370	0.0939	0.5370	0.0993	0.5670	0.0998	0.5696			0.2618	1.6078	0.2693	1.6481	0.2695	1.4693
21	0.1166	0.7039	0.1166	0.7039	0.1192	0.7151	0.1192	0.7168			0.2735	1.9121	0.2746	1.9144	0.2750	1.9192
22	0.1231	0.7635	0.1231	0.7635	0.1235	0.7635	0.1235	0.7677			0.2193	1.9057	0.2197	1.9217	0.2206	1.9251
23	0.1180	0.7066	0.1180	0.7066	0.1180	0.7189	0.1180	0.7206			0.1405	2.0109	0.1427	2.0410	0.1428	2.0442
24	0.1077	0.5415	0.1017	0.5415	0.1077	0.5776	0.1077	0.5754			0.0759	2.5269	0.0769	2.5301	0.0774	2.5350
25	0.0983	0.2932	0.0983	0.2932	0.9838	0.3212	0.9838	0.3322			0.0345	3.1219	0.0345	3.1303	0.0345	0.1304
26	0.0947	0.0000	0.0947	0.0000	0.0947	0.0000	0.0947	0.0000			0.0000	3.4680	0.0000	3.4910	0.0000	3.4950

1 ft = 0.305 m

be seen that the superposition of the first 5 modes is sufficiently accurate for this case.

The method of modal superposition assumes a linear relationship between forces and deformations. In its application to cable systems, therefore, the procedure is limited. A more rigorous approach would be to use an iterative scheme, or one in which the time interval is divided into increments of suitable size.

CHAPTER FIVE

Model Studies

5-1 GENERAL

In many structural engineering problems, an experimental study becomes necessary because either the problem is too complex for a theoretical treatment or the analysis is too expensive. Further, no theories can be called exact, because they are based on simplifying assumptions and approximations, and an experimental study may be essential to provide verification of a theory. In a majority of cases, a cable-roof system has a large degree of indeterminacy, and as pointed out in Chap. 3, a rigorous analysis could prove to be time-consuming and expensive, despite the availability of fast, large-capacity electronic digital computers. Moreover, in dealing with cable roofs, quite often dictated by functional and architectural requirements, one is faced with a structure of curvilinear plan as well as roof, and the problem may be both large and complex.[1] Although methods of static as well as dynamic analysis to handle problems of cable systems are reasonably well developed, as described in Chaps. 3 and 4, an experimental study may become necessary to analyze such structures which are not easily amenable to a theoretical treatment. Another advantage of a model study is the understanding gained with regard to constructional problems of the structure in question. In most cases the construction of a cable roof is a large project, and the experience

[1] As an example, consider the roof structures of the German pavilion in Expo '67 held in Montreal, Canada, and the roof of the main stadium for the 1972 Olympic Games held at Munich, West Germany.

gained will be helpful in planning the construction. For a majority of cable roofs it would be difficult to determine design wind-pressure coefficients from theoretical considerations. Though guidance can be taken from existing data for similar structures, such information is limited for cable roofs. Therefore a wind-tunnel study of the roof model seems to be the best solution for determination of wind pressure as well as for an assessment of aerodynamic behavior. It is for this purpose that roof models are used most. Another useful application is the evaluation of damping, in which a roof model permits a more realistic analysis for the dynamic response of the structure.

Important aspects of the design, fabrication, and measurement on models, with special reference to problems typical for experimentation on flexible systems of this type, are discussed in the following sections. Some aspects of wind-tunnel studies are also touched upon. However, for planning an aerodynamic study, detailed knowledge of fluid-flow principles and related phenomena is required. Since these matters are beyond the scope of this book, the reader is referred to detailed works on the subject. Elhaskar [23] has made a lucid presentation of different aspects of the problem.

5-2 DIMENSIONAL ANALYSIS

It is usual to make a model geometrically similar to the prototype. But, as is well known, this does not lead to a linear or direct ratio between all the parameters of the model and the prototype; only some of the parameters are accurately represented by the scale ratio. In order to predict the behavior of the prototype from the results obtained by the model study, dimensional analysis is necessary. This is a subject of extensive study and is dealt with in detail in several works. In this section only the important points which are relevant to static and dynamic models of cable systems are discussed.

Listed in Table 5-1 are the parameters related to cable systems. Parameters 1 to 12 relate to static as well as dynamic models, while the additional parameters 13 to 15 relate to dynamic models only. For a static model, therefore, the number of dimensionless pi-terms is 12 − 2, or 10, and for a dynamic model it is 15 − 3, or 12. The 10 pi-terms for a static model could be as follows:

$$N, M_c, \frac{F}{L}, \frac{u}{L}, \frac{x}{L}, \frac{EA}{H}, \frac{P}{H}, \frac{p_c L}{H}, \frac{p_a L^2}{H}, \frac{A}{L^2}$$

TABLE 5-1

Parameter	Notation	Dimension*
1. Area of cross section	A	L^2
2. Elastic constant	E	FL^{-2}
3. Span	L	L
4. Cable dip	f	L
5. Coordinates	x	L
6. Displacements	u	L
7. Horizontal component of cable tension	H	F
8. Concentrated load†	P	F
9. Load uniformly distributed along the cable†	p_c	FL^{-1}
10. Load uniformly distributed over surface area†	p_a	FL^{-2}
11. Number of cable members	M_c	
12. Number of nodes	N	
13. Mode shape	ψ	
14. Frequency of vibration	η	T^{-1}
15. Acceleration due to gravity	g	LT^{-2}

* F, L, T relate to basic dimensions of units of measurements, viz., force, length, and time.
† One or more of these as applicable.

The two additional terms for a dynamic model may be $\eta\sqrt{L/g}$ and ϕ.

Structure deformations term u/L can be expressed as having functional relationship with the remaining pi-terms:

$$\frac{u}{L} = F\left(\frac{x}{L}, N, M_c, \frac{f}{L}, \frac{EA}{H}, \frac{P}{H}, \frac{p_cL}{H}, \frac{p_aL^2}{H}, \frac{A}{L^2}\right) \qquad (5\text{-}1)$$

In order to predict results for the prototype structure from the observations on a model, the design and operating conditions of the model must be suitably scaled. In other words, each pi-term in the parentheses in Eq. 5-1 must be the same for both the model and its prototype. Choosing a model geometrically similar to the prototype, the design conditions corresponding to four dimensionless terms, N, M_c, f/L, and A/L^2, are satisfied automatically. Other conditions are worked out as follows:

1.
$$\left(\frac{EA}{H}\right)_m = \left(\frac{EA}{H}\right)_p$$

or
$$H_m = H_p \frac{E_m}{E_p} \frac{A_m}{A_p}$$

If $E_m = E_p$,

$$H_m = H_p \frac{A_m}{A_p} \quad \text{or} \quad H_p \frac{1}{\lambda^2}$$

where λ is the scale ratio and the subscripts p and m relate to the prototype and the model, respectively.

2.
$$\left(\frac{P}{H}\right)_m = \left(\frac{P}{H}\right)_p \qquad P_m = P_p \frac{H_m}{H_p}$$

or
$$P_m = P_p \frac{1}{\lambda^2}$$

3.
$$\left(\frac{p_c L}{H}\right)_m = \left(\frac{p_c L}{H}\right)_p \qquad (p_c)_m = (p_c)_p \frac{(L)_p}{(L)_m} \frac{H_m}{H_p}$$

or
$$(p_c)_m = (p_c)_p \lambda \frac{1}{\lambda^2} = (p_c)_p \frac{1}{\lambda}$$

4.
$$\left(\frac{p_a L^2}{H}\right)_m = \left(\frac{p_a L^2}{H}\right)_p \qquad (p_a)_m = (p_a)_p \frac{(L^2)_p}{(L^2)_m} \frac{H_m}{H_p}$$

or
$$(p_a)_m = (p_a)_p \lambda^2 \frac{1}{\lambda^2} = (p_a)_p$$

If the design and operating conditions (1) to (4) are satisfied for a geometrically similar model, then the prediction equation for deformation is

$$\left(\frac{u}{L}\right)_p = \left(\frac{u}{L}\right)_m$$

or
$$u_p = u_m \lambda$$

If the model, satisfying the above conditions, is used for dynamic measurements, the prediction equation for the frequency is

$$\left(\frac{\eta^2 L}{g}\right)_m = \left(\frac{\eta^2 L}{g}\right)_p$$

or
$$\eta_p = \frac{\eta_m}{\sqrt{\lambda}}$$

For models which have to be designed for wind-tunnel tests, certain other nondimensional quantities have also to be considered. Generally, in problems connected with fluid flow, parameters such as the Reynolds, Mach, Weber, and Froude numbers can be important, depending on the type of problem. In models to be tested for determination of wind effects under subsonic flow, only the Reynolds number is important, and is given as

$$\frac{\rho V D}{\mu}$$

where ρ = density
V = velocity
μ = dynamic viscosity
D = typical length dimension, usually width of structure

For true dynamic similarity the Reynolds number in the model and the prototype must be kept the same, though it is difficult to do so. Nevertheless it has been shown by experiments over a wide range of Reynolds number that the flow pattern is, to a large extent, independent of this parameter for sharp-edged bodies. Since a cable roof will usually have a well-defined parapet, scaling with respect to the Reynolds number is unimportant. Generally speaking, the Strouhal number is a function of the Reynolds number, and for flows in which the Reynolds number is known to become unimportant, the Strouhal number attains a constant value depending on the geometry of the structure.

For obvious reasons, it may not always be possible to satisfy all the design and operating conditions. For example, it may be particularly difficult, in a geometrically similar model, to satisfy the conditions for dead or permanent loads such as condition 3 or 4 above. The prediction equation in such a case has to be developed, taking into account the distortion with respect to the geometry, the loading, or the material properties, as the situation may be. It may be pointed out, however, that such a prediction equation can be developed only if the relationship among the relevant parameters is known. This, in some cases can be complicated, particularly in problems of cable systems, which are otherwise complicated on account of nonlinearity.

The conditions for dead loads such as condition 3 or 4 are usually achieved by suspending dead weights from wires composing the model. For dynamic models this practice may not be desirable, because the suspended weights behave like live loads and only a fraction of the suspended weight thus gets transferred to the system. Also, there may be some difficulty with such suspended weights when the model is vibrated. However, it may not be necessary to simulate masses for the prediction of frequencies and mode shapes, and a geometrically scaled model using the same material as the prototype may be adequate for this purpose. In this case, the prediction equation for the frequency of vibration is

$$\eta_p = \eta_m \frac{1}{\lambda}$$

It may also be realized that the prediction of dynamic stress or acceleration from the test on models cannot be easily made unless the cable material is specially designed. In most cases, a material with a lower Young's modulus than that of the prototype is required, and the possibility of using a soft rope (e.g., nylon) may be worth examining. For dynamic studies the common practice is to use the experimental results for the verification of the analytical approach, which may then be used to compute the response of the prototype.

5-3 CHOICE OF SCALE

While conducting experimental studies, an important decision to be taken is the choice of the model scale. There are many considerations that affect this decision. A large-sized model will suffer less from mechanical problems and has the advantage of greater proximity to prototype constructional problems and details. However, the construction of a large model will entail greater expense. The problem of measurement of deflections becomes increasingly easier with increase in model size, merely because it is easier to measure larger displacements. Also the size of wire used in the model may be adequately large to enable measurement of strain through direct application of electrical-resistance gauges.

Small-scale models, on the other hand, have the advantage of lesser cost and lesser involvement of physical effort, but can suffer large mechanical errors, as reported in some earlier works [42, 43] on the subject. Further, the procedures used in fabricating small models in the laboratory may not be suitable at all for use on a prototype and thus may be far removed from the reality of field problems. For example, in a small model the wires may be pretensioned by direct loading with small weights or by spring loading. Figure 5-1 is a photograph of a cable-truss model under fabrication, where vertical wires as well as the upper and lower main wires are loaded this way. The loads and the springs will be removed after the vertical wires are duly connected at their ends to the two main wires, and these are in turn clamped at the ends. This method cannot be used for pretensioning the prototype. Measurements of both forces and deflections tend to require greater precision for smaller-scale models. It may thus be seen that both large- and small-scale models have their advantages and disadvantages. A medium-size model may not, however, present the ideal answer.

For studying free-vibration behavior, the model frequency must be within the response range of the available measuring equipment. Also, if steady-state excitation is to be used for studying forced vibration response, the capacity of the mechanical oscillator and its operating frequencies should be such that resonant condition could be obtained and the resulting deformations could be monitored with reasonable accuracy by the available measuring instruments.

Models for wind-tunnel studies for both static and dynamic analysis have to be governed by the size of the tunnel. When the large-sized models are tested in a wind tunnel of relatively small size, the

FIGURE 5-1 Wire model of a pretensioned truss under construction. Note the direct and spring loading of the wires (*Ref. 42; photo by Joyce Gurr, Imperial College.*)

pattern of flow past the structure is considerably affected by the proximity of the tunnel ceiling. It is to be noted that atmospheric flow past the structure is of semi-infinite extent; the tunnel ceiling, however, causes appreciable *blockage effects* if it is too close to the structure, leading to distortion of the flow pattern. Since methods of correction of results for blockage effects in case of complicated structural forms are not available, it is recommended that the blockage be held at as low a value as possible, thereby enabling the model results to be interpreted without corrections.

5-4 WIRE MODELS

Models to represent cable-roof systems are usually made from steel wire. Mild, high-strength, as well as stainless steel, wires have been used for this purpose. While it would be more realistic to use wire ropes, there are two specific problems in doing so:

1. Rope of very small diameter may be difficult to obtain.
2. Strain, and thus force measurement by conventional procedures may not be accurate.

In certain cases steel strips have been used for models of pretensioned trusses [27] to facilitate use of electric strain gauges.

The connection of wires in laboratory models can be effected by the usual mechanical clamping [2, 46, 74] or even by spot (fusion) welding for small-size wires [42]. If spot welding is to be employed as the method of connection, an appropriate choice of steel is necessary. Fusion welding is, however, likely to lower the fatigue life of the joint and thus of the structure. If, therefore, the model is to be subjected to long-term cyclic loading, fusion welding may not be the ideal choice.

As pointed out earlier (Sec. 5-2), a geometrically similar model may not give the correct simulation of dead weights. The correct magnitude of dead weights can be obtained simply by hanging blocks of the required weight (not recommended for dynamic models). Care should be exercised, however, to ascertain that such blocks are not fixed to the wire but are rather hung from it. Fixing will introduce an undesirable flexural stiffness in the system.

Obtaining the correct geometry and introducing the required value of pretension and geometry is a tricky problem in both models and prototypes of cable systems. The problem occurs because of the interaction between these two parameters. A small change in pretension of any one wire of a truss or network model is bound to affect the pretension in all the other wires, as well as the geometric shape of the system.[2] The one accepted way seems to be to increase the prestress in small steps and adjust the geometry at the same time. For plane pretensioned trusses of the type shown in Fig. 5-1, or for hyperbolic paraboloid nets, the relationship between load and geometry is easily defined. In small-scale models of these structures, pretensioning can be done by hanging weights or through spring loading.

In small-scale models the problem of anchoring cables or wires can be solved easily by using rigid self-supporting frames. For larger models, anchoring into heavy reaction frames or floors may be the solution. Heavy concrete blocks may be used also to provide gravity anchors. The most common way of anchoring the wires is by mechanical clamping or with wedge and cone anchors, such as those used in prestressed concrete.

5-5 INSTRUMENTATION

Measuring instruments are required for force and displacement while dealing with both static and dynamic experiments. For measuring forces the type of gauges commonly used and the principles involved are briefly described in the following paragraphs.

Mechanical Gauges. The type of mechanical gauge generally used functions on the same basis as a spring balance. An illustration of such a gauge is shown in Fig. 5-2. Such an instrument will cover only a limited range of values of force and wire dimensions, but the great advantage is the ease with which it can be mounted and used. Indeed other mechanical gauges, of the type used for concrete specimens for measurement of strains, are available but may not prove suitable for applications on wire models. Similarly these may prove to be quite unsuitable for dynamic measurements.

[2] T. P. Agarwal [2] demonstrated how to determine the mode of this interaction and used results to arrive at the correct geometry and pretension values in constructing a hypar shape network model.

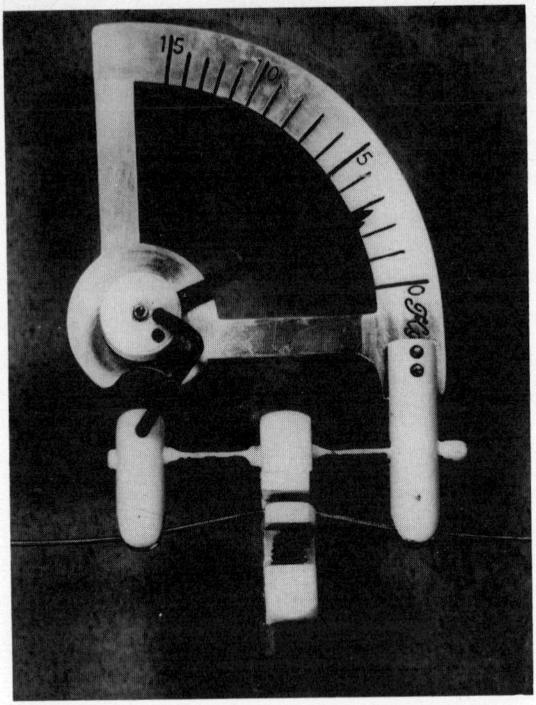

FIGURE 5-2 A mechanical gauge for measuring tension in wire models. (*Photo by Gopinath.*)

Electrical Gauges. The electrical-resistance strain gauges provide the widest range of applications for strain measurement. If the size of the wire is large enough to accommodate the gauge, it can be pasted directly; if that is not possible, load cells can be made and these can intercept the length of the wire at a suitable location. It will be obvious that direct application will be unsuitable for a model fabricated from twisted wire rope. Interception of the wire may introduce errors for these reasons:

1. The change in the overall stiffness of the wire
2. The undesirable freedom of movement of the different segments of the wire, thus made independent of each other

A load cell using electrical-foil-type gauges, illustrated in Fig. 5-3*a*, consists of two Duralumin strips (marked "S"), pinned down to anchor blocks (marked "B") at their ends. Four foil-type electrical-

FIGURE 5-3*a* Components of a *link* dynamometer for strain measurements in wires. (*Ref. 42; photo by Joyce Gurr, Imperial College.*)

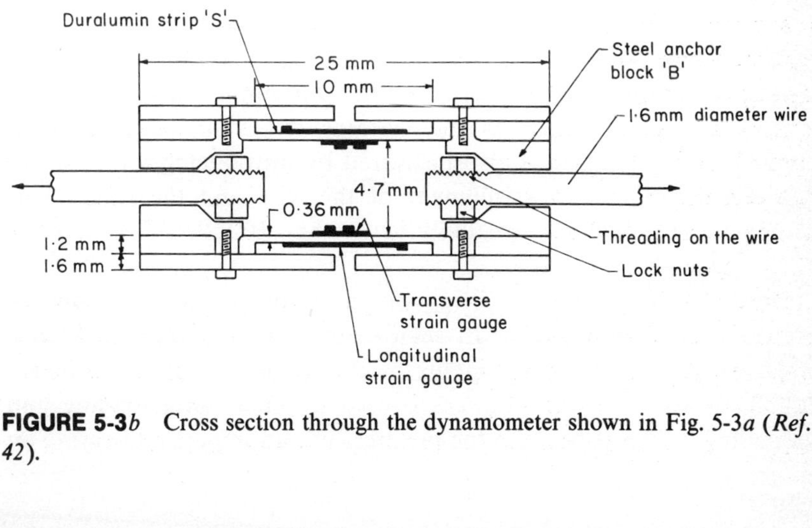

FIGURE 5-3*b* Cross section through the dynamometer shown in Fig. 5-3*a* (*Ref. 42*).

FIGURE 5-4 Link dynamometer using a transducer. (*Ref. 74; photo by Joyce Gurr, Imperial College.*)

resistance strain gauges, cemented to these strips, form part of a Wheatstone bridge circuit. The segments of the wire meeting at the dynamometer are threaded at their ends and are fastened at each end with nuts. These nuts sit inside the anchor block, within which they can rotate freely. Figure 5-3*b* shows a cross section through the load cell. The load cell was used on a small-scale model of a pretensioned plane truss and had a measurement range of 0 to 100 lb (0 to 445 N) of applied load and showed linear load-strain characteristics.

A ring-type load cell is illustrated in Fig. 5-4. Four electrical-resistance strain gauges, pasted on a Duralumin ring, provide a sensitivity of greater than 200 microstrains/lb (45 microstrains/N) of applied load. Tension is also measured by introducing a transducer between the points of attachment of the wire and the ring. This arrangement provides a sensitivity better than 100×10^{-6} in/lb (0.57×10^{-3} mm/N).

Here it is important to bring out a problem that is unique to pretensioned wire models. In such cases, it is important to keep a stress history of the components of the model; it is not sufficient simply to record strain changes caused by short-term loading and unloading. This is because the prestress has an important bearing on the stiffness of the structure. In applying electrical gauges, the problem of drift with time can present difficulties, since it is not easy to detect whether a change in reading after an interval of time is due to drift or due to change in prestress itself. A way has to be devised, therefore, to eliminate or to minimize electrical drift. For dynamic measurements, however, drift does not present a problem.

Sonic Gauges. The use of sonic gauges in long-term measurements of strains is well established and so is their property of stability. Such gauges can be embedded into structures to maintain the stress history of a particular part of the structure.[3] The principal component of the gauge is a wire under tension whose frequency of vibration is a function of the tension or of the change in it. This idea can be employed in a rather simple way for measurements of tension in a wire model, particularly for thin wires, where other methods present difficulties. In such an application, the wire itself forms part of the gauge. The wire is held at two points, a finite distance apart, and vibrated through electromagnetic excitation. The same electromagnet transmits the signal to an oscilloscope for measurement of its frequency of vibration. An application [42, 43] of this type is illustrated in Fig. 5-5. The wire in which force is required to be measured is 0.010 in (0.25 mm) in diameter and of nonmagnetic weldable stainless steel. In order that it can be excited by the electromagnets, it is magnetized locally by winding 0.002 in (0.05 mm) mild steel (m.s.) wire over a length of about 1 in (25.4 mm). Makowski et al. [59] report the application of the same principle in some of their experimental work.

Measurement of deflections presents no special difficulty. Dial gauges may be used for small- to medium-size models, as long as the spring load on the gauge spindle is small in comparison to the applied loading. This force can be as large as 0.5 lb (2.224 N). As an alternative, optical instruments—levels or cathetometers—can be used for deflection measurements. For large-size models and even in prototypes, the use of photogrammetry has been made.

The measurement of the effect of temperature change in cable structure models is of interest. It may be difficult to achieve a uniform temperature change in the model for this purpose. How-

[3] Though such an application is not possible or necessary for a cable structure.

FIGURE 5-5 A *sonic* gauge for measurement of wire tensions. (*Ref. 42; photo by Joyce Gurr, Imperial College.*)

ever, the effect can be assessed by introducing, at the end of each wire, the change in its length corresponding to a desired change in temperature.

Instrumentation for Dynamic Measurements. For measurement of dynamic strain, displacement, and/or frequency of vibration, three basic instruments—a pickup, an amplifier, and a recorder—will normally need to be used in series. The pickup can be either a strain gauge or a displacement, velocity, or acceleration transducer, depending upon convenience. The signal from this pickup is amplified and fed to a recorder unit, which could either be an ink-writing oscillograph or an oscilloscope fitted with a camera. For details of these instruments and their characteristics, the reader is advised to refer to a book on instrumentation technology.

The record of vibration on a continuous time basis is obtained on a paper which moves at a constant speed varying from approximately 0.05 to 20 in/s, and one can determine the frequency of vibration and the dynamic response parameter of interest in the model. The frequency of vibration may also be determined using a digital electronic counter.

For wind-tunnel studies where pressure measurements are involved, sensitive pressure pickups to monitor the dynamic pressures have to be used, in combination with the auxiliary instrumentation, for dynamic measurements, described earlier. Diaphragm-type pressure gauges employing solid-state devices, like the silicon strain gauge, exhibit high sensitivity and can be usefully employed. While the steady component of the wind velocity can be measured with the help of pitot tubes, the fluctuations may be picked up with a hot-wire anemometer or another instrument which works on the same principle.

5-6 RESULTS OF MODEL STUDIES—ILLUSTRATIONS

Experimental studies on models of existing or proposed roofs have been carried out with wide-ranging objectives such as verification of computed geometry; study of the static, dynamic, or aerodynamic behavior; determination of wind pressures; assessment of damping; and understanding of construction problems. Such experimentation has also been carried out in academic pursuits, generally with similar objectives. Published material of much value for anyone planning the design, the construction, or laboratory studies of a cable roof has resulted from these studies. Unfortunately, paucity of space does not permit reproduction of these works in detail as part of this chapter; only a brief description is included in the following paragraph. Results of certain wind-tunnel studies are also reproduced.

Swimming Pool for 1964 Tokyo Olympics, Japan [98]. This structure is shown in Fig. 6-9*a* and *b*. Two models of the roof, one with a 1/100 scale using nylon rope and another with a 1/30 scale using steel wire rope, both without cladding, were used for determination of forces and deflections under static loading as well as for determination of dynamic characteristics. Good agreement between computed and measured values for static loading has been shown. Frequencies of vibration obtained from the model in the first few modes were used to predict values for the prototype. However, the values obtained by vibrating the prototype do not agree well with the predicted values. This is perhaps explained by the presence of the rafters, cladding, ceiling, and central trusses in the real structure, which were not represented in the models. The frequency of vibration of the prototype is higher than that predicted by model tests. The frequency of vibration of the prototype for different modes is as follows:

Symmetric	2.38 Hz
Antisymmetric about the longitudinal axis of the roof	1.50 Hz
Antisymmetric about the transverse axis of the roof	5.54 Hz
Antisymmetric about both axes	4.18 Hz

A 1/300 scale model was placed in the wind tunnel and subjected to a steady wind with a maximum velocity of 22 m/s. The wind pressure distribution is shown in Fig. 5-6. It may be noted that most of the surface is subjected to suction of 60 to 90 kg/m^2 (589 to 884 N/m^2) but that at the steep portions near the poles (A, A') there is a high pressure of the order of 240 to 300 kg/m^2 (2355 to 2945 N/m^2). Further, a cut section of the roof with a 0.2-mm aluminum skin was

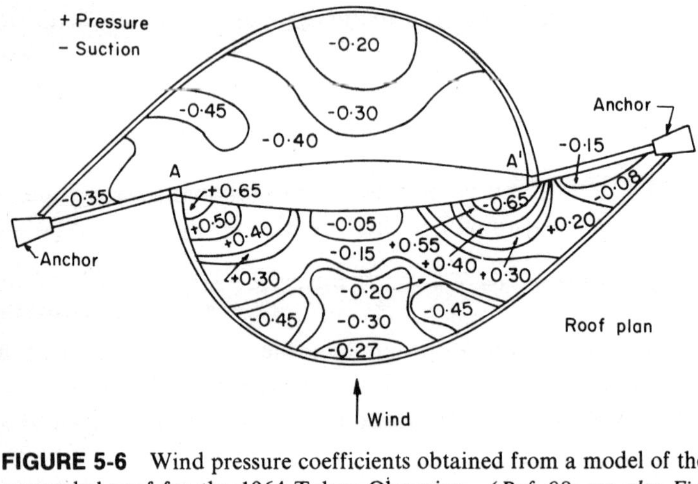

FIGURE 5-6 Wind pressure coefficients obtained from a model of the suspended roof for the 1964 Tokyo Olympics. (*Ref. 98; see also Fig. 6-9.*)

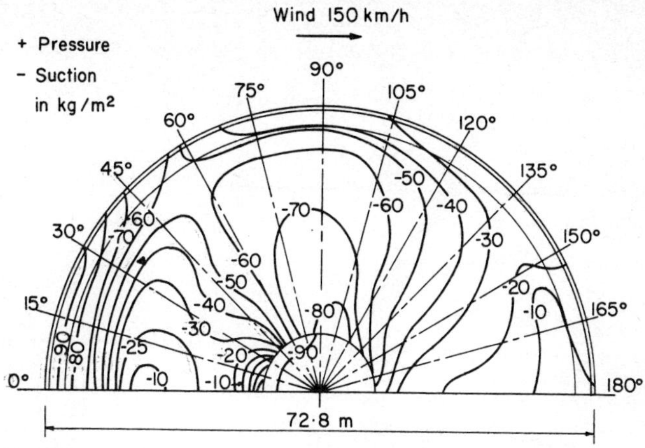

FIGURE 5-7 Wind pressure contours plotted over the roof plan for the Palasport Genova. (*Ref. 26; see also Fig. 6-10.*)

subjected to a fluctuating wind velocity of 22 ± 7 m/s. The dynamic pressure at resonance was noted to be about 50 percent more than the corresponding static pressure. The model did not indicate any tendency to flutter.

Palasport Genova, Italy [26]. A photograph of this structure is shown in Fig. 6-10, and a description of certain important features is included in Sec. 6-4. Various routine model studies were made. For one of these a 1/140 scale rigid model of the structure was placed in a wind tunnel; the pressure distribution obtained at the wind speed of 150 km/h is given in Fig. 5-7.

Farahabad Sports Stadium, Iran [59, 71]. Elaborate model studies are reported for the roof of this 145 × 50.5 m stadium near Teheran. The photograph in Fig. 5-8 is that of a 1/50 scale model, which was used for the determination of static behavior and temperature effects. Wind pressure distribution for different wind directions, shown in Fig. 5-9*a* to *c*, was obtained on a 1/500 scale rigid model. For the purpose of presentation of results the roof surface is divided into eight zones, and the average pressure coefficient in each zone is given.

Stadium for 1972 Munich Olympics, West Germany [55]. Figure 1-14 shows a model of the cable structure, nearly 440 m long and 80 m

FIGURE 5-8 A model of the Farahabad Sports Stadium roof at Teheran, Iran. (*Ref. 59; Space Structures Research Centre, University of Surrey.*)

wide. Some of its design features are also included in Chaps. 6 and 7. Extensive studies of the structural behavior were carried out for the determination of cutting patterns for the cables and also for the design of connection details. Studies are also reported for determining the dynamic and aerodynamic characteristics. One of the important conclusions from this study is that, while the logarithmic decrement for a network without cladding is of the order of 0.025, air damping enhances this value when the cladding is also present. A good value to use in such an event is 0.10. Tests were also carried out to determine the most favorable design of the edge parapet from the aerodynamic standpoint. The various shapes tried are shown in Fig. 5-10*a* to *f*, and the one given in *f* was seen to cause the least aerodynamic disturbance.

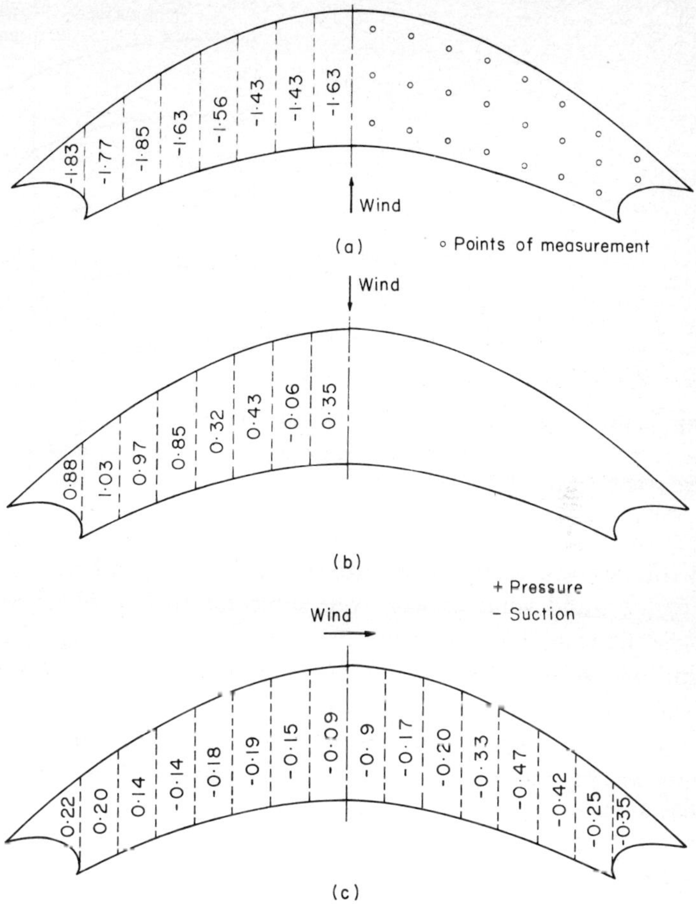

FIGURE 5-9 Wind pressure coefficients obtained from a model of the Farahabad stadium roof, shown on the roof plan for three different wind directions. (*Ref. 59; see also Fig. 5-8.*)

Oklahoma State Fair Arena Roof, United States. This elliptical paraboloid roof, formed of 3.67-in (9.34-cm) precast concrete panels, supported on posttensioned cables, is placed over a 70-ft-high (21.38 m) boundary wall and has a circular plan 346 ft (105.4 m) in diameter. Elhaskar [23] tested a 12-in-diameter (30.48 cm) aeroelastic model with a 2.5-in-high (6.35 cm) plexiglass drum wall. The roof was represented by a neoprene rubber membrane approximately 0.01 in (0.25 mm) thick. There were 10 holes 1 in (2.54 cm) in diameter in the drum wall to represent the openings. The model, designed to

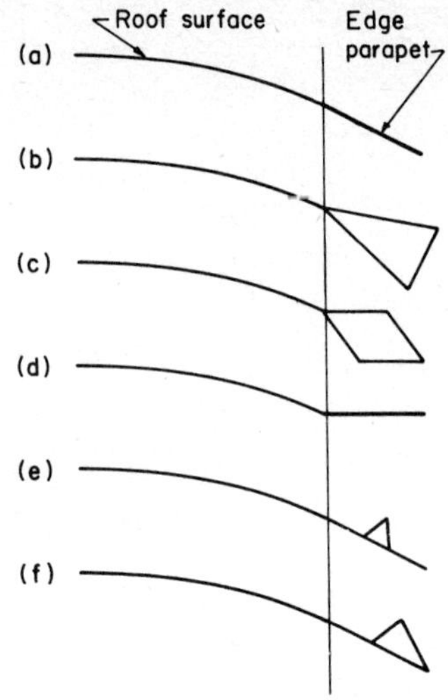

FIGURE 5-10 Alternative designs examined for the edge parapet of the 1972 Munich Olympic stadium roof (*Ref. 55*).

satisfy the scaling requirements for the Strouhal number, stiffness, mass density, and wind velocity, was subjected to free-vibration and aerodynamic tests. Results were obtained for three modes of vibration and are shown in Table 5-2. It may be seen that the results in the first mode, in which the air underneath the membrane is actually displaced, were affected by the number of openings. Aerodynamic damping was high in the first mode and was dependent upon the wind velocity, while for the second mode it was virtually independent of the same.

Mode		*Resonant frequency, Hz*	*Critical damping ratio,* %*	*Aerodynamic damping, %*
1			0.24–1.00†	7.00–11.00
2		12.00	2.10	0.50
3		17.00	2.65	Not reported

* Structural and acoustical.

† Values vary with openings in the wall.

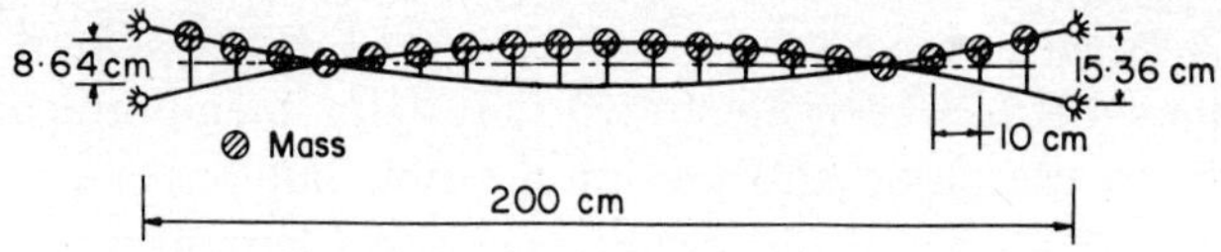

FIGURE 5-11 Cable truss model for dynamic measurements (*Ref. 27; see results in Fig. 5-12*).

New Palasport in Milan, Italy. Gallo-Curcio and Piccarreta [29] have reported studies on a 1/100 scale geometric model of the 142-m-diameter anticlastic roof with a reinforced concrete (R.C.) ring beam. The cables are modeled by 0.4 × 0.5 mm high-tension steel bands at 100 mm c/c, and the ring beam is represented by a 16 × 37 mm rectangular aluminum casting. The force scale is 1/5000.

Other studies. Jensen has reported extensive static as well as dynamic experimentation [37–40] on both cable trusses and networks.

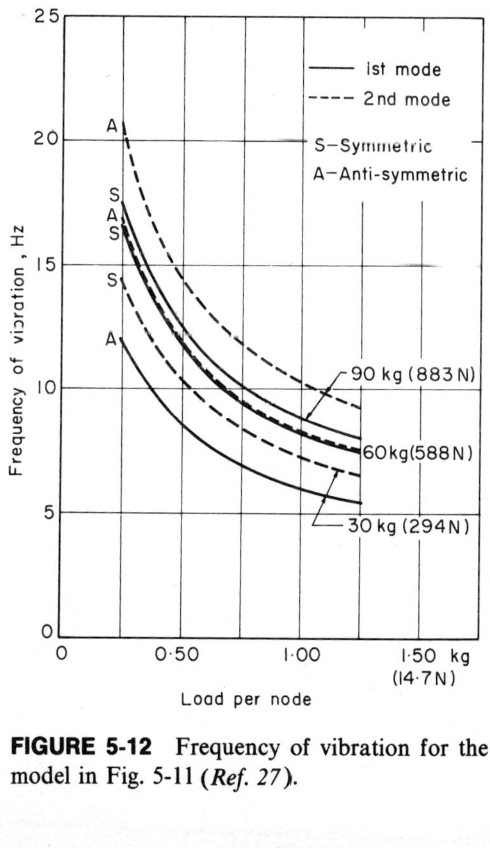

FIGURE 5-12 Frequency of vibration for the model in Fig. 5-11 (*Ref. 27*).

Measured values show good correlation with the corresponding computed values. It has been illustrated that nylon membrane adds only marginally to the static stiffness of the net, but has a pronounced effect on the damping of the system (see also Chap. 4).

Franchi and Scirocco [27] have carried out dynamic tests on the truss shown in Fig. 5-11. The values of the resonant frequency in the symmetric and the antisymmetric mode for different values of prestress and applied loads are shown in Fig. 5-12. It is interesting to note that the lowest frequency for the prestress of 30 kg is obtained in the antisymmetric mode, while for the case of 90 kg prestress, it is obtained in the symmetric mode. (A similar theoretical result is obtained by Buchholdt and Chana [8].) Experimental studies for the verification of theoretical values of pretensioned trusses and networks subjected to static loads are also reported by the author [42, 46].

CHAPTER SIX

Design

6-1 GENERAL

A brief discussion of general aspects of the design of cable roofs, such as the types of cable-roof systems and their supports, concepts of planning and economy, serviceability, safety, fire protection, durability, and materials is included in Chap. 1. Design aids in the form of curves and approximate formulas for typical systems are given in Chaps. 2 to 4, to help in the analysis and design investigations. This chapter further elaborates on general aspects and deals with design loads and specifications. Also included is a summary of several existing designs with respect to their major dimensions, design loads, structure type, cost, and response to loads, to the extent the information was available.

6-1.1 Proportions

The dip/span ratio in cable roofs usually varies on an average between 1:15 and 1:20, notwithstanding exceptional forms of structures. It is of interest to note that this ratio varies between 1:8 and 1:12 in suspension bridges. Minimum clearance requirements remaining the same, while the height of the supporting structure increases with the increase in cable sag, so also does its cost. In lightly loaded structures such as roofs, therefore, it is advisable to keep the sag as small as possible. It should at the same time be realized that an increased sag affords greater stiffness to the cable. While maintaining the clearances required, a compromise is often possible in cable trusses by employing the form shown in Fig. 1-9*d*. Further, as established in Chap. 3, the upper cable in trusses is the primary load-bearing member, and for this reason, the sag of the upper cables is usually kept larger than that for the lower cables. For example, in a total sag of approximately 8 percent of the span, 5 percent may be assigned to the upper-cable sag. In doubly curved networks, curvatures for the hogging and the sagging cables are kept nearly the same. It may be noted that the load-bearing character of a net is a function, among other parameters, of the curvature of the surface generated, rather than of individual cable sags, as is the case in a truss.

6-1.2 Choice of Structure

For temporary cable roofs, such as those used in expositions, the tent type or an air-inflated type of system is most often employed. However, for the more conventional and permanent roofs the choice falls among the simply suspended truss types or the doubly curved (anticlastic) types of cable system. Whereas it is difficult to specify any definite criteria, it is obvious that the plan shape, the span, and, to some extent, the loading on the structure will govern this choice. For narrow rectangular or circular forms it is the simply suspended or truss type of system that will usually be employed, although the use of networks over circular plans is also a possibility. For squarish plans the hypar type of arrangement will perhaps be more suitable. For oblong plans each of these types has been successfully employed. A cable truss is used in the same fashion as a girder in single or continuous spans. For covering a rectangular plan, trusses are placed parallel to each other, while for a squarish plan these may be used to form a grid. For circular plans, trusses are placed radially in most cases, but formation of an orthogonal truss grid provides an alternative worth examining. The latter will eliminate the need for a central tension ring, essentially required in a radial arrangement. The grid will also generate better two-way action. The use of tridiagonal truss grids is uncommon, although such an arrangement may sometimes be considered in single-layer nets.

As pointed out in Chap. 1, the simply suspended type of roof is suitable only for small spans, on account of its low rigidity, unless it is preloaded heavily.[1] For larger spans the stiffer pretensioned systems

[1] An example of a heavily preloaded cable roof is the 404-ft-diameter (123 m) Madison Square Garden roof, in New York City.

have to be used. In structures where design loads are heavy or spans are large, deflection limits are important, and cable trusses prove to be superior to simply suspended cables, on account of their greater static stiffness. Cable truss systems also provide more effective damping. It may be pointed out, however, that in trusses with diagonals (Fig. 1-9*a*) almost half the diagonals become slack under uniformly applied loading, unless they are heavily prestressed, and also that such trusses are comparatively more difficult to erect. But these trusses exhibit better damping characteristics than do trusses with vertical web members (Fig. 1-9*b* and *c*). For the same reasons as pointed out for trusses, networks of anticlastic curvature are more satisfactory than synclastic two-way systems. In erecting networks, accuracy in cable lengths is a very important factor. This is particularly true for complex networks. Generally, trusses are easier to erect than nets.

Anchoring and supporting arrangements have been described in Chap. 1. In brief it may be said that where space around the structure is easily available, various types of supporting arrangements are possible, for example, a *column-and-backstay* combination for roofs with cable trusses. This arrangement is more commonly used in structures of rectangular plan. However, in crowded locations this is not possible and the anchoring and supporting systems have to be combined. As already mentioned for networks and for structures on a circular (or near-circular) plan, a system such as a ring beam or an arch, which supports cables by developing primarily direct forces, is preferable to a straight-edge beam, which is a flexural member.

Cost Comparisons. One of the most important considerations governing the choice of the structural system is the cost factor. Generalization of this factor is most difficult, however, because it is dependent on a large number of components which are quite often local in character. Results of some cost studies are given in this section, and cost breakdowns for certain existing structures are given in Sec. 6-4 for guidance. In evaluating the cost of a structure, due consideration ought to be given to its maintenance and air-conditioning costs.

Valuable work on theoretical cost analysis of cable systems has been carried out by White et al. [100]. Taking the basic data given in Tables 6-1 and 6-2 and assuming that the minimum height of the roof is 32.8 ft (10 m) and that the maximum permissible deflection under

TABLE 6-1 Design Criteria

1. Loads	Decking Insulation Ceiling finish	20 kgf/m^2
	Roof services	25 kgf/m^2
	Superimposed loads	75 kgf/m^2
	Design wind pressure	100 kgf/m^2
	Wind uplift	80 kgf/m^2
2. Steel	BS449-grade 50. Allowable stress	230 N/mm^2
3. Cables	Minimum breaking stress	1800 N/mm^2
4. Soil	Clay to at least 10 m hardening with depth; allowable bearing pressure	0.2 N/mm^2
5. Safety factors	Parabolic cables	2.00
	Straight tie cables	2.25
	Dead-weight anchorages	2.00
	Tension piles	3.00

1 $kgf/m^2 = 9.806\ N/m^2 = 0.205\ lb/ft^2$

1 $N/mm^2 = 145\ lb/in^2$

SOURCE: Ref. 100, by courtesy of the authors.

load is not to exceed $\frac{1}{300}$ of span for rigid trusses and $\frac{1}{100}$ for cable trusses, cost comparisons are carried out for cases I and II as described below.

CASE I Costs are evaluated for single-curvature roofs consisting of simply suspended cables or pretensioned trusses at a spacing of (13.1 ft) 4 m, and they are compared with a conventional steel truss roof. The variable parameters are as follows:

Sag/span ratio	6 to 12 percent
Depth/span ratio	1:18 to 1:12
Anchorage	Dead weight or tension piles
Span	100 to 425 ft (30 to 130 m)

Figure 6-1 shows the variation of costs with the different combination of parameters. It is seen that the roof with trusses of variable sag, using tension piles, has the minimum cost for spans over 130 to 165 ft (40 to 50 m) and the steel truss system is more economical for the smaller spans. The cable truss costs only about 15 to 20 percent of the total cost of the structure.

CASE II Evaluation of costs for circular roofs consisting of radially arranged cable trusses of the type shown in Fig. 6-2 is made in this case. Trusses are anchored into circular compression rings at the outer circumference and are held by tension rings at the center. The variable parameters are the same as for case I except that the need for tension piles or dead-weight anchorages is eliminated. Cost comparisons are given in Fig. 6-2. Costs are nearly the same and

TABLE 6-2 Schedule of Notional Rates

	Unit	*Rate, £**
1. Excavate in normal ground including planking and strutting	m^3	8.00
2. Fill excavated earth or remove	m^3	2.00
3. Mass concrete	m^3	15.00
4. Reinforced concrete including formwork and reinforcement (normal uses)	m^3	60.00
5. Same as item 4, heavily reinforced or precast or with excessive strutting	m^3	75.00
6. Mild-steel normal sections, fabricated, assembled, and fixed	tonne	300.00
7. Mild-steel heavy sections, fabricated, assembled, and fixed	tonne	380.00
8. Mild-steel welded columns and tension rings	tonne	400.00
9. Aluminum roof decking with insulation (self-supporting)	m^2	8.00
10. Wood-wool screed and felt roofing (on purlins, not included)	m^2	6.00
11. Tension piles—bored underreamed piles	tonne stay force	5.00
12. Inclusive cost of cables, ties, connections, and anchorages per tonne of cable erected	tonne	1300.00

* British Pound Sterling. 1 tonne = 9.8 kN, 1 m^2 = 10.76 ft^2, 1 m^3 = 35.3 ft^3

SOURCE: Ref. 100, by courtesy of the authors.

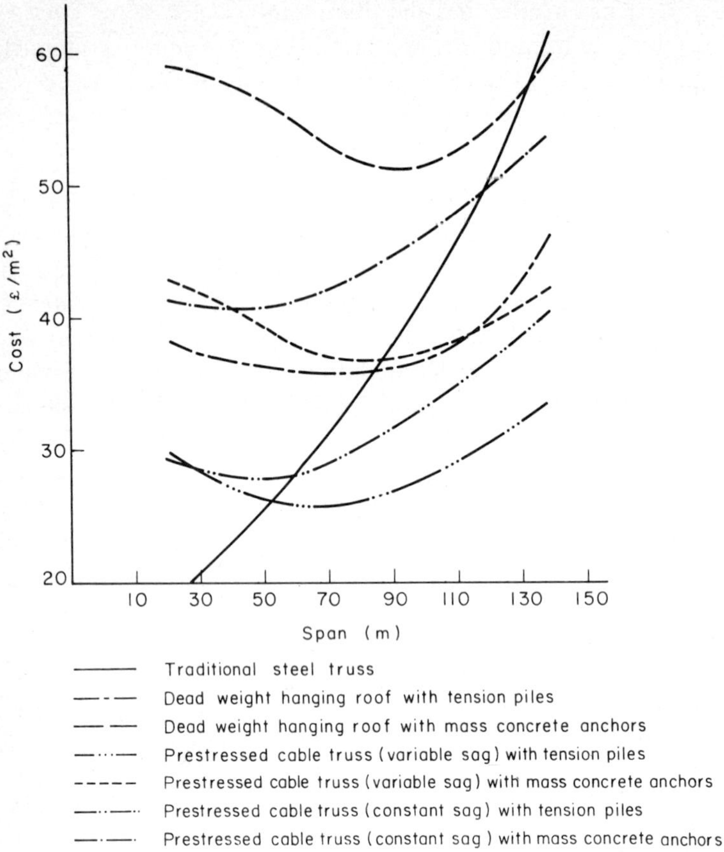

FIGURE 6-1 Cost comparisons for different forms of cable roofs. Effect of the mode of pretensioning, geometry, and anchoring arrangement is studied. (1 £/m^2 = 0.093 £/ft^2, 1 m = 3.28 ft) (*Ref. 100; courtesy of the authors.*)

minimum for all the alternatives at spans of about 80 ft (25 m). Figure 6-3 gives the breakdown of costs for a roof typical of case II. It will be seen that a sizable proportion of costs is taken up by the compressions rings. If ring beams other than circular, i.e., elliptic or parabolic, are used, bending is likely to increase and raise the cost of the rings.

Magnusson [57] has reported the cost of 198.8 × 213.2 ft (60 × 65 m) and 246 × 262.4 ft (75 × 80 m) double-curved network roofs built in Sweden as 14.9 Kr/ft^2 (160 Kr/m^2).[2] Butler [13] gives a compari-

[2] Kr represents the Swedish Kroner. It is currently equivalent to 0.23 U.S. dollars.

son of the quantity of cable used in cable trusses, bicycle wheels, and dished shells for different spans. On that basis the following figures are obtained:

Trusses and dished shells	0.0042 lb/ft^2 per foot of span (0.65 N/m^2 per metre of span)
Bicycle wheel	0.0064 lb/ft^2 per foot of span (1.0 N/m^2 per metre of span)

Interesting cost comparisons are reported for the 310-ft-diameter (94.5-m) Montevideo stadium which employs a bowl-shaped, post-tensioned, precast concrete-shell roof supported on a single layer of cables. The edge compression ring into which the cables are anchored is supported on a thin cylindrical R.C. shell. It is reported

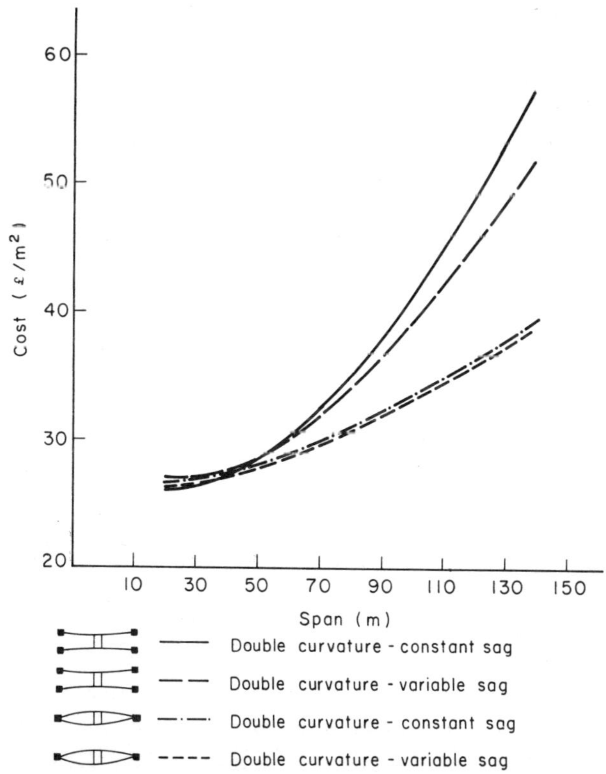

FIGURE 6-2 Cost comparisons for circular roofs using pretensioned trusses. (*Ref. 100; courtesy of the authors.*)

that a structure using a steel-frame roof with aluminum sheeting was bid at approximately 3 times the cost of the present structure, and one with a thin R.C. dome roof was estimated to cost approximately 4 times.

A useful comparison of costs for two large post-tensioned precast concrete shells supported on doubly curved cable nets is reported by Lin and Young [56]. The cost breakdown in U.S. dollars per square foot (exclusive of the supporting columns and foundations) is as follows.

	380-ft-diameter (116 m) hyperbolic paraboloid roof, Arizona State Fair Grounds	*400 × 320 ft (124 × 97 m) elliptical roof, Oklahoma State Fair Grounds*
Post-tensioning tendons (cables) Placed and stressed	1.41	0.83
Precast panels in place	0.99	
Grouting between panels	0.22	2.25 (for precast panels, grouting and scaffolding combined)
Scaffolding	1.63	
Steel struts (provided between cables)	0.22	Nil
Concrete ring beam	0.62	0.85
	5.09	3.93

6-1.3 Prestressing

Cable roofs must be pretensioned almost invariably and the need for this has already been established. Pretensioning is achieved either by preloading the cables, or by employing *counterstressed* systems. A simply suspended roof may also be stiffened by designing the cladding as a prestressed shell. This is a simple process in which precast concrete slabs are placed on the cables and loaded with predetermined weights. The spaces between the slabs are then grouted with concrete, and the applied loads are removed when the concrete sets. Two examples of this are the sports stadium in Montevideo, Uruguay,

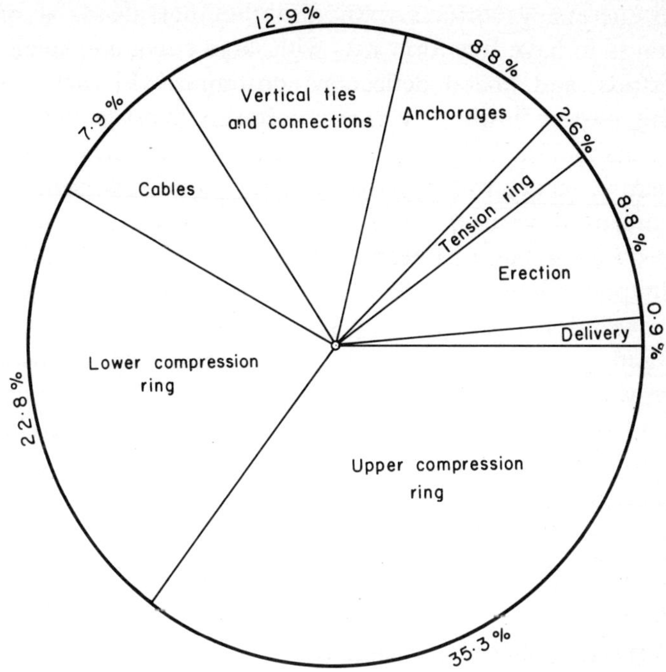

FIGURE 6-3 Cost breakdown for a circular roof of the type studied in Fig. 6-2. (*Ref. 100; courtesy of the authors.*)

and a water reservoir roof in Durham, South Africa (see also Chap. 1). In pretensioned systems such as nets or trusses, the cables are pretensioned by means of jacks or by a suitable introduction of turnbuckles. The extent to which the structure has to be pretensioned depends upon factors such as the type of structure, the ratio of dead to live loads, and the limitation on deflections. The degree of pretensioning required can be determined by trial and error, on the basis of approximate calculations, and can be considered satisfactory if, on application of applied loads, the deflections are within the permissible limits and the cable elements do not become slack. A large amount of prestressing may become necessary in networks with flat zones, to avoid excessive deformations and slackening of members. A designer may further consider whether it would be advisable to allow certain members in flat zones to slacken under occasional loads, rather than to increase the overall prestress in the network. Increased curvature is another effective and perhaps more economical way of increasing stiffness. An ideal but not always a practical situation is to have free-form nets with large curvature, small degree of prestress, and liberal deflection constraints. In such cases the cladding must be flexible. It is also necessary to assign pretensioning values such that the fundamental frequency of the system is in a range such that resonance with applied dynamic loads is avoided.

As already described, cables in a roof system are in many cases anchored into arches, ring beams, or closed frames. When the cables are fully pretensioned and the roof is loaded with uniform symmetric loads, a well-designed supporting system has a predominance of axial forces and the value of moments is low. However, during pretensioning, while the axial forces are built up gradually, large deformations and bending moments can occur. Prestressing, therefore, has to be carried out in a carefully determined sequence of operations to avoid the occurrence of large moments. Quite often it becomes necessary to introduce the entire value of prestress in a cable, or a set of cables, not in a single step but rather in a number of steps.

6-1.4 Drainage and Watertightness

It was mentioned in Chap. 1 that drainage from cylindrical roof surfaces (single curvature) presents little problem if the roof is appropriately sloped in the longitudinal direction (perpendicular to cable span). From anticlastic surfaces, too, the water drains easily. In synclastic (saucer-shaped) roofs, water collects in the concavity. Therefore, the water has to be pumped into an exterior gravity flow system or the design has to provide an interior system of pipes for drainage.

Large movements, which are possible in a cable system on account of dead and live loads and because of temperature changes, create a probability of water leakage through the roof. In this respect, pretensioned systems, which are more rigid than simply suspended systems, are better. An ideal way to provide a leakproof roof is to have a prestressed concrete decking; although from other considerations in cable-roof design, this is not commonplace. It would be good practice to seal the joints between decking elements after all dead loads are on the roof. Joint movements are then limited only to those caused by subsequent loads and temperature changes. Sealing may be done with pitch, synthetic plastics, or any other suitable material.

6-2 LOADS

The loads to be considered in the design of cable roofs are the same as in any other type of roof, apart from the prestress, which cannot strictly be considered as an applied load. These loads are now enumerated.

Dead Load. The dead load consists of the weight of cladding, insulating material, cables and fittings, etc. The value of dead weight may be as low as 5 to 15 lb/ft^2 (240 to 720 N/m^2) if cloth, plastics, or corrugated metal sheets are used and may be anywhere from 15 to 30 lb/ft^2 (720 to 1440 N/m^2) if concrete or timber is used. Of necessity, these values have to be higher for simply suspended systems.

Live Load. Most specifications specify the intensity of a uniformly distributed live load to be taken into account. In making use of these specifications, however, the designer should consider that a cable roof generally has a large span and has a curved surface. It can thus be considered inaccessible to people except for maintenance purposes. Therefore, it may be quite justified to design the cladding units for the normal live load, but a lighter design live load may be considered in designing the cable system and the supporting structure.

Snow Load. In areas where snowfall occurs, the load corresponding to the expected intensity of snow has to be considered. The intensity to be taken into account will vary with the geographic location of the structure. In this context it is worth mentioning that snow accompanied by wind gusts can cause unequal snow intensity over the roof. Similarly, alternate thawing and freezing may lead to nonuniform accumulation. Greater amounts of snow may collect in the flatter parts of the roof. Since unequal loading can result in critical deflections in cable systems, this point should be borne in mind. When there is a possibility of ponding of melted snow or rainwater, this should be treated as an additional load.

Wind Load. Wind forces often govern the design of cable-roof systems. However, there are no special specifications which set out the design wind loads to be considered in designing such systems. It is usual to design cable roofs for a uniform wind pressure or suction, despite the established evidence from experimental studies (see Chap. 5) that this practice is far from satisfactory. Whereas the practice is in many cases unavoidable in view of the lack of information on the subject, it is recommended that wherever feasible wind loads on cable-roof surfaces must be determined through wind-tunnel testing. If aeroelastic models are considered too expensive, rigid models may be used and will usually afford fairly dependable data. The basic wind pressure depends upon the location of the structure, the values of which can be picked up from relevant code specifications for buildings. In addition to the static effects of wind, due cognizance ought to be given to the time-dependent nature of wind and its dynamic effects on a roof (see discussion in Sec. 4-5.1).

Dynamic Loads. Apart from the dynamic loading that wind can cause on a roof, blast and earthquake loads ought to be considered, when applicable.

Prestress. Prestress can be treated as a permanent load similar to dead load. As mentioned in Sec. 6-1, the magnitude of prestress will vary from structure to structure and will be governed by factors already discussed.

Erection Load. Loads to which the structure is subjected during erection are temporary but can be quite heavy. Erection loads and their effect need to be worked out at every important stage during construction of the roof. Erection stresses can be severe in supporting structures like rings unless the prestressing procedure is carefully planned.

Creep and Temperature Effects. In addition to the above-mentioned loads, the effect of creep in the supporting structure (if it is of concrete) or in the cables has to be taken into account. In both areas creep will result in relaxing the cable tension and thus reducing the stiffness of the system. The slippage of cable at the anchors and an increase in temperature will effect the stiffness in a similar manner. It stands to reason that loss of prestress due to creep will be less seriously felt for cables using higher-tension steels. Also, *stress relax-*

ation is more pronounced for flexible ropes than for strands which are stiffer. For example, the loss of prestress due to creep in the roof of the German pavilion at the Montreal exposition (1967) was of the order of 25 percent, while in the Munich Olympics (1972) stadium roof it was estimated as 10 percent. The former structure used softer rope. This loss can be taken care of by suitably overstressing the cables during pretensioning.

6-3 LOAD COMBINATIONS

The design of a cable roof needs to be checked for adequate strength and serviceability, just as the design of any other structure. The section of a cable is designed on the basis of its tensile breaking strength, and it is imperative that this strength be adequate to sustain the worst combination of loads, amplified by the specified load factors. The stipulations of the AISI *Manual for Structural Applications of Steel Cables for Buildings* [60], regarding the choice of load combinations, are as follows:

The effective design breaking strength of the cable should be equal to or greater than the largest value produced by the following cable tension conditions:

(*a*) $2.2T_1$

(*b*) $1.6T_1 + 2.7T_2$

(*c*) $2.2T_3$

(*d*) $2.0T_4$

(*e*) $2.0T_5$

(*f*) $2.0T_6$

where T_1 = net tension in cable due to dead load and prestress
T_2 = change in cable tension due to application of live load
T_3 = net tension in cable due to dead load, prestress, and live load
T_4 = net tension in cable due to dead load, prestress, live load, and wind or earthquake
T_5 = net tension in cable during erection of structure
T_6 = net tension in cable due to dead load, prestress, and wind

Further, the effect of decrease in temperature is to cause an increase in the cable tension, and this should be taken into account along with T_1.

For the structure to be serviceable under different loading conditions, the following requirements need to be satisfied:

1. The maximum deflection limit for the roof should not be exceeded, or in cases where it is critical for the cladding, the maximum change in slope should be within limits. Whereas no specific limitations governing cable roofs have been established by building codes, it would seem reasonable to accept less stringent limitations than those specified for conventional roofs. Designers may use their discretion, keeping in view the local and overall serviceability requirements.
2. None of the cable elements should become slack, and, preferably under all conditions of loading, all elements should have a residual prestress,[3] whose minimum value can be specified.
3. When the roof is subjected to dynamic loads (wind, blast or earthquake), conditions of resonance or quasi-resonance leading to large deformations have to be avoided.

In order to check for the first requirement, an appropriate combination of dead load, prestress, live/snow load, and wind/blast/seismic[4] load should be considered. It is well worth noting that partial or concentrated loading can often create the critical condition. If the actual distribution of wind pressure is not known, it is recommended that the following wind load distribution be combined with other loads:

1. Wind pressure on full span
2. Wind pressure on one half of span and suction on the other half

Creep, cable slip at the anchorage, and an increase in the temperature all relax cable tension and lead to a greater possibility of slackness in members. This also reduces the overall stiffness of the system and results in greater deformations.

[3] See the discussion in Secs. 1-4.3 and 6-1.3.

[4] Static equivalents.

For structures in strongly seismic zones, or subjected to blasts or gusty winds, a dynamic analysis is recommended to check for dynamic safety.

6-4 DETAILS OF SOME EXISTING STRUCTURES—CASE STUDIES

This section gives the details of dimensions, structural features, loads, structural response, and costs of some of the existing cable-roof structures. The details are given only to the extent to which the author could obtain information and therefore, are not complete in all respects. Yet it is hoped that these examples will help the designer in making preliminary decisions.

Madison Square Garden Building, New York, New York. The Madison Square Garden complex [91] is covered with a circular cable roof. This is an outstanding example of a suspended roof built without pretensioning the cables in the conventional manner. The cables are, however, loaded with equipment to provide sufficient stiffness in the system. The main dimensions of the roof are shown in Fig. 6-4. The roof structure consists of a 1.5-in (37-mm) metal deck covered with 2-in-thick (50 mm) insulation and built-up roofing. The loads considered in design[5] are as follows:

Dead load of roof	15 lb/ft^2 (720 N/m^2)
Dead load of equipment	36 lb/ft^2 (1730 N/m^2)
Live load (equipment)	75 lb/ft^2 (3600 N/m^2)
Snow load	40 lb/ft^2 (1920 N/m^2) in the flat portion 30 lb/ft^2 (1440 N/m^2) in the sloping portion
Wind load	20 lb/ft^2 (960 N/m^2)

Forum Sports Arena, Inglewood, California. Figure 6-5 shows the main features[6] of a cable roof of a type and overall dimensions similar to those of the Madison Square Garden roof. The loading considered is as follows:

[5] Information provided by Severud and Partners, N.Y.

[6] Based on information provided by Johnson and Nielsen Associates, California.

Dead load	30 lb/ft² (1440 N/m²)
Live/snow load	20 to 12 lb/ft² (960 to 580 N/m²)
Wind load	20 lb/ft²
Unbalanced live load	6 lb/ft² (290 N/m²)
Seismic coefficient for dead load only	0.133

The structure is claimed to be 20 percent more economical than the alternatives available. The cost breakdown is as follows:

Item	*Cost* dollars/ft² (dollars/m²)
Cables	1.04 (11.0)
Fittings for cables	0.79 (8.5)
Supported roof on top of cables	7.00 (75.0)
Columns supporting the roof	7.00 (75.0)
Total roof	15.83 (170.0)

Utica Auditorium, Utica, New York. This 240-ft-span (73.2 m) circular roof [17, 34] employs counterstressed cable trusses at 5° intervals.

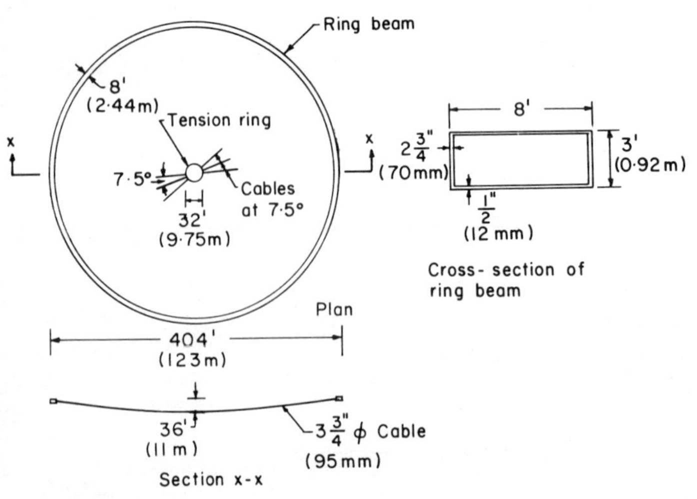

FIGURE 6-4 Details of the Madison Square Garden roof.

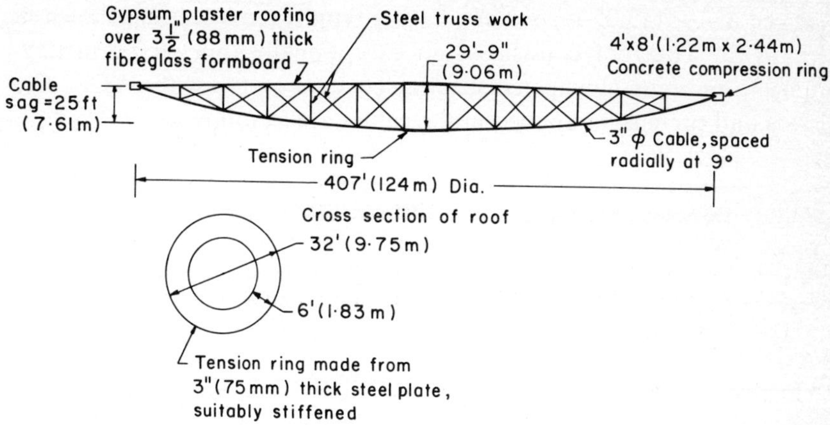

FIGURE 6-5 Forum sports arena, Inglewood (see Sec. 6-4).

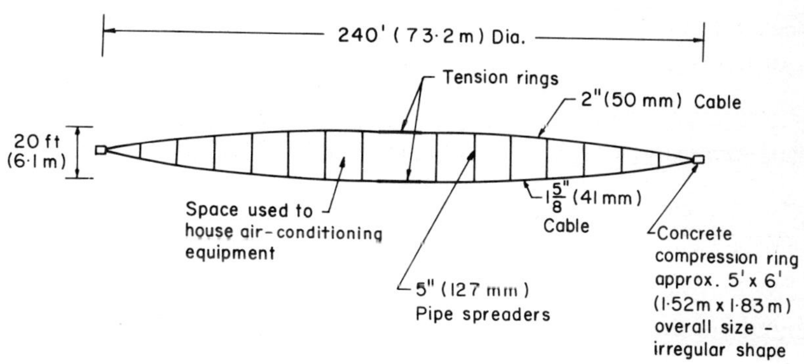

FIGURE 6-6 Cross section through Utica auditorium roof.

Some of the features of the arrangement are shown in the cross section presented in Fig. 6-6. The roof consists of corrugated metal deck with insulation boards. The roof employs 72 tons (692 kN) of cable, and the total dead load is 14 lb/ft^2 (670 N/m^2). The live load considered is 30 lb/ft^2 (1440 N/m^2). The minimum prestress in the cables was 135,000 lb (600 kN). The deflection caused by the roof and air-conditioning equipment amounts to 4 in (102 mm).

Whorley Reservoir, Durham County, England. The Whorley reservoir roof[7] covers an area of 185 × 123 ft (56.3 × 37.5 m). The trusses,[8]

[7] Information provided by White-Young and Partners, United Kingdom.

[8] The trusses were designed and studied at the Cable Research Centre, Polytechnic of Central London.

spaced at 10-ft (3.05-m) centers, have a typical cross section, shown in Fig. 6-7. The roof consists of an exceptionally thin—0.035-in (0.9-mm)—aluminum corrugated sheeting covered on top with PVC. The loads and pretension applied in the cables are as follows:

Dead load + snow load	30 lb/ft^2 (1440 N/m^2)
Wind load	8 lb/ft^2 (385 N/m^2)
Pretension in each cable	30,000 lb, equivalent to a load of 10 lb/ft^2 (480 N/m^2) approx.

Deflection expected under full dead and snow load is 18 in (460 mm). The computed time period of vibration for the unclad truss is 0.25 s and for the truss with the roofing is 0.31 s.

The roof costs approximately £2.56 per square foot (£27.6 per square metre) of covered area.[9] The cost breakdown for some of the major items is as follows:

Cable girders	£0.34 per square foot (£3.7 per square metre)
Cladding	£0.60 per square foot (£6.5 per square metre)
Anchors	£0.70 per square foot (£7.5 per square metre)

In considering the total cost, which may appear to be high, it must be pointed out that the access conditions at site were extremely difficult.

Sports Hall, Billingham, England; Transit Store, Schiphol, Holland; and Ice Stadium Johanneshov, Stockholm, Sweden. These are some of the outstanding illustrations of the use of counterstressed trusses,

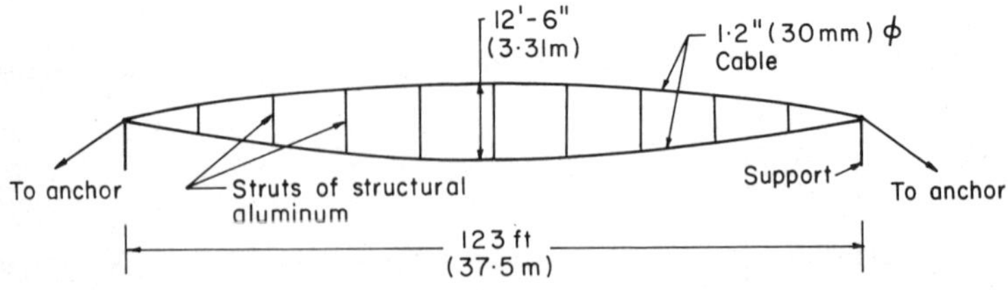

FIGURE 6-7 Cross section through Whorley reservoir roof.

[9] £ represents the British pound sterling. It is currently equivalent to 1.7 U.S. dollars.

TABLE 6-3*

	Name of building		
	Sports Hall *Billingham, England*	*Transit Store* *Schiphol, Holland*	*Ice Stadium Johanneshov,* *Stockholm, Sweden*
Trusses Section	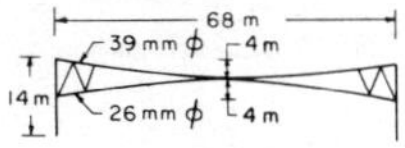	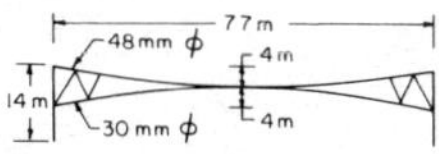	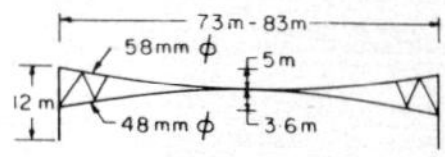
Type: Spacing:	Jawerth 3.35 m	Jawerth 4.7 m	Jawerth 5.28 m
Length of building	40 m	140 m	120 m
Type of supports	Inverted arch	Steel column and backstay	Steel column and backstay
Prestress (equivalent)	28 kg/m²	20 kg/m²	39 kg/m²
Design loads			
Dead	40 kg/m²	50 kg/m²	42 kg/m²
Snow	75 kg/m²	50 kg/m²	100 kg/m²
Wind	− 60 kg/m²	− 42 kg/m²	− 40 kg/m²
Temperature range	− 20 to + 20°C	− 10 to + 25°C	− 20 to + 20°C
Deflections			
Snow	530 mm	400 mm	430 mm
Wind	− 280 mm	− 310 mm	− 180 mm
Natural period of vibration, under dead load only	0.31 s	0.5 s	0.37 s
Cost			
Cables and fittings erected		40 to 60 Kr†/m²	
Cladding		90 to 100 Kr/m²	

* Based on information provided by David Jawerth of JAWERTH INTERSTATIK.

† Swedish kroner.

1 kg/m² = 9.806 N/m² = 0.205 lb/ft²

1 m = 3.28 ft

with diagonal web members, developed by Jawerth. The main features of the roof trusses used in these three structures and their structural response are tabulated in Table 6-3.

State Fair Arena, Raleigh, North Carolina. This roof could well be said to have ushered in the present era of cable-roof construction. As shown in Fig. 6-8, the roof employs a doubly curved network of

FIGURE 6-8*a* Raleigh arena. (*Bethlehem Steel.*)

cables[10] anchored into two parabolic arches [62, 87]. The network is stabilized against vibrations by means of *spring-loaded* guy ropes which act as dampers. The roof consists of corrugated 22-gage asbestos-protected metal deck, covered with $1\frac{1}{2}$-in-thick (38 mm) insulation. The design loading is

Dead load	6 lb/ft^2 (290 N/m^2)
Live load	25 lb/ft^2 (1200 N/m^2)
Wind suction	16 lb/ft^2 (770 N/m^2)

The cost of the building, excluding seating, etc., was $16.21 per square foot ($174.5 per square metre) including the cost of services and $13.15 per square foot ($141.5 per square metre) if services are excluded.

Swimming Pool, Tokyo, Japan. This roof (see Fig. 6-9) consists of doubly curved networks [98] supported on main cables and concrete

[10] The maximum sag of the main cable is 31 ft $3\frac{5}{8}$ in (9.55 m) and the maximum rise of the secondary cable is 28 ft 8 in (8.75 m).

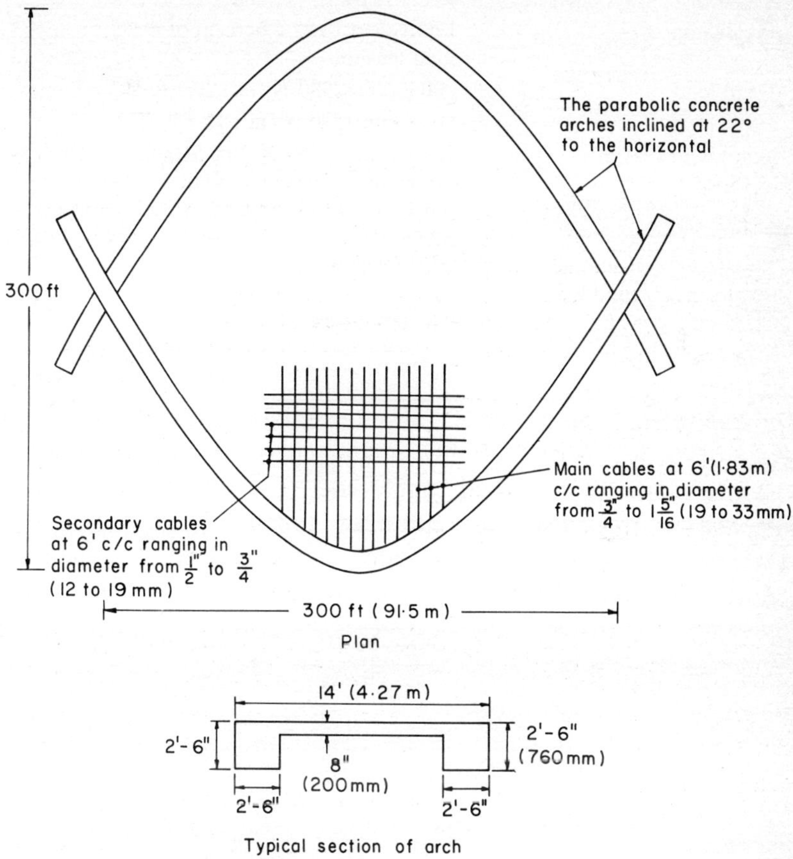

FIGURE 6-8*b* Details of Raleigh arena.

ring beams. The network has sharp changes of slope near the pylons, and to accommodate this geometry, the main members are built-up I sections, bent to shape, instead of cables. The deck for the roof consists of 0.18 in (4.5-mm) steel plates welded together. A special feature of the anchoring system is that the gravity anchors for the cables support only the vertical component of the tensile force. The horizontal component is transmitted between the anchor and the pylon by means of 4.9×9.4 ft (1.5×3.0 m) concrete struts, and between the pylons by the inspecting gallery under the pool, acting as a huge strut. The main cables are installed with an oil damper system at each of the two pylons, to safeguard against the possibility of large amplitudes of vibration. The loads considered are as follows:

Pretension	Equivalent to 22.4 percent of dead load tension
Dead load	100 kg/m^2 (980 N/m^2, 20.5 lb/ft^2)
Snow load	60 kg/m^2 (590 N/m^2, 12.3 lb/ft^2)
Wind load (basic)	250 kg/m^2 (2450 N/m^2, 51.2 lb/ft^2) (Wind-tunnel tests showed a variation of 0.65 to −0.45 for wind-pressure coefficient over roof surface. See Fig. 5-6.)
Maximum deflections due to snow/wind load	200 mm (7.9 m)
Maximum period of vibration	0.67 in unsymmetric mode

The Palasport, Genova, Italy.[11] The central 236.2-ft-diameter (72 m) portion (see Fig. 6-10*a*) of this building is a tension structure [26] which is supported on radial cantilevers. The cables anchor into a compression ring made in reinforced concrete at the outer circumference and into a tubular frame at the center. The main dimensions and features of the system are shown in Fig. 6-10*b*. The

FIGURE 6-9*a* Swimming pool roof for the 1964 Tokyo Olympics. (*Courtesy of RETORIA Yukio Futagawa and associate photographers.*)

[11] Based on Ref. 26 and a personal communication from G. Maeir.

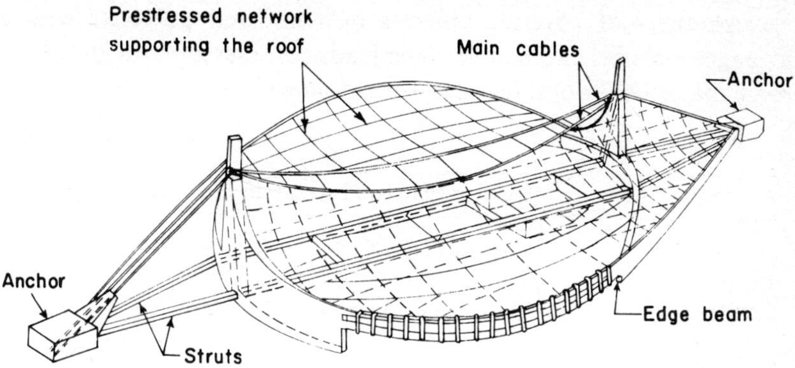

FIGURE 6-9*b* Structural scheme for the structure in Fig. 6-9*a*.

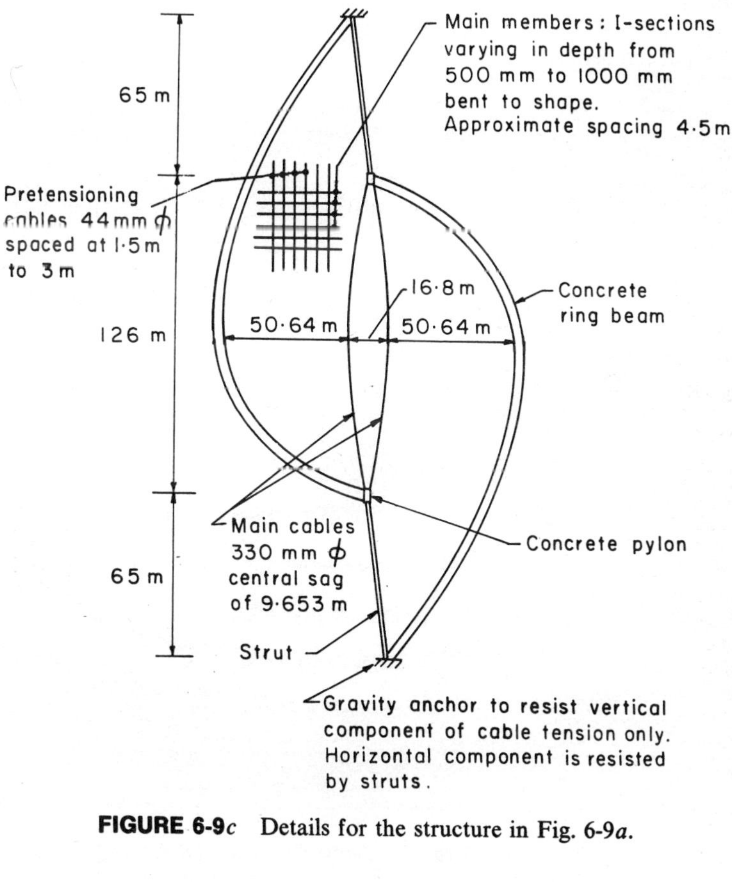

FIGURE 6-9*c* Details for the structure in Fig. 6-9*a*.

transparent roof covering consists of reinforced polyester with an average thickness of 3.5 mm. The loads considered in the design and some major behavioral features are as follows:

Pretension	10 tonnes (98 kN, 22.4 kips) in each upper cable 27 tonnes (265 kN, 59.6 kips) in each lower cable
Dead load	23 kg/m² (1105 N/m², 4.7 lb/ft²)
Live load	10 tonnes suspended from the central frame
Snow load	100 kg/m² (980 N/m², 20.5 lb/ft²)
Basic wind pressure	150 kg/m² (1470 N/m², 30.75 lb/ft²) (wind coefficients as obtained from wind-tunnel studies are shown in Fig. 5-7)
Maximum calculated deflection due to snow load	250 mm (9.84 in)
Fundamental period of vibration	< 0.5 s

The structure employs a total of 21.5 kg/m² (211 N/m², 4.4 lb/ft²) steel out of which 10.5 kg/m² (103 N/m², 2.15 lb/ft²) is used in cables and fittings.

FIGURE 6-10*a* Palasport Genova. (*Ref. 26.*)

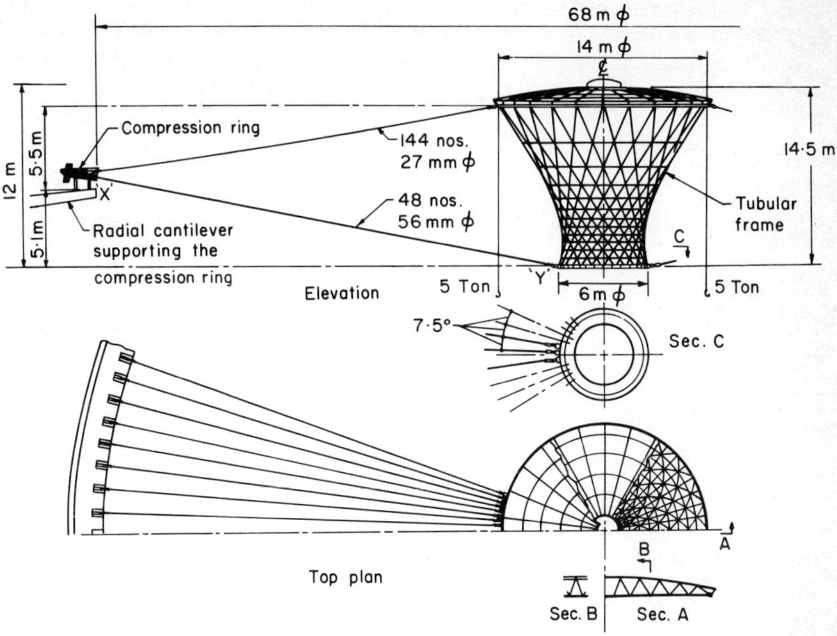

FIGURE 6-10*b* Structural scheme for the suspended roof structure in Fig. 6-10*a*. Details at *X* and *Y* may be seen in Fig. 7-9.

Sports Hall, Herning, Denmark.[12] This is a bicycle wheel type of cable roof 262.4 ft (80 m) in diameter with the typical cross section as shown in Fig. 6-11. The system consists of 42 trusses with equal angular spacing, anchored into two 13.1-ft-diameter (4 m) steel tension rings at the center and two precast concrete compression rings at the outer circumference. The supporting columns also consist of precast concrete elements, with in situ concrete foundation. The roof cladding consists of galvanized steel sheeting with an 0.31-in-thick (8 mm) insulation. The sheeting is supported on a radial system of beams, which are in turn suspended from the cable ties (see Fig. 6-12).

A special feature of the design is that the cables are connected at the center into two separate tension rings, contrary to the common practice in such structures of connecting all cables into a single tension ring. The advantage is a considerable reduction in the col-

[12] The information is based on Ref. 31, and personal communications from V. L. Gill, and from H. Mollmann who acted as consultant for the analysis of the structure to the consulting engineer R. Christiansen I/S.

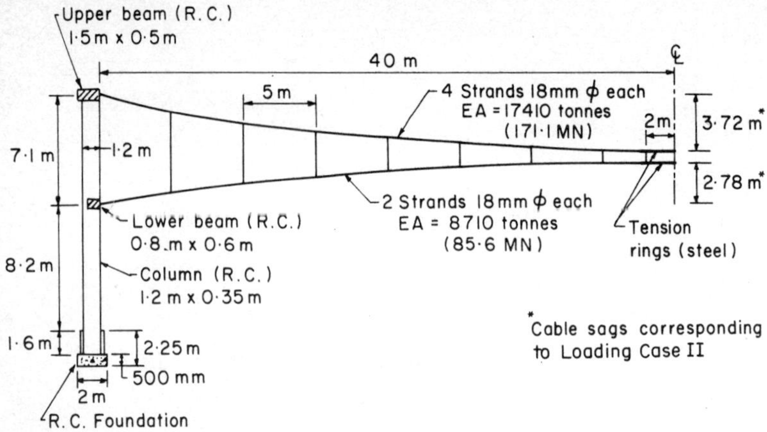

FIGURE 6-11 Cross section through the roof of sports hall, Herning (see Sec. 6-4).

umn and foundation moments when the roof is subjected to unsymmetric loading. The idea is explained qualitatively by the sketches in Fig. 6-13. In the present case, as envisaged in the section in Fig. 6-13*a*, unsymmetric loading causes very small changes in cable tension, and the load is resisted mainly by a change in geometry, indicated by the broken line. The moments in the columns and foundations are therefore small in this case. In the case of a single

FIGURE 6-12*a* Sports hall, Herning: a view of the roof. (*Courtesy of V. L. Gill.*)

FIGURE 6-12*b* Sports hall, Herning: suspender and bottom cable connection. (*Courtesy of V. L. Gill.*)

tension ring, cable forces develop as shown in Fig. 6-13*b*. The ring beams are pushed or pulled in the radial direction as rigid bodies. The forces ΔH have therefore to be borne by the columns, causing heavy moments in these members and their foundation.

The design loads considered for the roof structure are

Self-weight of the roof (including cables)	42 kg/m^2 (412 N/m^2, 8.6 lb/ft^2)
Superimposed snow load	75 kg/m^2 (735 N/m^2, 15.4 lb/ft^2)
Wind suction	100 kg/m^2 (980 N/m^2, 20.5 lb/ft^2)

Loading cases I to VI are given below. The structure was also analyzed for the effects of change in temperature of $\pm 20°C$, shrinkage, and creep in concrete supporting structures.

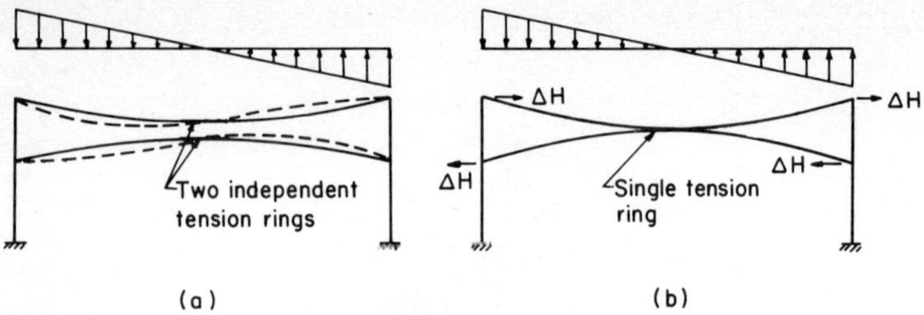

FIGURE 6-13 A schematic representation of the effect of antisymmetrical load on a truss with (*a*) two independent tension rings, and (*b*) single tension ring, at the center.

TABLE 6-4 Sports Hall, Herning, Denmark. Results of Analysis for Different Load Combinations*

		Loading case					
		I	*II*	*III*	*IV*	*V*	*VI*
Maximum force in upper cables		23.7 (229)	37.7 (370)	61.8 (606)	56.6 (556)	6.0 (59)	36.9 (363)
Maximum force in lower cables		24.2 (238)	18.2 (179)	10.2 (100)	18.9 (186)	33.3 (327)	17.4 (171)
Maximum vertical displacement of roof		− 234	0	320	525	− 531	13
Upper ring beam	Direct compressive force	176 (1730)	259 (2545)	401 (3945)	341 (3350)	71 (698)	252 (2475)
	Maximum bending moment	0	0	0	52.3 (514)	0	0
Lower ring beam	Direct compressive force	111 (1090)	94 (925)	75 (737)	91 (895)	139 (1365)	91 (895)
	Maximum bending moment	0	0	0	1.6 (16)	0	0
Column	Bending moment at level of lower compression ring	− 18.2 (− 179)	− 8.4 (− 82)	− 4.2 (− 41)	19.7†, 18.3† (194) (180)		
	Bending moment at foundation level	35.1 (345)	22.4 (220)	7.5 (73)	4.4†, − 18.4† (43) (− 181)		

* Units used for forces are tonnes (kN); for moments, tonne-metres (kN · m); for deflections, mm.

† On opposite columns.

1 tonne = 2.2 kips, 1 tonne-metre = 7.22 kip-ft, 1 mm = 0.04 in

LOADING CASES

Loading case I: Cable prestress + weight of cables

Loading case II: Cable prestress + weight of cables and roof decking

Loading case III: Loading case II + uniformly distributed snow load over entire surface

Loading case IV: Loading case II + uniformly distributed snow load over half of surface

Loading case V: Loading case II + wind suction over entire surface + temperature rise of 20°C

Loading case VI: Loading case II + creep of concrete ring

In analyzing the structure, the flexibility of the ring beams was accounted for, and a summary of the results is given in Table 6-4. It is worth noting from Table 6-4, loading case I, that the thrust in the two outer ring beams is in the ratio 176:111, differing from the ratio in the corresponding cable tension, which is 23.7:24.2. This difference is caused by readjustments due to the bending of the column.

COSTS The structure, covering an area of 5000 m^2, cost £57 per square metre. The cost breakdown is as follows:

Item		*Cost, %*
Ground beam	Precast concrete	3.8
Lower ring beam		6.3
Upper ring beam		13.4
Columns		20.7
Erection of the concrete items		7.3
Cables		5.5
Tensioning, anchorages, ties, and fittings		31.6
Erection of cables		11.4
		100.0

PRETENSIONING AND ERECTION The roof construction was started after the supporting structure had been erected. The tension ring was initially placed at the level of the top compression ring, and the top cables were connected to it. The ring was then freely suspended from the top cables, and the lower cables were connected to the ring, the vertical aluminum tie bars being connected to the cables by aluminum clamps. The bottom cables were pulled at the center in a predetermined sequence consisting of eight complete sets of operations, leading to a total pull of 7.9 in (200 mm) and a pretension of 23.8 tonnes (234 kN) in each bottom cable. In each step, cables were pulled in diametrically opposite pairs. Also in each step, the rings were subjected to an axisymmetric load equal to the pretension in the cables at the end of the preceding step, plus an unsymmetric load, which caused bending of the ring. Eight steps were necessary to make sure that stresses in the ring resulting from axial forces and bending moments were within safe limits at all stages of erection.

The top cables consisted of four strands 0.6 in (18 mm) in diameter; the lower cables, two strands. All strands were provided with a screwed end terminal for adjustment. The top cable strands were cut to a tolerance of ± 0.12 in (3 mm).

6-5 DESIGN FOR PRELIMINARY COSTING

In designing large-span roofs, the choice of the system is indeed most important. As discussed in other parts of this book, an assessment of factors in favor and against the suitability of cable roofs points to the general conclusion that these roofs are competitive with other systems for the medium-span range and are distinctly better for large spans. The various other forms that the designer may consider for comparison are shells, domes, and latticed structures of steel or concrete. Cable-roof systems may need to be compared among themselves too. One of the prime factors for comparison is cost. Notwithstanding the cost of land, services, design, and financing, each of which may not be too sensitive to structural form, the estimation of structural material and labor is extremely important. With the help of general discussions in Secs. 1-3, 1-4, and 6-2 to 6-4, analytical aids in Secs. 3-4 and 3-5, and cable properties in Sec. 7-2, it is attempted herein to get a preliminary material estimate for a proposed cable roof. Once an estimate of materials is available, the cost of the materials and the cost of labor component are determined, depending upon the rates applicable where the structure is to be built.

For the purpose of illustration, the roof of an aircraft hangar is considered. The length of the building is 560 ft (170.7 m) and its width is 460 ft (140.2 m). The minimum clearance required at the center is 80 ft (24.4 m). The roof is to slope longitudinally from the center outwards to allow for drainage. To allow for this and vertical deflections under load, the elevation of the central truss is raised by an additional 5 ft (1.5 m). This is a building with a rectangular plan with a minimum span of 460 ft (140.2 m). Material and cost breakdown for the structure designed herein is given based on Indian specifications for permissible stresses in steel and reinforced concrete construction, and rates for Indian conditions.

6-5.1 Choice of Structural Form

For clear spans of the order under consideration, the suitability of cable roofs is well established. Also a comparison of costs for the Jumbo aircraft hangars at the London and Frankfurt airports reveals that the cost per unit area of the latter, which uses a cable truss roof, is nearly 40 percent of the former. The structure at the London airport uses a steel space system. Among cable systems the choice is between trusses and networks. Provision of a doubly curved network may not be very satisfactory from the viewpoint of clearances and also for the reason that there should be a fully openable gate on one of the 460-ft (140.2-m) sides. Cable trusses shown in Fig. 1-9 are therefore considered. Trusses in Fig. 1-9*b* provide greater central clearance than those in 1-9*c*, and are therefore adopted. In order to reduce deformations under asymmetric loading due to wind, the cables are connected at the center. *It may be mentioned here that the choice in this case is further encouraged by the need to use a simple illustrative example*. The proposed truss cross section is shown in Fig. 6-14. The sag of the upper cable is taken as 0.075 times the span and equals 34.5 ft (10.5 m). The spacing of trusses is taken as 25 ft 5 in (7.75 m). The roof cladding is carried by the upper cable.

6-5.2 Loads

The loading cases considered are the following:

$$\text{Dead weight} = 8\ \text{lb/ft}^2\ (385\ \text{N/m}^2)$$

$$\text{Live load} = 8\ \text{lb/ft}^2$$

As mentioned repeatedly in this book, estimation of wind load is indeed a difficult task. For this example two cases of wind are considered.

1. 10 lb/ft^2 (480 kN/m^2) suction over the entire span: to check the safety of the bottom cable.

2. Same intensity of suction over one half of the span and pressure over the other: to check for deformations.

The value of initial prestress (equivalent to 8 lb/ft^2 of uniform load)	= 156 kips (693.9 kN) in the upper cable
	= 233 kips (1036.4 kN) in the lower cable

6-5.3 Estimation of Cable Size

Estimates Based on Strength. Cable sizes are estimated first on the basis of minimum required strength, using the criteria given in Sec. 6-3. In calculating the tension increment for applied loads, it is assumed that the two cables share the load in proportion to the inverse of the cable sag. For example,

An applied load of 8 lb/ft^2 will cause a uniformly distributed load of $\frac{8 \times 25.4}{1000}$ = 0.203 kip/ft (2.96 kN/m) on a truss.

The tension change in the upper cable

$$= \frac{0.203 \times 460^2}{8 \times 34.5} \times \frac{23}{34.5 + 23}$$

$$= 62.3 \text{ kips } (277.1 \text{ kN})$$

Tension change in the lower cable

$$= 62.3 \times \frac{34.5}{23}$$

$$= 93.4 \text{ kips } (415.4 \text{ kN})$$

Tension values as per the combinations given in Sec. 6-3 are tabulated in Table 6-5. The design tension for the upper cable is 617.3 kips (2745.7 kN), and for the lower cable, 512.7 kips (2280.5 kN), and

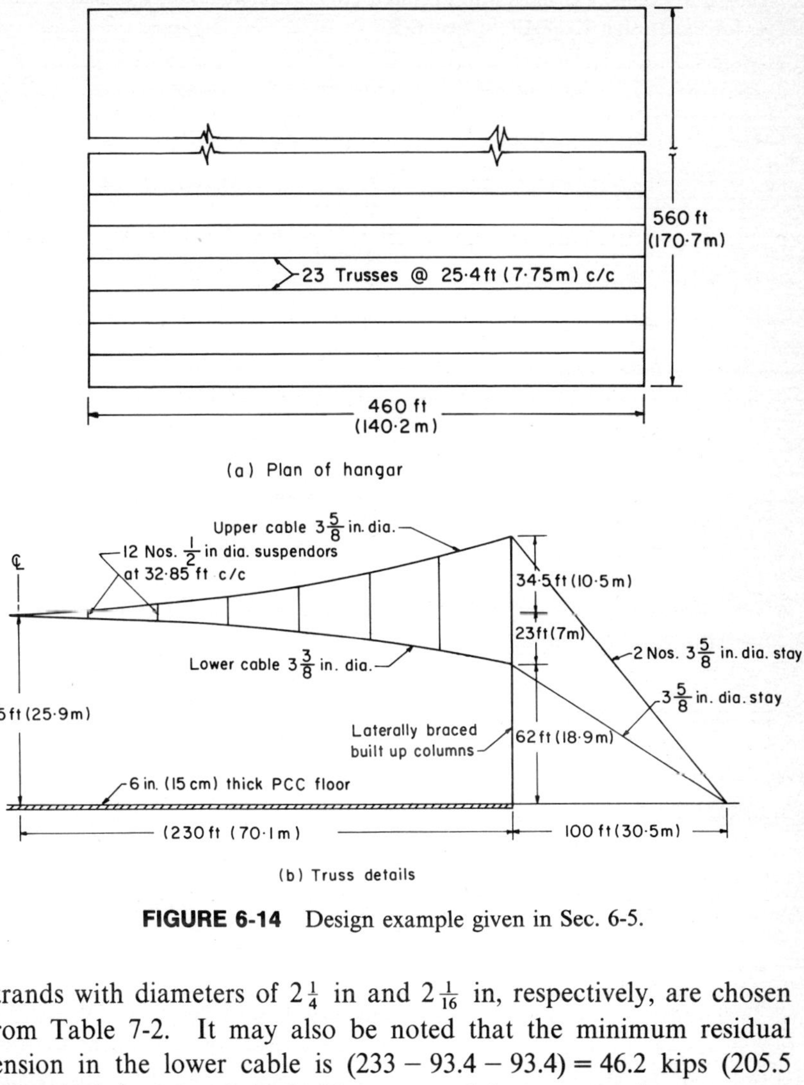

FIGURE 6-14 Design example given in Sec. 6-5.

strands with diameters of $2\frac{1}{4}$ in and $2\frac{1}{16}$ in, respectively, are chosen from Table 7-2. It may also be noted that the minimum residual tension in the lower cable is (233 – 93.4 – 93.4) = 46.2 kips (205.5 kN), which is approximately 20 percent of the prestress force.

Estimates Based on Deformations. It is next attempted to obtain an estimate of cable sizes required to limit deformations, with the use of design curves given in Sec. 3-4.1. It is necessary first to indicate the conditions on which the estimate is to be based.

1. It is stipulated that the dead load of the roof will be trans-

TABLE 6-5 Cable Tensions with Factored Loads, Computed for Estimating Cable Size, for the Design Example in Sec. 6-5

*Combination**	*Upper-cable tension, kips (kN)*		*Lower-cable tension, kips (kN)*	
$2.2T_1$	$2.2(156 + 62.3)$	$= 480.3$ (2136.4)	$2.2(233 - 93.4)$	$= 307.1$ (1366)
$1.6T_1 + 2.7T_2$	$1.6 \times 218.3 + 2.7 \times 62.3$	$= 517.5$ (2301.8)	$1.6 \times 139.6 - 2.7 \times 93.4$	$= -28.8$ (128.1)
$2.2T_3$	$2.2(156 + 62.3 + 62.3)$	$= 617.3$ (2745.7)	$2.2(233 - 93.4 - 93.4)$	$= 101.6$ (451.9)
$2.0T_4$	$2.0(156 + 62.3 + 62.3 - 77.9)$	$= 405.4$ (1803.2)	$2.0(233 - 93.4 - 93.4) + 116.75$	$= 325.9$ (1449.6)
$2.0T_6$	$2(156 + 62.3 - 77.9)$	$= 280.8$ (1249)	$2.0(233 - 93.4 + 116.75)$	$= 512.7$ (2280.5)

* T_1 = net tension in cable due to prestress and dead load
T_2 = change in cable tension due to application of live load
T_3 = net tension in cable due to prestress, dead load, and live load
T_4 = net tension in cable due to prestress, dead load, live load, and wind
T_6 = net tension in cable due to prestress, dead load, and wind.
(Only the effect of wind suction on the whole span is considered. For the purpose of this preliminary exercise, the antisymmetric wind effect is ignored; it is not likely to be large.)

ferred completely to the trusses before these are adjusted to the final geometry.

2. The critical condition for the deflection check due to condition 1 is, therefore, the live load ($= q$) + antisymmetric wind load ($= 1.25q$)
3. It is assumed that:
   a. The effect of variation in f_l/L and $(EA)_l$ is negligible.
   b. The limiting deflection $w = L/250$, or $w/L = 4 \times 10^{-3}$.
4. Kinematic linearity is assumed.

The cable areas must now be estimated with the help of various curves given in Fig. 3-12. It is obvious that neither these loading cases, nor the structure in question, is directly covered by the results in Fig. 3-12. However, a suitable manipulation can yield reasonably good results. This is attempted through the following steps:

1. The curves for maximum positive deflection given in Fig. 3-12*a* are reproduced in Fig. 6-15 and demarcated as curves I and II respectively. The loading cases corresponding to the curves are also shown on the same figure.

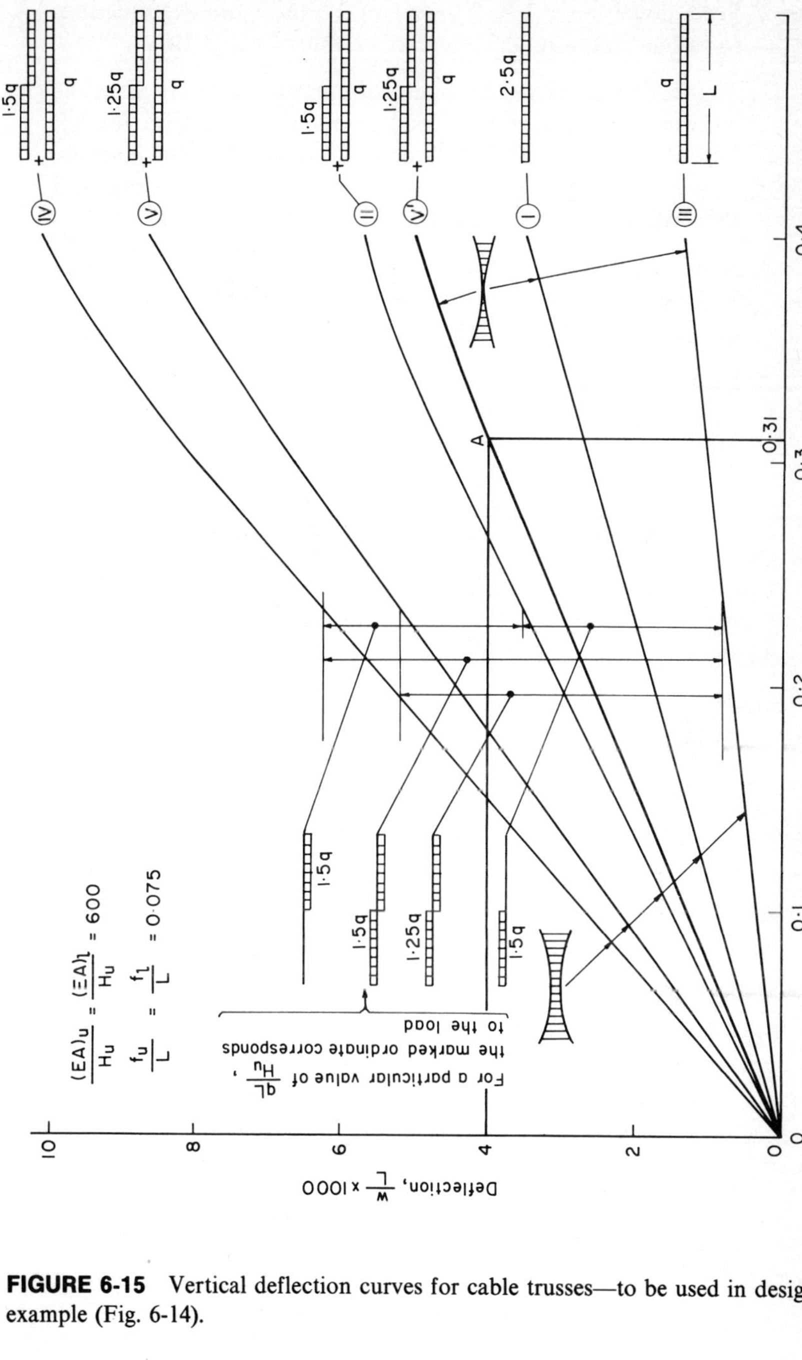

FIGURE 6-15 Vertical deflection curves for cable trusses—to be used in design example (Fig. 6-14).

2. Scale down curve I by a factor of 2.5 to obtain deformations for a load intensity q. Demarcate this as curve III.

3. The difference in the ordinates of curves II and III gives the deformation for a positive load of intensity $1.5q$ over the left half of the span. If, in addition, a negative load of intensity $1.5q$ is now applied to the right half of the span, assume that this ordinate will be doubled.[13] Thus draw curve IV.

4. Scale the ordinate between curves IV and III by a factor of $1.25/1.5$ because deformations are sought for an antisymmetric load of $1.25q$. Thus draw curve V.

5. Since in this case the cables of the truss are connected at the center, deformations under antisymmetrical loading will be smaller as compared to a truss with unconnected cables. On the basis of results discussed in Sec. 3-4.1, further reduce the difference between curves V and III by[14] 50 percent and draw the final curve V′.

6. Draw a abscissa through the point $w/L = 4 \times 10^{-3}$, intersecting the curve V′ at A. Drop a vertical from A to obtain a value of 0.31 for qL/H_u.

7. Now,

$$q = 0.203 \text{ kip/ft}$$

$$L = 460 \text{ ft}$$

$$H_u = \frac{(EA)_u}{600}$$

Therefore,

$$\frac{qL}{(EA)_u/600} = 0.31$$

or

$$(EA)_u = \frac{600qL}{0.31} = 18.07 \times 10^4 \text{ kips}$$

Accordingly, a strand $3\frac{5}{8}$ in (92 mm) in diameter is chosen for the upper cable. The lower cable is proportioned as 512.7/617.3 (see Table 6-5) and taken as a $3\frac{3}{8}$-in-diameter (85.7-mm) strand.

[13] Factorization made in steps 3 and 5 is at best an approximation based on judgment derived from only a limited number of results. There could be a fair degree of variation in these values.

[14] See note 13 above.

6-5.4 Check on Design

Use of the approximate formulas of Sec. 3-5.1 has been illustrated by Example 3-7. Now, taking the following data for the cables and loads, and using the formulas for cases II and VII in Table 3-12*b*, deflections and forces in the truss are computed and summarized in Tables 6-6*a* and 6-6*b*.

$$(EA)_u = 18.9 \times 10^4 \text{ kips } (84 \times 10^4 \text{ kN})$$

TABLE 6-6*a* Maximum Deflections for the Design Example in Sec. 6-5

			Cable deflections, ft (m)	
Serial no.	*Loading, kips/ft (kN/m)*		*At x = 0*	*At $x = \mp L/4$*
1.	Uniformly distributed dead load	= 0.203(2.96)	0.686(0.209)	0.514(0.157)
2.	Uniformly distributed live load	= 0.203(2.96)	0.686(0.209)	0.514(0.157)
3.	Uniform wind suction	= 0.254(3.7)	− 0.858(0.261)	− 0.644(0.196)
4.	Uniform wind suction on right half and uniform pressure on left half	= 0.254(3.7)	0	± 0.787(0.24)

The maximum deflection will be at $x = \mp L/4$, due to live load + antisymmetric wind load and equal to 1.3 ft (0.396 m). This is equal to $L/354$.

TABLE 6-6*b* Maximum Values of Horizontal Component of Cable Tension for Different Load Combinations for the Design Example in Sec. 6-5

Serial no.	*Type of loading*		*Upper-cable tension, kips (kN)*	*Lower-cable tension, kips (kN)*
1.	Initial prestress		156 (693.9)	233 (1036.4)
2.	Uniformly distributed dead load	= 0.203 K/ft (2.96 kN/m)	107.1 (476.4)	− 61.9 (275.3)
3.	Uniformly distributed live load	= 0.203 K/ft (2.96 kN/m)	107.1 (476.4)	− 61.9 (275.3)
4.	Uniform wind suction	= 0.254 K/ft (3.7 kN/m)	− 134.0 (596)	+ 77.4 (344.3)
5.	Uniform wind suction on right half and uniform pressure on left half	= 0.254 K/ft (3.7 kN/m)	± 95.3 (423.9)	∓ 95.3 (423.9)
DIFFERENT LOAD COMBINATIONS				
a.	1 + 2 + 3		370.2 (1646.6)	109.2 (485.7)
b.	1 + 2 + 4		129.1 (574.3)	248.5 (1105.3)
c.	1 + 2 + 5		358.4 (1594.2)	266.4 (1184.9)

$$(EA)_l = 16.38 \times 10^4 \text{ kips } (72.8 \times 10^4 \text{ kN})$$

$$\text{Live load, } p = 0.203 \text{ kips/ft } (2.96 \text{ kN/m}) \text{ of truss span}$$

$$\text{Wind load (suction)} = 0.254 \text{ kip/ft } (3.7 \text{ kN/m})$$

$$\text{Wind load (antisymmetric)} = 0.254 \text{ kip/ft } (3.7 \text{ kN/m})$$

It may be noted from Table 6-6*a* that the maximum deflection is well within the permissible limit assumed. However, the design is not revised at this stage, keeping in view the possibility of increased deformations on account of support flexibility. It is worth mentioning that, while maintaining the designed cable sizes previously given, it became necessary to design the cable stays from the consideration of limiting deformations rather than strength. Table 6-6*b* gives maximum forces in the cables and as such the design forces on the supporting system. In addition, the columns will be subjected to forces of wind. The details of the design for the columns, backstays, foundations, and bracing, etc., which is an exercise in conventional design work, are omitted to save space.

TABLE 6-7 Material and Cost Estimate for the Design Example in Sec. 6-5

Serial no.	*Item*	*Quantity*	*Rate*	*Cost Rs.*	*Percentage of total cost*
1.	Roof				
	a. Roof cables inclusive of suspendors, connectors, and end fittings	551 kips	Rs. 2,400/-kip	13,22,400/-	14.6
	b. Purlins	339 kips	Rs. 1,100/- kip	3,72,900/-	4.1
	c. Roofing material C.G.I. sheets	2,61,52 ft^2		11,00,000/-	12.1
2.	Supporting structure	505 kips	Rs. 1,820/-kip	9,19,100/-	10.1
	a. Stay—cables				
	b. Tension piles	46 Nos		18,40,000/-	20.3
	c. Steel column inclusive of bracing and side rails	1105 kips	Rs. 1,100/-kip	12,15,500/-	13.4
	d. Column foundations (R.C.)	46 Nos		4,40,000/-	4.6
	e. Side cladding 6 mm A.C. sheeting	1,24,880 ft^2		3,70,000/-	3.0
3.	Flooring				
	6 in P.C.C. over 6-in rammed hard core	1,28,800 t^3		11,60,000/-	12.8
				86,39,900/-	
4.	Miscellaneous (5% of above)			4,31,995/-	
				90,71,895/-	Say Rs. 90,72,000/-

6-5.5 Costs

A breakdown of materials and costs is given in Table 6-7. The estimate excludes the entrance and exit structure on the 460-ft sides. For a floor area of 257,600 ft^2 (24,100 m^2) the cost is Rs.25.82[15] per square foot ($35 per square metre) for the items included.

Further refinements are indeed possible in the work presented for improving upon the estimate of materials and costs. For example, a more accurate analysis may be carried out by using the computer program contained in the Appendix or otherwise, duly accounting for the flexibility of the supporting members.

[15] Rs. represents the Indian rupee. It is currently equivalent to 0.113 U.S. dollars.

CHAPTER SEVEN

Construction Materials and Details

7-1 GENERAL

Information regarding construction details and available material is as important to a designer as the analysis procedures or information on design loads. It is particularly important when suspended roofs are concerned, because some of the construction material employed for these roofs is not as conventional as reinforced concrete or rolled steel. This chapter gives such information on cables, fittings, anchorages, roof decking, and certain other related aspects.

7-2 CABLES

The strength of cables under static as well as fatigue loading is dependent upon cable composition, clamps, fittings, and changes in curvature. The basic component of a cable is wire drawn from high-strength steel rods. The wire is then galvanized by the hot-dip or the electrolytic process. In some cases wires are drawn further after galvanizing and are known as *drawn galvanized.* A number of such wires are then spun to form a *strand.* Wire *ropes* are formed by assembling a number of strands. Figure 7-1 shows some of the numerous[1] arrangements possible. The usual practice is to arrange six strands over a central core which is itself a strand or a rope. Both strands and ropes are used in suspended roofs. High-strength steel rods, rolled or built-up sections, have also been used in certain roof

[1] May be seen from rope manufacturers' catalogues.

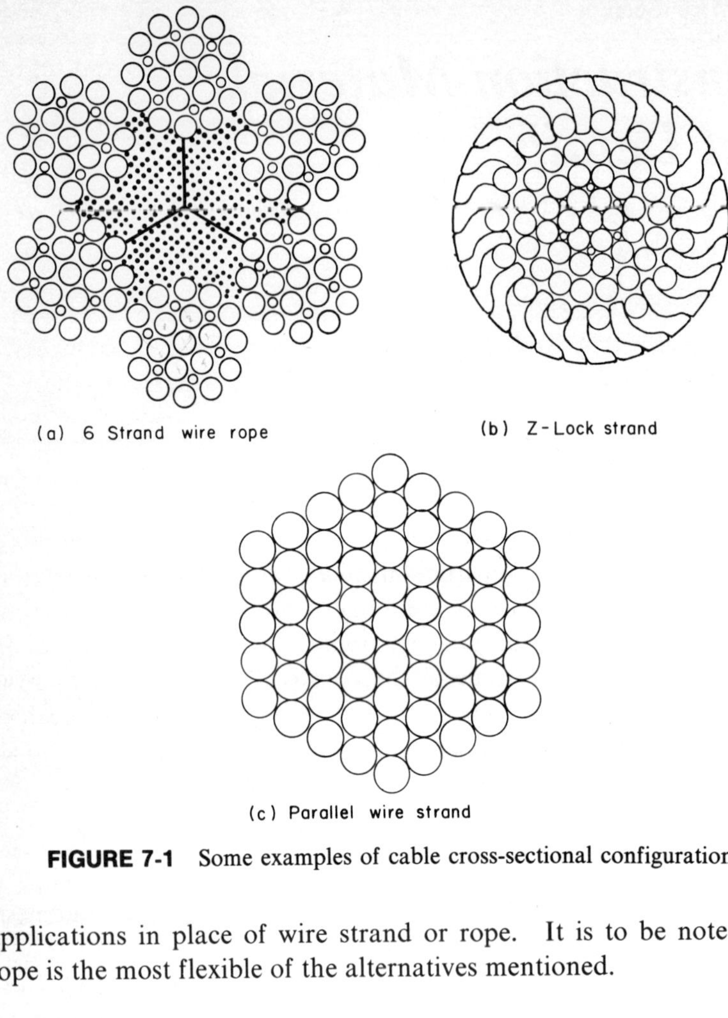

FIGURE 7-1 Some examples of cable cross-sectional configurations.

applications in place of wire strand or rope. It is to be noted that rope is the most flexible of the alternatives mentioned.

7-2.1 Classification[2]

Types of strand commonly used are described as follows:

1. *Spiral strand.* As already described, it consists of a number of round wires spun together.
2. *Locked coil strand.* In this type, one or more of the outer layers consists of shaped wires. Most commonly the wire is

[2] Parts of this section are condensed from a personal communication from J. A. Boyle of British Ropes.

shaped as a "Z." These wires lock together to form a seal, which prevents the entry of moisture into the inner layers. The "Z" wires cannot be drawn to the same strength as round wires. Thus spiral strands have a higher strength/weight ratio than locked-coil strands. The advantage gained through the protective seal in the latter is therefore partly lost. Similar protection can be gained in spiral strands through a protective coating, as described later in this section. These cables suffer a greater loss of strength due to transverse clamp pressure and changes in curvature [28].

3. *Parallel-Wire Strand.* This strand consists of a set of wires assembled parallel to each other. The obvious advantage gained is the greater length of strand for the same material and also a greater value of Young's modulus of elasticity, which could be as much as 28×10^6 lb/in^2 (193 kN/mm^2). Parallel-wire strand has been used in several suspension bridges and suspended structures. Instead of this a parallel-strand cable can also be used. The parallel-strand cable has less bending stiffness as compared to the parallel-wire cables.

4. *Prestressing Strand.* Such a strand is obtained by grouping together concrete prestressing wire, and has the advantage of being a readily available standard material. The terminal fittings arc also standard.

7-2.2 Protective Coating

For permanent structural applications, such as in suspended roofs, it is necessary to protect the wire with anticorrosive material. This protection is provided by galvanizing the wire by the *electrolytic zinc* or the *hot-dip* processes. The life of the zinc coating thus obtained is directly proportional to its weight (or thickness). The ASTM standards A586-68 specify the weight of zinc, and on that basis the wires fall into different classes, such as A, B, and C. Among these, class A refers to the lightest coating of zinc.

Steel rope companies have also developed special protective measures, which are likely to differ with the organization. As an example, for severely corrosive conditions Bethlehem Steel Corporation uses the Bethlume hot-dip aluminizing process. British Ropes, Ltd., recommends the application of their Metalcoat A during the manufacture of the strand as well as at site. Metalcoat A is a compound which has a suspension of aluminum particles in a hydrocarbon resin, diluted by a solvent for application.

Concrete prestressing strands, which are usually not galvanized, also have to be protected if used in suspended structures. Protection can be provided through an extruded plastic coat. The coating has to be removed at anchorages, and after stressing, the strand is reprotected by the application of an air-drying plastic.

7-2.3 Mechanical Properties and Strength

The mechanical properties for strand and rope and the minimum acceptable breaking strengths as specified by the ASTM Standards A586-68 and A603-70 are included in this section to facilitate their design. The mechanical properties expected of the wire, with different classes of zinc coating, are given in Table 7-1, and the minimum breaking strengths and other properties for strands of particular diameter are given in Table 7-2. The corresponding values for wire rope are given in Table 7-3. It may be noted that cables with class B or C coating, which provide heavier coats of zinc for better corrosion protection, have lesser strength. When a cable is required to pass over a saddle, bending is introduced in the cable to a larger or smaller degree, based on the saddle/cable radius ratio. The AISI *Manual for Structural Applications of Steel Cables for Buildings* [60] specifies the reduced strengths as given in Table 7-4.

The minimum elastic moduli for strand and rope fabricated from wire with grade A coating are given in Table 7-5; it is specified that the value may be reduced by 1×10^6 lb/in^2 (6.9 kN/mm^2) if outer wires have a class B or C coating. It may be noted too that the modulus values for locked-coil strands are lower than those for spiral strands to the extent of about 2×10^6 lb/in^2 (13.8 kN/mm^2).

Greenberg [32] has developed expressions to describe the stress-strain relationship through the elastic as well as inelastic range to the ultimate failure of a cable corresponding to stress σ_{ult} and strain ϵ_{ult}. Accordingly, the behavior in the elastic range, i.e., up to a stress of σ_e and strain ϵ_e, is treated as linear, and in the inelastic range,

Stress: $$\sigma = \sigma_e + \sigma_0\left\{1 - \left[1 - \left(\frac{\epsilon - \epsilon_e}{\epsilon_0}\right)\right]^j\right\}$$

Tangent modulus: $$E_t = \sigma_0\left[\frac{j}{\epsilon_0}\left(1 - \frac{\epsilon - \epsilon_e}{\epsilon_0}\right)^{j-1}\right]$$

Strain:
$$\epsilon = \epsilon_e - \epsilon_0\left[\left(1 - \frac{\sigma - \sigma_e}{\sigma_0}\right)^{1/j} - 1\right]$$

where $j = \epsilon_0 E/\sigma_0$

$\sigma_0 = \sigma_{\text{ult}} - \sigma_e$

$\epsilon_0 = \epsilon_{\text{ult}} - \epsilon_e$

For use of these expressions it may be assumed that:

1. The prestretch load of the cable is one-half of the breaking load.
2. Elastic behavior occurs up to three-fourths of the prestretch load.
3. Ultimate strain ϵ_{ult} is 3 percent for wire strand and 2 percent for wire rope.

7-2.4 Prestretching and Cable Length

Cables, as manufactured, are not truly elastic, and although they cannot ever be made perfectly elastic, repetitive prestretching to a certain proportion of its breaking load makes them behave nearly elastically. This enables wires to readjust into the final position and the cable to assume a well defined elastic character. The inelastic constructional stretch is thus almost removed.

For application in suspended roofs, achievement of a dependable stress-strain law (preferably linear) for the cables is significant for the following reasons:

1. It enables the markings on the cable, for erection, to be appropriately spaced.
2. It makes possible the determination of accurate values for cable lengths when subjected to a prescribed force of tension.
3. It permits accurate computation of deflections.

Prestretching is usually the responsibility of the cable manufacturer, and major rope companies have the facilities to provide cables cut to the required length with great accuracy. Such accuracy is most essential for a smooth erection of the structure. Sources of inaccuracy in cable lengths may be the following:

1. Insufficient accuracy in computing cutting patterns

TABLE 7-1 Mechanical Properties

Zinc coating class	*Nominal diameter, in (mm)*	*Minimum stress at 0.7% extension under load, lb/in² (N/mm²)*	*Minimum ultimate tensile strength, lb/in² (N/mm²)*	*Total percentage elongation in 10 in (254 mm)*
A	0.040–0.110 (1.016–2.794)	150,000 (1060)	220,000 (1520)	2.0
	0.111 and larger (2.820 and larger)*	160,000 (1100)	220,000 (1520)	4.0
B	0.090† and larger (2.286 and larger)*	150,000 (1060)	210,000 (1450)	4.0
C	0.090† and larger (2.286 and larger)*	140,000 (960)	200,000 (1390)	4.0

* This is not to imply that larger wire will be manufactured to any unlimited diameter. It implies only that the wire sizes chosen by the strand manufacturer must meet the requirements of this specification.

† This value is 0.040 in (1.016 mm) for wire rope.

2. Wrong marking and cutting of cables
3. Nonuniformity of prestretching or other conditions during cutting of cables
4. Wrong positioning of edge clamps
5. Incorrect clamping and subsequent dislocation of cables during erection or pretensioning
6. Incorrect formation of edge loops
7. Incorrect fixing of turnbuckles

Cable systems, particularly networks, are extremely sensitive to errors in lengths and can suffer large over- and understressing on this account. The net may also distort as a result of these inaccuracies. The requirement in the accuracy of cut lengths of cables is to some extent dependent on the procedure of erection. A net may be assembled fully on the ground and then pulled up at one or more points into position. It may then be tensioned by straining any one element, preferably one with a small tension force. The requirements of accuracy in such cases are extremely stringent, say, of the order of 1/10,000 to 1/20,000. In a particular case [33], cables of average length 98.4 ft (30 m), having errors of ±0.3 in (7.5 mm) (±0.025 percent) are reported to have suffered an average error of ±20 percent and a peak error of up to 60 percent in force. This poses a serious problem of local overstressing but may not affect the ultimate load-carrying capacity of the system, or the overall response characteristics, quite so seriously. It is difficult, however, to obtain

TABLE 7-2 Properties of Zinc-Coated Steel Structural Strand (ASTM-A586-68)

Nominal diameter		Minimum breaking strength: Class A coating throughout		Minimum breaking strength: Class A coating inner wires; Class B coating outer wires		Minimum breaking strength: Class A coating inner wires; Class C coating outer wires		Approx. gross metallic area		Approx. weight	
in	mm	kips	kN	kips	kN	kips	kN	in^2	mm^2	lb/ft	N/m
1/2	12.7	30.0	133.4	29.0	129.4	28.4	126.5	0.150	96.8	0.52	7.55
9/16	14.3	38.0	168.6	36.8	163.7	36.0	159.8	0.190	122.6	0.66	9.60
5/8	15.9	48.0	213.7	46.6	206.9	45.6	202.9	0.234	151.0	0.82	11.76
11/16	17.5	58.0	257.9	56.2	250.0	55.0	245.1	0.284	183.2	0.99	14.70
3/4	19.1	68.0	302.0	66.0	293.2	64.6	287.3	0.338	218.1	1.18	17.64
13/16	20.6	80.0	355.9	77.6	345.2	76.0	338.3	0.396	255.5	1.39	20.58
7/8	22.2	92.0	408.9	89.2	397.1	87.4	388.3	0.459	296.1	1.61	23.52
15/16	23.8	108.0	480.5	104.8	465.8	102.6	455.9	0.527	340.0	1.85	27.44
1	25.4	122.0	542.3	118.4	526.6	115.8	514.8	0.600	387.1	2.10	30.38
1 1/16	27.0	138.0	613.8	133.8	595.2	131.0	582.4	0.677	436.8	2.37	34.30
1 1/8	28.6	156.0	694.3	151.4	673.7	148.2	658.9	0.759	489.7	2.66	39.20
1 3/16	30.2	172.0	764.8	166.8	742.3	163.4	726.6	0.846	545.8	2.96	43.12
1 1/4	31.8	192.0	854.1	188.2	837.4	184.4	819.8	0.938	605.6	3.28	48.02
1 5/16	33.3	212.0	943.3	208.0	925.7	204.0	907.0	1.03	664.5	3.62	52.92
1 3/8	34.9	232.0	1031.0	228.0	1014.0	222.0	987.4	1.13	729.0	3.97	57.82
1 7/16	36.5	252.0	1121.0	246.0	1088.0	242.0	1076.0	1.24	800.0	4.34	63.70
1 1/2	38.1	276.0	1228.0	270.0	1201.0	264.0	1174.0	1.35	871.0	4.73	68.60
1 9/16	39.7	300.0	1334.0	294.0	1308.0	288.0	1280.0	1.47	948.4	5.13	74.48
1 5/8	41.3	324.0	1441.0	318.0	1414.0	310.0	1378.0	1.59	1025.8	5.55	81.34
1 11/16	42.9	352.0	1566.0	344.0	1529.0	338.0	1503.0	1.71	1103.2	5.98	87.22
1 3/4	44.5	376.0	1673.0	368.0	1636.0	360.0	1601.0	1.84	1187.1	6.43	94.08
1 13/16	46.0	404.0	1797.0	396.0	1761.0	388.0	1725.0	1.97	1271.0	6.90	100.94
1 7/8	47.6	432.0	1931.0	424.0	1885.0	414.0	1841.0	2.11	1361.3	7.39	107.80
1 15/16	49.2	460.0	2046.0	452.0	2010.0	442.0	1966.0	2.25	1451.6	7.89	114.66
2	50.8	490.0	2180.0	482.0	2143.0	476.0	2117.0	2.40	1548.4	8.40	122.50
2 1/16	51.1	522.0	2322.0	514.0	2286.0	506.0	2250.0	2.55	1645.2	8.94	130.34
2 1/8	54.0	554.0	2464.0	546.0	2429.0	538.0	2392.0	2.71	1748.4	9.49	138.18
2 3/16	55.6	586.0	2606.0	578.0	2571.0	568.0	2526.0	2.87	1851.6	10.05	147.00
2 1/4	57.2	620.0	2757.0	610.0	2713.0	602.0	2678.0	3.04	1961.3	10.64	154.84
2 5/16	58.7	654.0	2909.0	644.0	2863.0	634.0	2820.0	3.21	2071.0	11.24	163.66
2 3/8	60.3	688.0	3060.0	678.0	3015.0	668.0	2961.0	3.38	2180.7	11.85	172.48
2 7/16	61.9	720.0	3202.0	710.0	3158.0	698.0	3104.0	3.57	2303.2	12.48	182.28
2 1/2	63.5	752.0	3345.0	740.0	3292.0	730.0	3246.0	3.75	2419.4	13.13	191.10
2 9/16	65.1	784.0	3487.0	772.0	3434.0	760.0	3380.0	3.94	2541.9	13.80	200.90
2 5/8	66.7	834.0	3709.0	822.0	3656.0	808.0	3593.0	4.13	2664.5	14.47	210.70
2 11/16	68.3	864.0	3843.0	850.0	3781.0	838.0	3727.0	4.33	2793.6	15.16	221.48
2 3/4	69.9	904.0	4021.0	890.0	3958.0	876.0	3897.0	4.54	2929.0	15.88	231.28
2 7/8	73.0	988.0	4395.0	972.0	4323.0	958.0	4260.0	4.96	3200.0	17.36	252.84

TABLE 7-2 (Continued)

Nominal diameter		Minimum breaking strength: Class A coating throughout		Minimum breaking strength: Class A coating inner wires; Class B coating outer wires		Minimum breaking strength: Class A coating inner wires; Class C coating outer wires		Approx. gross metallic area		Approx. weight	
in	mm	kips	kN	kips	kN	kips	kN	in^2	mm^2	lb/ft	N/m
3	76.2	1076.0	4786.0	1060.0	4714.0	1044.0	4644.0	5.40	3483.9	18.90	275.38
$3\frac{1}{8}$	79.4	1168.0	5195.0	1150.0	5114.0	1132.0	5035.0	5.86	3780.7	20.51	298.90
$3\frac{1}{4}$	82.6	1250.0	5560.0	1232.0	5479.0	1212.0	5391.0	6.34	4090.3	22.18	323.40
$3\frac{3}{8}$	85.7	1346.0	5986.0	1326.0	5898.0	1306.0	5809.0	6.83	4406.5	23.92	348.88
$3\frac{1}{2}$	88.9	1448.0	6440.0	1428.0	6351.0	1404.0	6244.0	7.35	4742.0	25.73	375.34
$3\frac{5}{8}$	92.01	1536.0	6832.0	1514.0	6733.0	1490.0	6627.0	7.88	5083.9	27.60	402.78
$3\frac{3}{4}$	95.3	1644.0	7312.0	1620.0	7205.0	1594.0	7089.0	8.43	5438.7	29.50	430.22
$3\frac{7}{8}$	98.4	1756.0	7810.0	1730.0	7694.0	1704.0	7579.0	9.00	5806.5	31.50	459.62
4	101.6	1850.0	8229.0	1822.0	8104.0	1794.0	7979.0	9.60	6193.6	33.60	490.00

such a high degree of accuracy, and errors in pretension are inadvertent.

It may further be observed that length measurements computed by an accurate analytical procedure are more dependable than those obtained from small-scale models (this is one of the techniques sometimes adopted). By another procedure of erection, one which is used more often, the cables composing the network are positioned and secured to the supporting structure independently of each other, and pretensioned in carefully planned steps. For this procedure accuracy required in cut lengths may be less strict. Adjustments in the length of a cable may be made at site by introducing turnbuckles or similar devices at the end or along the length of the cable. This method indeed has its limitations where a net has a large number of cable elements. Accuracy in cable length is comparatively less important for simply suspended roofs or trusses. It would be simpler to pretension trusses in comparison to nets.

7-2.5 Rope vs. Strand

Ropes are more flexible than strands. This means that ropes are easier to handle, and that, where cables have to pass over saddles, smaller radii can be used for a rope than for a strand. Owing to their greater bending stiffness, strands develop bending stresses, particularly at clamps and terminal fittings. Strands have a greater value

TABLE 7-3 Properties of Zinc-Coated Steel Structural Wire Rope (ASTM-A603-70)

Nominal diameter		Minimum breaking strength: Class A coating throughout		Minimum breaking strength: Class B coating throughout		Minimum breaking strength: Class C coating throughout		Approx. gross metallic area		Approx. weight	
in	mm	kips	kN	kips	kN	kips	kN	in^2	mm^2	lb/ft	N/m
3/8	9.53	13.0	57.8	12.4	54.9	11.8	52.9	0.065	41.9	0.24	3.53
7/16	11.11	17.6	78.4	16.8	74.5	16.0	71.6	0.091	58.7	0.32	4.70
1/2	12.70	23.0	102.0	22.0	98.0	21.0	93.2	0.119	76.8	0.42	6.08
9/16	14.29	29.0	129.4	27.6	122.6	26.4	117.7	0.147	94.8	0.53	7.74
5/8	15.88	36.0	159.8	34.4	152.9	32.8	146.1	0.182	117.4	0.65	9.51
11/16	17.46	43.0	191.2	41.0	182.4	39.0	173.6	0.221	142.6	0.79	11.56
3/4	19.05	52.0	231.4	49.6	220.6	47.2	209.8	0.268	172.9	0.95	13.82
13/16	20.64	60.0	266.7	57.2	253.9	54.6	243.2	0.311	200.7	1.10	16.07
7/8	22.23	70.0	311.8	66.8	297.1	63.6	282.4	0.361	232.9	1.28	18.62
15/16	23.81	80.0	355.9	76.4	340.3	72.8	323.6	0.414	267.1	1.47	21.46
1	25.40	91.4	406.9	87.2	388.3	83.0	368.7	0.471	303.9	1.67	23.52
1 1/8	28.58	115.6	513.8	110.2	490.3	105.0	466.8	0.596	384.5	2.11	30.77
1 1/4	31.75	144.4	642.3	137.8	612.9	131.2	583.4	0.745	480.7	2.64	38.51
1 3/8	34.93	175.6	781.5	167.6	745.2	159.6	709.9	0.906	584.6	3.21	46.84
1 1/2	38.10	208.0	924.7	198.4	882.5	189.0	840.4	1.076	694.2	3.82	55.66
1 5/8	41.28	246.0	1098.0	234.0	1039.0	224.0	990.4	1.270	819.4	4.51	65.76
1 3/4	44.45	286.0	1275.0	272.0	1216.0	260.0	1157.0	1.470	948.4	5.24	76.44
1 7/8	47.63	328.0	1461.0	312.0	1392.0	298.0	1324.0	1.690	1090.4	6.03	87.91
2	50.80	372.0	1657.0	354.0	1579.0	338.0	1500.0	1.920	1238.8	6.85	99.86
2 1/8	53.98	420.0	1863.0	400.0	1785.0	382.0	1696.0	2.170	1400.1	7.73	112.70
2 1/4	57.15	470.0	2089.0	448.0	1991.0	428.0	1902.0	2.420	1561.4	8.66	126.32
2 3/8	60.33	522.0	2324.0	498.0	2216.0	474.0	2108.0	2.690	1735.6	9.61	140.14
2 1/2	63.50	576.0	2559.0	550.0	2442.0	524.0	2334.0	2.970	1916.2	10.60	154.55
2 5/8	66.68	634.0	2824.0	604.0	2687.0	576.0	2559.0	3.270	2109.8	11.62	169.44
2 3/4	69.85	694.0	3089.0	662.0	2942.0	630.0	2804.0	3.580	2309.8	12.74	185.81
2 7/8	73.03	758.0	3373.0	724.0	3216.0	688.0	3059.0	3.910	2522.7	13.90	202.66
3	76.20	824.0	3667.0	786.0	3491.0	748.0	3334.0	4.250	2742.1	15.11	220.30
3 1/4	82.55	950.0	4226.0	906.0	4030.0	864.0	3844.0	5.040	3251.8	18.00	262.44
3 1/2	88.90	1110.0	4942.0	1058.0	4707.0	1008.0	4491.0	5.830	3761.5	21.00	306.25
3 3/4	95.25	1280.0	5697.0	1222.0	5432.0	1164.0	5177.0	6.670	4303.5	24.00	349.96
4	101.60	1460.0	6491.0	1392.0	6197.0	1328.0	5903.0	7.590	4897.1	27.00	393.76

for the modulus of elasticity (see Table 7-5), and so cable roofs made from strands deflect less. However, this also means that extensions are smaller and, therefore, that greater accuracy is required on length tolerances. Strands have a better strength/weight ratio. Ropes are easier to grip than strands.

TABLE 7-4 Effect of Saddle Radius on Breaking Strength

Ratio of saddle radius to strand diameter	*Ratio of saddle radius to wire rope diameter*	*Effective design breaking strength as % of specified minimum breaking strength*
20 and over	15 and over	100
19	14	95
18	13	90
17	12	85
16	11	80
15 min.	10 min.	75

TABLE 7-5 Modulus Values for Strand and Rope

Nominal diameter for strand or rope, in (mm)	*Minimum modulus with class A coating, lb/in² (kN/mm²)*
$\frac{1}{2}$ to $2\frac{9}{16}$ (strand) (12 to 65)	24,000,000 (165)
$2\frac{5}{8}$ and larger (strand) (67)	23,000,000 (158)
$\frac{3}{8}$ to 4 (rope) (10 to 102)	20,000,000 (138)

7-2.6 Fatigue

It is well established that a tension member with a mean tensile stress superimposed repetitively with a fluctuating stress can fail from fatigue. The fatigue life is dependent on the mean stress level and the amplitude of the fluctuating stress. In suspended roofs, cables are subjected to a mean tensile stress, which will change as a result of the application of wind and other live loads. Gabriel [28] has observed that the tension due to prestress may be in the range of 0.15 to 0.35 of the ultimate strength of the cable, while addition of the fluctuating stress due to wind may increase this to as much as one-half of the ultimate. Whereas the amplitude of fluctuation is substantial in comparison to the mean stress, the number of cycles of this extreme value of stress is unlikely to be as large as in certain other structures, such as bridges. It is of course likely that the structure will be subjected to a much larger number of cycles of smaller stress fluctuation on account of vibrations or application of loads, whose effect may not be so serious from the point of view of fatigue.

Fatigue study of strands with and without the influence of fittings and clamps [28, 61] has shown that the fatigue life of strands can be reduced considerably by their presence. Studies [15] have also shown that strands comprising a greater number of wires for the same diameter have better fatigue strength. The current state of the art, however, does not permit a quantitative assessment which would enable consideration of fatigue in designing a cable. It has been shown further that strands with cold-poured sockets (using a casting material consisting of steel balls, zinc dust, and epoxy resin) have a better fatigue life than those with metal-poured sockets. The strength of the latter strands is affected by the heating of wires during pouring. A study to determine fatigue strength of cables when subjected to bending moments (e.g., cables passing over pulleys) has also been made, by the British Ropes Group [86].

7-3 FITTINGS

Fittings are attachments used to grip the cable at the ends or along its length. They can be classified, in accordance with the type of application, as the friction or clamp type, the pressed or swaged type, or the socketed type. Clamps usually provide intermediate fittings. Pressed fittings are used either at the ends or along the length of small-size strands or rope, and sockets are always terminal fittings used for larger sizes of cables. Pressed or socket fittings, which are made from forged or cast steel, are provided by the manufacturer along with the cable, and are expected to be fully tested to withstand the required forces. The process of making a zinc[3] socket is illustrated in Fig. 7-2. Figure 7-2*a* shows the end of the strand pushed into the socket and opened like a broom. After the wires have been cleaned, molten zinc is poured into the socket. On cooling, the joint looks as seen in Fig. 7-2*b*. A typical application for a swaged strand eye fitting is shown in Fig. 7-3. Applications of socket fittings are shown in Fig. 7-4. When prestressing strands are used, standard fittings commonly available for these strands can be employed.

[3] White metal can also be used for this purpose.

FIGURE 7-2*a* Cable and socket joint: ready for pouring molten zinc. (*Bethlehem Steel.*)

Whereas pressed or socket fittings are standard appliances, friction grips or clamps are usually not. These, therefore, have to be designed for particular applications. The clamp should be designed so that the pressure it exercises on the cable is as uniform as possible. An important problem connected with clamps is to know the coefficient of friction. The value of the coefficient of friction can vary widely[4] in practice, and great care is needed in determining it. The AISI *Manual for Structural Applications of Steel Cables in Buildings* [60] recommends 0.07 as the maximum value to be used. It is further recommended that bearing pressures on the clamp should not exceed 4000 lb/in^2 (27.6 N/mm^2) for cable diameters greater than or equal to 3 in (76 mm) and 6000 lb/in^2 (41.4 N/m^2) for a diameter of 1 in (25 mm).

For purposes of clamping, mild steel or high-tension bolts can be

[4] Rice [82] reports values between 0.10 and 0.30 for sliding friction in experiments carried out by him.

FIGURE 7-2*b* Cable and socket joint: after the zinc has cooled. (*Bethlehem Steel.*)

used. Some designers prefer mild steel bolts, since they have greater tolerance for distortion effects on the clamp. When the clamped cable is tensioned, it decreases in diameter and there is a consequent decrease in clamp compression. This can be readjusted by tightening the bolts. A typical clamp is illustrated in Fig. 7-5. A typical U clip for gripping ropes and its applications are shown in Fig. 7-6. Figure 7-7 shows a typical detail for a connection between the top cable in a truss (see Fig. 1-9*c* or *d*) and a compression spreader. Also shown, in Figs. 7-8 to 7-14, are certain connection, terminal, and roof details for some existing structures.

7-4 ROOF CLADDING

The components of roof cladding, as has already been explained in Sec. 1-4.2, are the roofing, deck, auxiliary framing, and insulation. It has also been explained that auxiliary frames are not always required if the deck is sufficiently strong to span the spacing between support-

FIGURE 7-3 Swaged strand eye—a typical application. (*Bethlehem Steel.*)

ing purlins or cables. Lightweight metallic roofs are preferable for pretensioned cable structures, while concrete or timber may be advantageous in simply suspended systems in which gravity stiffness is important. In a very large number of cases corrugated metal decking is employed. Translucent or transparent plastic or glass may be considered, generally for temporary or semipermanent construction, where admittance of natural lighting through the roof is an important factor. It has also been pointed out, in Sec. 1-4.2, that the use of a separate insulating layer may not always be necessary when using timber decking. Practical details of some roof cladding arrangements are shown in Figs. 7-10 to 7-13. Some details of materials[5] used for cladding now follow.

Roofing. A large number of materials and their products are marketed for use as the waterproofing roof skin. There are different

[5] Condensed in part from Chapters 8, 15, and 17 of Ref. 90.

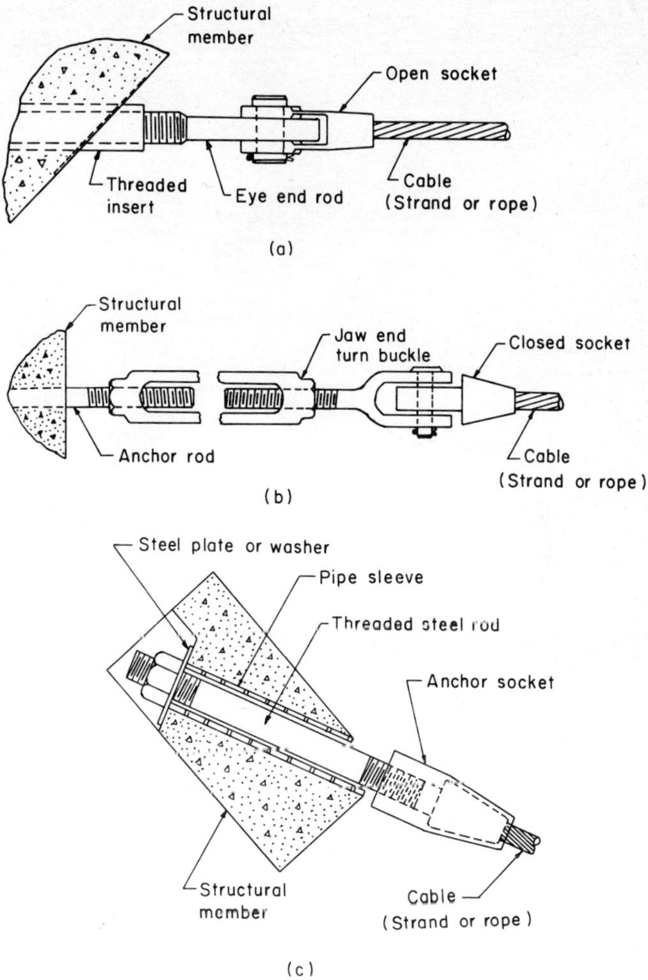

FIGURE 7-4 (*a*) Open socket application, (*b*) closed socket application, (*c*) anchor socket application. (*Bethlehem Steel.*)

types of wood, asphalt, aluminum, or asbestos shingles that can be used apart from tiles, sheet metal, glass, or plastics [90]. Sheet-metal roofing can be made from steel, aluminum, lead, zinc, copper, etc. Shingles and tiles will generally be considered satisfactory for large slopes and may not be applicable for most cable roofs, which quite often have varying curvature and flat zones. Asbestos-cement corrugated sheeting is popularly used for roofing as well as decking.

FIGURE 7-5 A typical cable clamp. (*Courtesy of Leonhardt und Andra. Photo by Dressler.*)

Galvanized-steel sheeting is available[6] in 2 to 3 ft (600 to 900 mm) widths and 6 to 10 ft (1.8 to 3 m) lengths. Thickness varies from 0.018 to 0.138 in (0.46 to 3.5 mm) and the weight from 0.78 to 5.78 lb/ft^2 (37 to 277 N/m^2). Steel coated with a mixture of lead and tin is called *terneplate*, and is used to make plates in small sizes, as well as strips that are 20 in (500 mm) wide and 50 ft (15 m) long. Stainless steel or monel metal are also usable in roofing, but their use is limited because both are rather expensive. Application of zinc is uncommon in roof sheeting, but its use in galvanizing steel sheeting is well known. Lead sheeting may be used in particular where the roof surface is warping sharply, since it is a soft, pliable material. Hard lead sheets are also available in sizes limited to 2×4 ft (60×120 cm). When a lead surface is expected to come into contact with fresh concrete, it should be painted with asphalt paint. Copper sheets are produced in 2 to 3 ft (600 to 900 mm) widths and 8 to 10 ft (2.4 to 3 m) lengths and are also available in strips. Where aesthetic beauty and long life are a consideration, copper is a good material to use. Aluminum sheets

[6] Size and measures given here are for general guidance only and are liable to variation.

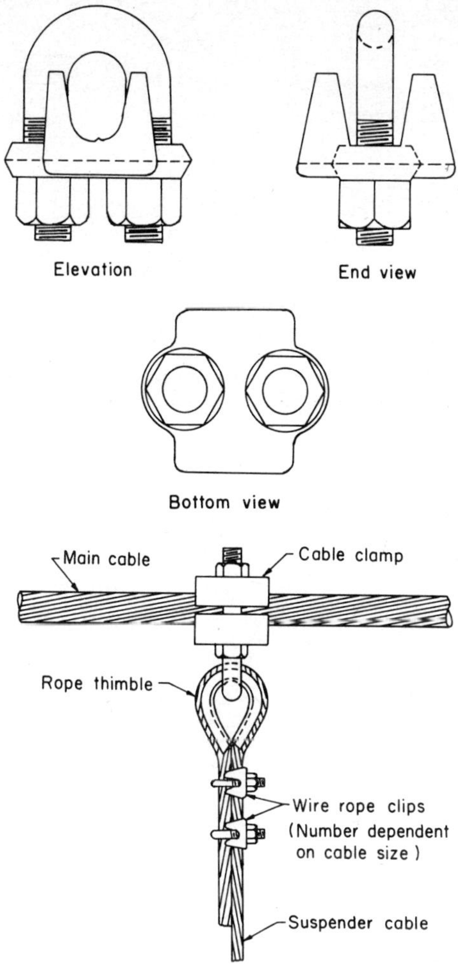

FIGURE 7-6 (*a*) A U clip and (*b*) application of clips for cable-to-cable connections. (*Courtesy of the Crosby Group.*)

are generally available in different finishes, and vary in width from 2 to 4 ft (600 to 1200 mm) and may be 4 to 12 ft (1.2 to 3.6 m) long. Aluminum is well known for its durability, lightness, and good resistance to corrosion. However, both aluminum and copper have a large coefficient of thermal expansion.

Seams or slip joints are provided between roof sheets to affect waterproofing and permit thermal expansion or other movement.

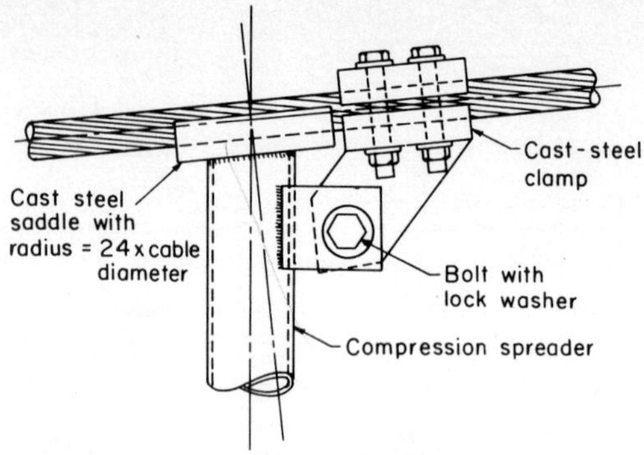

FIGURE 7-7 A typical connection detail between a cable and a compression spreader. (*Ref. 30, p. 22–14.*)

Flat seams may be provided in flat zones of the roof with solders as an additional waterproofing measure, while standing seams are suitable for large slopes. Standing or ribbed seams are more suitable for jointing in expansive material or for permitting large movements. Welding is also possible for waterproof jointing between sheets.

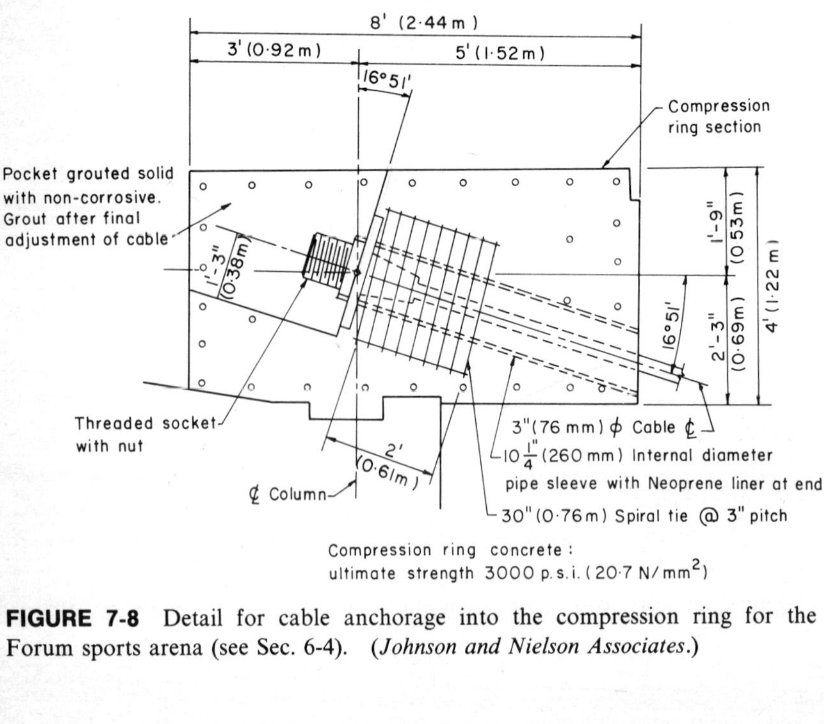

FIGURE 7-8 Detail for cable anchorage into the compression ring for the Forum sports arena (see Sec. 6-4). (*Johnson and Nielson Associates.*)

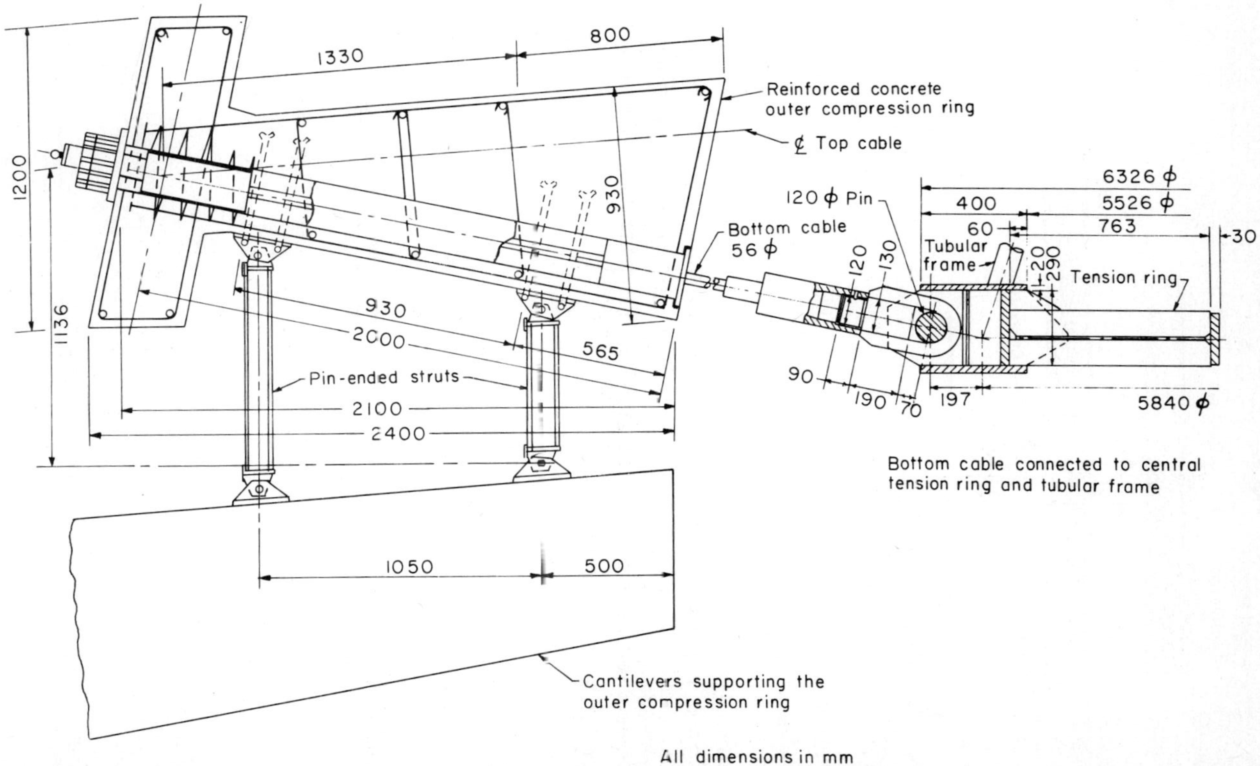

FIGURE 7-9 Terminal details for the bottom cable in the roof for Palasport Genova (see Fig. 6-10*b*).

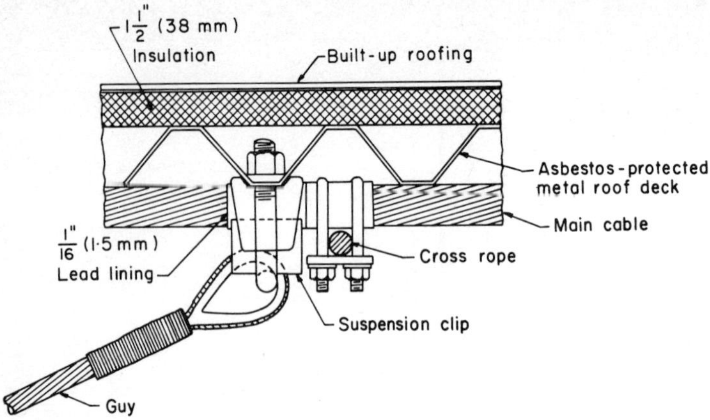

FIGURE 7-10 Connection detail between cables and guy rope for the Raleigh arena (see Sec. 6-4). (*Bethlehem Steel.*)

Another form of roofing is possible with glass, which is available in two forms: corrugated sheets of 3/8 in (10 mm) thickness with a maximum size of 50 × 144 in (1.25 × 36 m); and 1/4-in-thick (6-mm) glass sheets reinforced with wire mesh of not more than 7/8 in (22 mm) opening, and a maximum size of 60 × 132 in (1.5 × 3.3 m). The advantage in using glass is the entry of natural light through the roof.

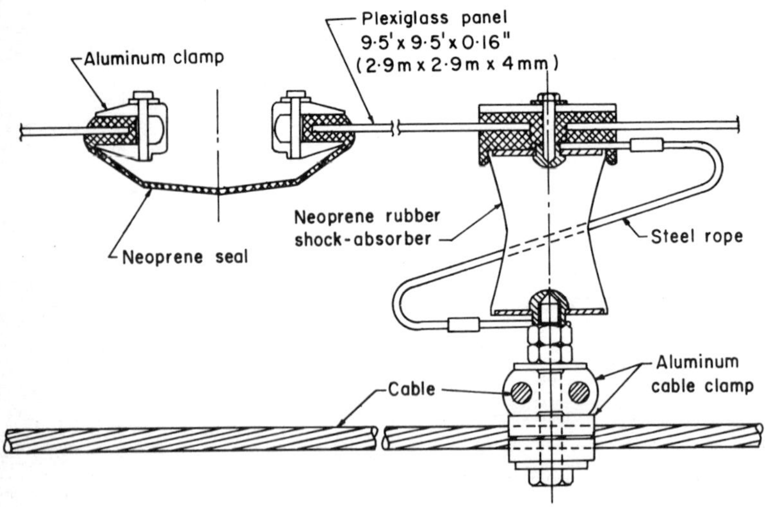

FIGURE 7-11 Connection detail between the cables and the plexiglass panel cladding for the Munich Olympic stadium roof (Fig. 1-14). (*Ref. 54.*)

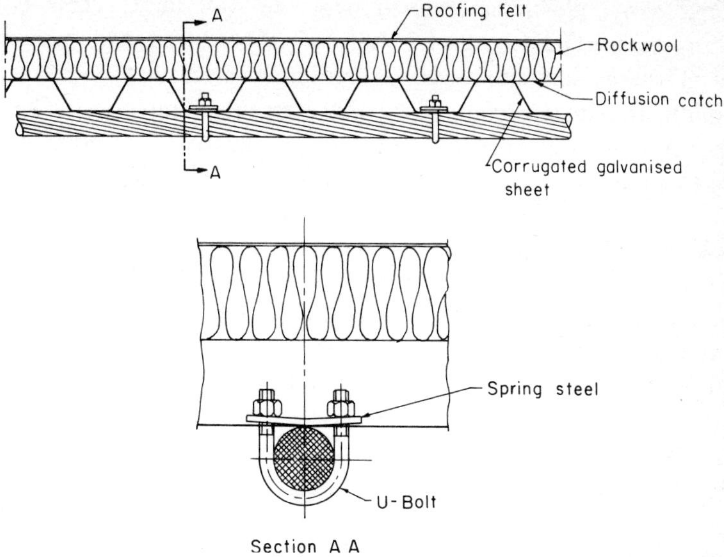

FIGURE 7-12 Cable-to-cladding connection. (*Courtesy of Jawerth Interstatik.*)

Alternatively, corrugated sheets of glass-fiber-reinforced plastic are also available, in 8 to 12 ft (2.4 to 3.6 m) lengths and 34 in (850 mm) widths. These sheets may be translucent or colored. Another way in which a plastic material may be used for roofing is to spray it onto the roof deck. This is known as *envelope roofing*, in which a 30 to 40×10^{-6} in thick (750 to 1000×10^{-6} mm) continuous film of

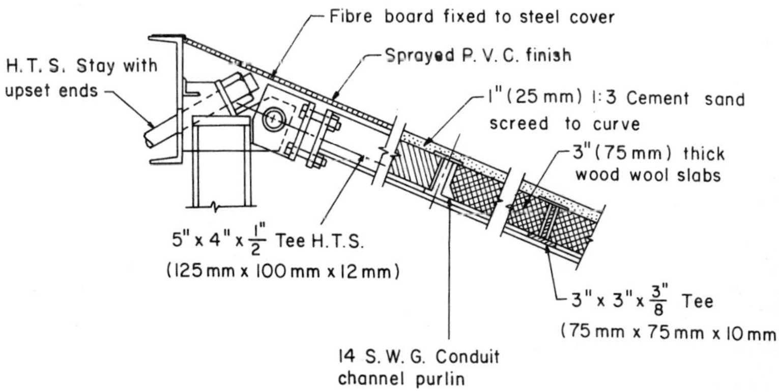

FIGURE 7-13 Roof detail for the structure in Fig. 1-18. (*Ref. 25.*)

opaque vinyl plastic is sprayed over the roof deck. This type of roofing is particularly useful for curved surfaces and may be of special interest in cable roofs. The film has a reasonably good tensile strength and may permit up to 200 percent elongation. The vinyl film has high resistance to deterioration caused by prolonged exposure to sunlight and a mildly acidic or dirty atmosphere.

Decking. The function of roof decking is to transmit loads to the supporting cable system. Corrugated sheets of galvanized steel, asbestos-protected steel (Figs. 7-10 and 7-12), or aluminum are a popular choice and provide light, rigid decking. Concrete, both lightweight and ordinary, can be employed as well in a variety of ways. In the application of concrete it is a desirable practice to avoid in situ work (at the roof level) as far as possible and use precast units, in order to minimize the use of shuttering and centering. Precast roof decking made from gypsum is either reinforced with wire mesh or rods, or both, and is available in widths of 15 in (375 mm) and lengths varying between 4 to 10 ft (1.2 to 3 m) depending upon the type. Gypsum decking may be about half as heavy as concrete but is still much heavier than metallic decks. Insulation boards made generally from organic fiber such as wood, straw, cane, or cork are also used as structural decking.

Insulating Layer. The purpose of using insulating material in a roof, as already explained in Sec. 1-4.2, is to prevent the transfer of heat through the roof. The heat may get transferred by conduction, convection, or radiation. Poor conductors (insulators) are used to prevent heat loss by conduction. To avoid transfer by convection, one should try to prevent air currents from being set up. To prevent loss of heat by radiation, it is important that materials which reflect the heat be used.

Loose fill; blankets; batts; structural insulation board; slab or block insulators; reflective insulators; and foamed-in-place, sprayed on, and corrugated insulation sheets are the basic kinds of thermal insulators. For roofs with large curvatures, the softer form of insulations which are comparatively more pliable are preferable. For flatter roofs, the choice is greater.

The loose-fill insulation is usually bulky. It is either fibrous, made from mineral wool or wood fiber, or granular, made from expanded minerals or ground vegetable matter. The mineral wool fill insulation is suitable for conditions of moisture, vermin, and combustion. The fibrous loose fill may have a tendency to settle over a period of time in the vertical spaces, or where slopes are large. Blanket insulations are manufactured generally in thicknesses from 1 to 4 in (25 to 100 mm) and widths from 16 to 24 in (40 to 60 cm). Rolls up to 100 ft (30 m) in length are marketed. Blanket insulation is made from fibrous material in the form of a mat, sometimes with a paper cover. Batts are similar in basic manufacture to blankets. The rigid slab insulation is relatively stiff and inelastic, and made of inorganic materials like mineral wool with binder, foamed plastic, foamed concrete, cellular glass, hard rubber, etc. Slab insulation with thicknesses varying from $1\frac{1}{2}$ to 4 in (37 to 100 mm) is available in 12×18 in (300×450 mm) or 24×48 in (600×1200 mm) sizes. Reflective insulations are metallic or other insulation surfaces which rely for their performance on their surface characteristics. They also act as a vapor barrier. Foamed-in-place insulation is made from liquid resin. Two ingredients are used which, when mixed, foam up and solidify to fill the space into which they were introduced. The average density of this material when set is about 2 lb/ft^3 (320 N/m^3). Sprayed on insulations are produced by mixing some fibrous or cellular material with an adhesive and blowing the mixture onto the surface to be insulated. This method is suitable for areas that are otherwise difficult to insulate because of their shape, location, etc. Corrugated insulation is made from paper, corrugated and cemented into multiple layers. Some types are sprayed with an adhesive, which hardens, for extra stiffness, while others are faced with foil for extra insulative value. The thickness of corrugated sheet generally varies from 1 to 4 in (25 to 100 mm).

To know the relative suitability of a particular material as an insulator, the thermal conductivity of the material may be compared. Based on this type of comparison [90], it would appear that insulation boards made from cork, foamed-in-place insulation, blankets, and sprayed-on asbestos fiber have lower conductivity than some of the other available material.

7-5 ANCHORAGES

In self-supporting systems, cables are anchored into structural members, such as a concrete or a steel ring, edge beam or arch, or an edge cable. Certain possible details of this type are shown in Figs. 7-6 to

7-9. In other systems, cables are to be anchored into the ground and tension anchors are required. These can be classified as

1. Gravity anchors
2. Rock anchors
3. Plate anchors
4. Tension piles

7-5.1 Gravity Anchors

This is the type of anchor most commonly selected and can be considered applicable universally. However, it is not the most economical. The reason is easily understood if one considers the magnitude of forces that must be resisted in cable roofs. It may be noted that for cables consisting of several strands or ropes it is common practice to secure these separately into the anchor.

7-5.2 Rock Anchors

In locations where firm rock is available, it is usually feasible to use the rock anchor. A dependable assessment of the strength of rock in mass is naturally a prerequisite, and would generally require a site study. If there are no serious accessibility problems, rock anchors should prove to be more economical than gravity anchors. Cost estimates [11] range from £8 to £9 per metre in Great Britain, provided extensive grouting is not required. Rock anchors can be classified as mechanical, bonded, and rock sockets [11, 94].

7-5.3 Plate Anchors

Ultimate pull-out loads for plate, mushroom, or other footings can be estimated approximately by existing theories. The theory considered most dependable is by Meyerhof [63]; this is based on a semitheoretical approach involving a large number of model and field tests. The results described herein (Secs. 7-5.3 and 7-5.4) relate to vertical anchors under vertical pull. Meyerhof [65] has extended this work to inclined anchors under axial pull, but it has been shown that for inclinations of less than about 45° the *pull-out* capacity is not affected much by change in inclination. The effect is particularly small for deep anchors, the type most commonly expected to be used for anchoring cables.

Strip Footing. The ultimate pull-out force P_u (per unit length of strip for this case) for shallow footings is given as

$$P_u = 2cD + \gamma D^2 B_u \tan \psi + Q \tag{7-1}$$

where c = unit cohesion

B_u = nominal uplift coefficient for earth pressure on a vertical plane through edge of footing (for all practical purposes this value may be taken as 0.95)

γ = unit weight of soil

ψ = angle of internal friction for soil

D = depth of footing

Q = weight of footing and soil mass lifted with it at failure

For deep footings, expression 7-1 is modified as

$$P_u = 2cD' + \gamma(2D - D')D'B_u \tan \psi + Q \tag{7-2}$$

The value of D' as determined from tests are given in Table 7-6. The upper limit on P_u for a plate of width b is given as

$$P_u = b(cN_c + \gamma DN_q) + \mu_s A_s + Q \tag{7-3}$$

where A_s = surface area of anchor shaft

μ_s = average unit skin friction of soil on shaft

N_c, N_q = bearing-capacity factors for footing under downward load [63]

Circular Footings. For footing of shallow depth [64],

$$P_u = \pi cbD + \chi \frac{\pi}{2} \gamma bD^2 B_u \tan \psi + Q \tag{7-4}$$

where χ is a shape factor. For depths $D > D'$,

$$P_u = \pi cbD' + \chi \frac{\pi}{2} \gamma b(2D - D')D'B_u \tan \psi + Q \tag{7-5}$$

TABLE 7-6 Values of D′

ψ	20°	25°	30°	35°	40°	45°	48°
D'/b	2.5	3	4	5	7	9	11

SOURCE: After Ref. 64, Table 1, p. 233.

TABLE 7-7

ψ	20°	25°	30°	35°	40°	45°	48°
j	0.05	0.10	0.15	0.25	0.35	0.50	0.60
Max. value of χ	1.12	1.30	1.60	2.25	3.45	5.50	7.60

SOURCE: After Ref. 64, Table 2, p. 234.

with an upper limit of P_u as given by an equation similar to Eq. 7-3.

The values of shape factors χ are given by Eq. 7-6*a* with the maximum value as given by Eq. 7-6*b*. The values of j are indicated in Table 7-7.

$$\chi = 1 + j\frac{D}{b} \tag{7-6a}$$

with a maximum of,

$$\chi = 1 + j\frac{D'}{b} \tag{7-6b}$$

Rectangular Footing. For footings with length a and width b, the value of P_u for shallow and deep footings is given by Eqs. 7-7 and 7-8, respectively:

$$P_u = 2cD(a + b) + \gamma D^2(2\chi b + a - b)B_u \tan\psi + Q \tag{7-7}$$

$$P_u = 2cD'(a + b) + \gamma(2D - D')D'(2\chi b + a - b)B_u \tan\psi + Q \tag{7-8}$$

On the basis of tests on shallow anchors in clay, it has been recommended [64] that only half the value obtained from the theory (Eqs. 7-1 to 7-5, 7-7, and 7-8) be used.

7-5.4 Tension Piles

Use of tension piles for anchoring cables is another versatile alternative. Piles may be underreamed for greater resistance to pull. Like a plate anchor, a pile is provided at an inclination such that the cable pulls at it axially. Little work has been done to establish the effect of inclination on the pull-out strength of a pile. However, G. Ranjan [81] has observed that the capacity of tension piles in sandy soils, inclined at an angle not greater than 30° to the vertical, is closely given by vertical tension piles under similar conditions.

Sand. The pull-out capacity of a vertical bored pile in sandy soils can be assessed as follows [81]:

$$P_u = Q_p + \tfrac{1}{2}\gamma D^2 \pi b B_s \tan \xi \tag{7-9}$$

where Q_p = weight of pile
D = depth of pile
b = diameter of pile
B_s = coefficient of earth pressure
ξ = angle of wall friction, in this case equal to ψ

The value of B_s varies from 0.5 for loose sand to 1.0 for dense sand, but much higher values, which could approach values for the passive state, are reported for driven piles [63].

Cohesive Soil. The pull-out capacity is given as

$$P_u = \pi b D j c \tag{7-10}$$

where j = adhesion factor, depending upon type of clay and varying from 0.4 for stiff clays to 1.0 for soft clays

It is worth noting that Eq. 7-10 assumes the contribution of skin friction against a *pull-out* force to be the same as against a *push-in* force.

Pile Group. The total uplift resistance of a vertical pile group is taken as the smaller of the following two values:

1. Uplift capacity of a single pile times the number of piles in the group
2. Uplift capacity of the entire group considered as a block

Cost. Buchholdt and Vadgama [11] have given an estimate of tension-pile costs as

450-mm ϕ piles	£7 to £10 per metre length of pile
600-mm ϕ piles	£9 to £12 per metre length of pile
Larger piles (including underreamed piles)	£22 per tonne of concrete

1 £/m = 0.305 £/ft

7-6 ERECTION

As mentioned earlier, in Chaps. 1 and 6, the erection of cable roofs requires careful consideration. It is always important for the erector to fully understand the sequence of construction envisaged by the designer. This has a special significance in cable roofs because erection stresses can be substantially large. The design follows a particular sequence of construction, and the safety of different elements is checked at various stages. A change in this sequence can cause sharp changes in the force pattern of the structure and may lead to serious consequences. In general, structures supported on closed rings, arches, or frames show greater sensitivity to the erection and pretensioning procedure. Also large movements of cables are possible during erection, on account of the gradual placement of decking and other roof loads, which cannot indeed be placed all at once. Checks against computed coordinates are made at different stages to ensure that the correct configuration is achieved.

The order of operations in erecting a cable roof cannot be generalized but usually the supporting structure, including the anchors, columns, rings, or frames, is first constructed. This is followed by the stringing of cables to a preliminary position and placement of decking units thereon. Next, the cables are adjusted as necessary to the final geometry and decking is interlocked; the roof is then finished. The deflections of the roof because of live, snow, or wind load, or because of creep and temperature changes, need to be computed with respect to the final geometry. Schematic sketches showing the erection sequence for a typical structure of the type shown in Fig. 6-14 are given in Fig. 7-15. For this particular structure, trusses would have no interaction with each other during erection, and can therefore be pretensioned independently.

For radial forms (see Fig. 1-11), however, the trusses interact through the compression ring. The erection sequence for one such structure (see the plan in Fig. 7-16*a*) is shown in Fig. 7-16*b*. If the cable tensions cause an unsymmetric loading of the rings, these will develop bending moments. Whereas there can be no control over this situation insofar as the effect of live or wind loads is concerned, an attempt should be made to minimize the unbalance of loading and consequent bending by choosing a suitable sequence of erection. It may be a good practice to design the rings for the worst combination of (1) thrust due to pretension + dead load and (2) bending moment

FIGURE 7-14 Turnbuckles incorporated into cables at Munich stadium. (*Courtesy of Leonhardt und Andra, Photo by Dressler*.)

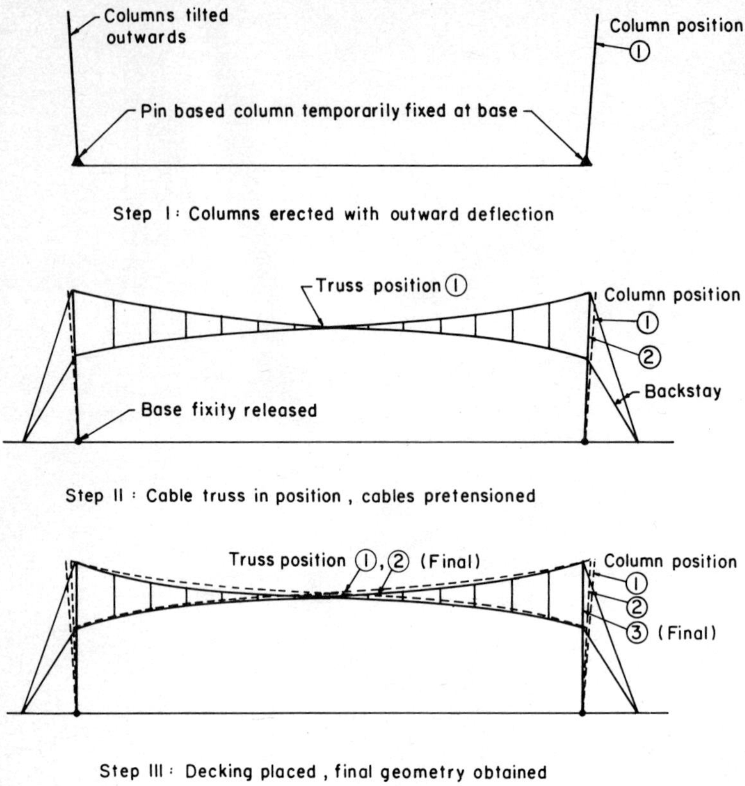

FIGURE 7-15 Schematic sketches to indicate the sequence of erection of the roof shown in Fig. 6-14.

TABLE 7-8 Sequence of Pretensioning Operations Suggested for the Structure in Fig. 7-16

Operation	*Truss numbers*
1	1, 8, 16, 24
2	4, 12, 20, 28
3	2, 14, 18, 30
4	6, 10, 22, 26
5	3, 15, 17, 29
6	5, 13, 19, 27
7	7, 11, 21, 25
8	9, 23

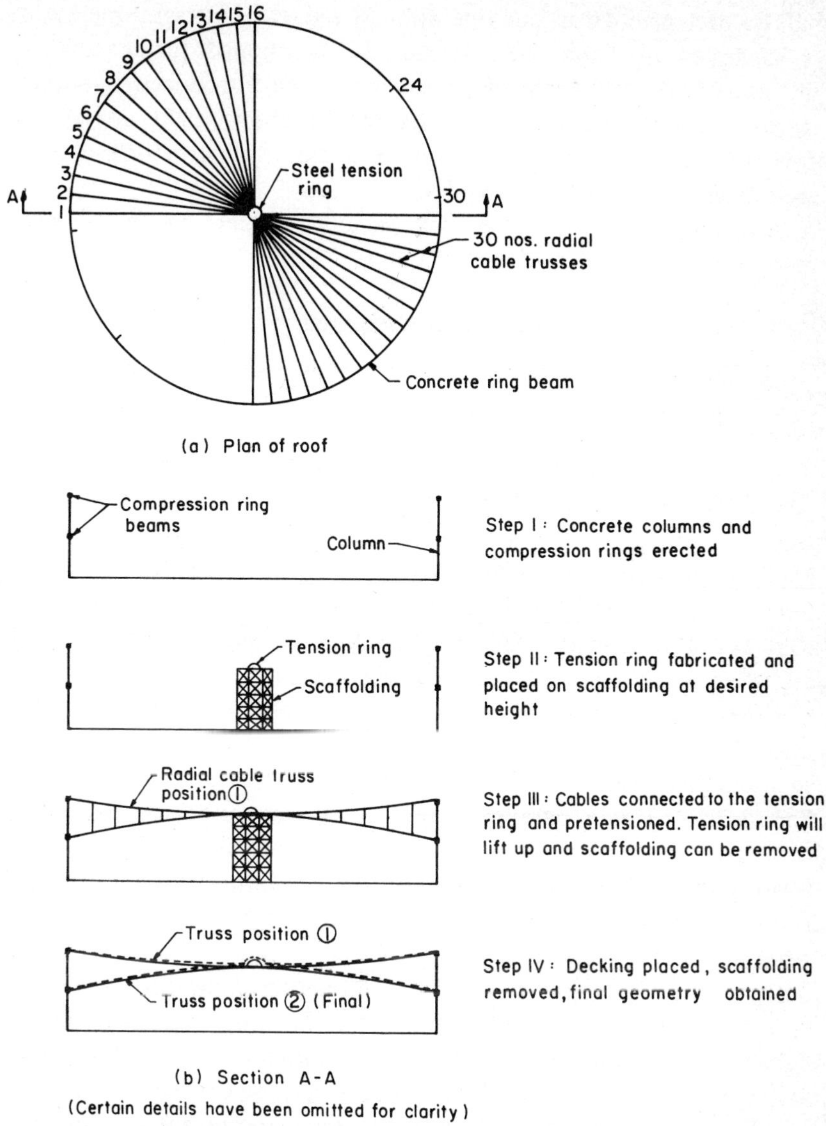

FIGURE 7-16 A scheme of erecting a circular cable roof with radial trusses.

due to live or wind loads and thrust. The erection sequence should then ensure that the rings are not overstressed at any stage during erection. The ideal situation will be to pretension all the 30 trusses simultaneously and thus cause only direct thrust in the ring. However, this may not be practicable. Assuming that only four trusses can be pretensioned in one operation, a sequence of eight operations is suggested in Table 7-8. It may be mentioned further that, if necessary, the total force of prestress may be introduced in several stages—each stage involving a set of the above eight operations. Description of erection procedures for some other structures may be seen in Refs. 10, 34, 62, 85, and 92.

In regard to erection, cable roofs have two distinct advantages over many other forms of roof construction:

1. Very little or no scaffolding is required.
2. Erection can be achieved quite rapidly; completion within a period of 4 to 6 weeks is not uncommon.

As already mentioned, cables are prestretched in factories and duly marked for the position of the fittings. These are then transported to site [15] on reels having a large enough diameter to avoid excessive bending and loss of the effect of prestretching. After the cables are placed, adjusted into a preliminary position, and secured at their ends, they are pretensioned. This may be done by means of jacks, or by adjusting lengths, controlling the cable sag, or both. It is also possible that the cable assembly is prepared on the ground and then lifted into position and tensioned through any one element. The latter requires the cables to be cut to extreme precision in length (see also Sec. 7-2.4). For a parallel system of trusses, either of the above procedures may be adopted. However, for radial forms where uniformity of cable tension due to prestress has utmost importance, the former method could be advantageous in that better control is possible at various prestressing steps. For anticlastic nets, too, some designers find it simpler to position the sagging cables accurately into their preliminary configuration, place over them the hogging (or cross) cables, and tension the system in a predetermined sequence by pulling at the ends of the hogging cables. This may be carried out in small steps, as also suggested for the radial systems, to prevent overstressing of the supporting structure.

CHAPTER EIGHT

Air-Supported Roofs

8-1 INTRODUCTION

It has been mentioned, in Sec. 1-3, that building roofs can be supported by air pressure. Air pressure stretches the roof skin to form a tensioned membrane and thus sustains the loads. Some of these roofs consist of a membrane stabilized by a system of cables or ropes. As the air pressure stretches the membrane which is attached to the cables, the cables are tensioned too. This chapter is intended to introduce the subject and discuss briefly the different aspects of air-supported roofs. For more detailed information the reader's attention is drawn to Refs. 3, 22, 75, 78, and 103.

8-1.1 Development

Lanchester introduced the idea of using the pneumatic principle for buildings as early as in 1917. During the second world war, considerable research and development took place; by 1946, *radomes*, air-supported weather covers for radar installations, were constructed for the U.S. Air Force, and lightweight portable structures were developed further for military and commercial applications. The Pan American Airways Pavilion at the 1958 Brussels World Fair further popularized the use of such structures. In the middle sixties, sufficient interest had been generated by these developments for the organization of the first International Symposium on pneumatic structures in Stuttgart, Germany, in 1967. The greatest exploitation

FIGURE 8-1 U.S. Pavilion at the International Exposition, Osaka, Japan, 1970. (*Photo by Taiyo Kogyo.*)

of these structures was made during Expo '70 at Japan, and this form of construction is continuing to engage the interest of the architect as well as the engineer with increasing intensity, particularly in the United States, Europe, and Japan. The air-supported roof constructed for the Pontiac Metropolitan Stadium in Michigan has the capacity to seat as many as 80,000 persons. The stadium covers an area of 10 acres (0.0384 km^2). A roof to cover a 3 km^2 city in the Arctic for housing 45,000 people is being mooted. Figure 8-1 shows the elegant U.S. Pavilion at Expo '70. The roof covered an elliptical plan area [35] of 1,025,000 ft^2 (95,000 m^2).

8-1.2 Classification

Pneumatic roofs can be classified as air-inflated or air-supported. An *air-inflated* structure is of a tubular or cellular construction which is capable of transmitting applied loads to the points of support. If leakage through the membrane is eliminated, these do not require constant pumping, but such roofs can be comparatively expensive because they require a larger membrane area and higher air pressures. An *air-supported* structure provides a single-wall enclosure. The membrane is attached to the support along the periphery. The membrane is stretched and elevated by a slight increase in the internal

(a) Exterior view of an air supported roof

(b) Transparent air supported roof – an interior view

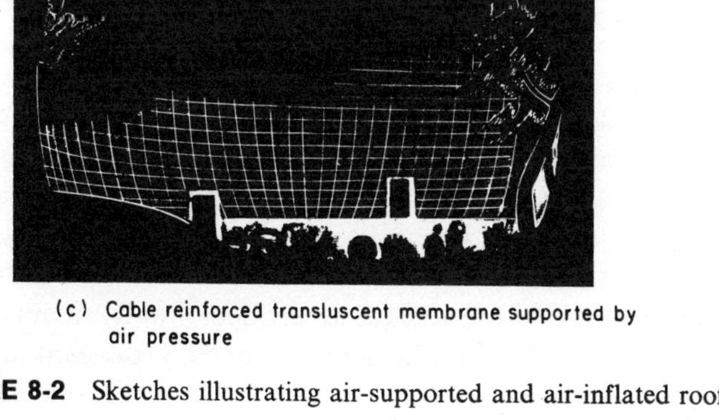

(c) Cable reinforced transluscent membrane supported by air pressure

FIGURE 8-2 Sketches illustrating air-supported and air-inflated roofs.

air pressure so that it can support applied loads. Greater attention is given herein to the latter type, since it is often supported on cables or ropes. Typical forms of such roof structures are shown in Fig. 8-2. The basic elements of this structure are the structural membrane, cables, inflation equipments, anchorage, entrances, or exits.

8-1.3 Applications

Air-supported structures or air structures are lightweight and pliable. They have a small packaging ratio, can be easily transported and quickly erected almost anywhere, and are likely to involve smaller initial cost compared to conventional buildings. As with cable roofs, large spans can be achieved and the unobstructed building space efficiently utilized. Since the membrane is in tension, the material use is most efficient. A unique feature of this form of roof is that it is possible to adjust the rigidity of the system by increasing or decreasing the internal air pressure in accordance with the variation in external loads. For example, in the event of a wind storm of high velocity, the internal pressure may be increased for greater resistance.

The use of air structures has so far been made for a variety of roof applications, such as swimming pools, sports halls, pavilions, warehouses, coverage for military vehicles and equipment, and for production or assembly lines. Increasing applications are likely for space-exploration programs where light weight and compactness are imperative. Air structures provide an excellent answer for temporary construction. Limitations, though not insurmountable, are (1) the need to maintain the pressure differential across the membrane by constant pumping, (2) thermal and acoustic insulation of the thin membrane, and (3) susceptibility to vandalism.

Cable-reinforced Envelope. For structures with large spans or radii of curvature, cables and nets are required to divide the membrane into a number of smaller elements, thus reducing the membrane stresses. Lower fabric loads are thus achieved. The main load-bearing elements are the cables. Circumferential cables are suitable for high domes and radial cables for shallower forms. Nets of uniform gauge are most efficient as the surface is divided into a number of equal-sized elements of uniform curvature. The U.S. pavilion at Expo '70, the Pontiac Metropolitan Stadium at Pontiac, Michigan, and a dome

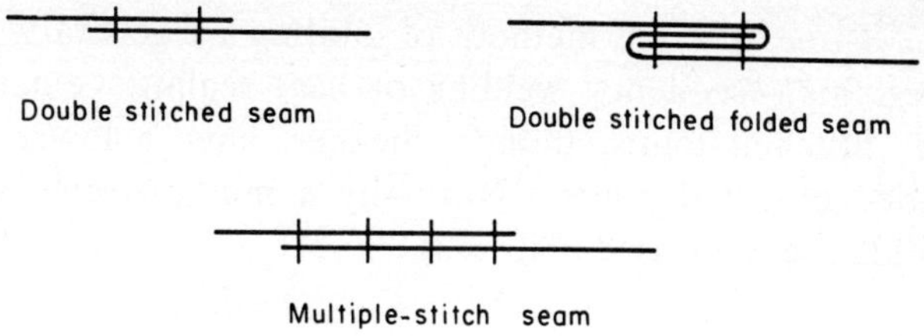

FIGURE 8-3 Joints in membrane material.

at the University of Northern Iowa prove the application of cable networks for air-supported roofs. A large number of free forms suggested by soap-bubble conglomeration are possible pneumatically. Care is required, however, to prevent the chafing that can occur on the membrane due to large movements in gusty winds.

8-2 MATERIAL AND EQUIPMENT

The material used in air-supported roofs would mainly be cables (or synthetic ropes) and their connections, and the membrane skin, the latter being a special feature. Some of the equipment required is also unique to this type of roof. The provision of appropriate material for the membrane and suitable equipment is important for the successful construction of air-supported roofs and their maintenance. These two aspects are now discussed.

8-2.1 Material

In selecting the material for the membrane, requirements such as toughness, strength, impermeability, flame resistance, light weight, and flexibility have to be taken into account. Synthetic fabrics are generally coated with plastics to increase their resistance to gaseous diffusion. Polyesters coated with PVC, neoprene, or hypalon are generally used. The life of PVC-coated fabric is estimated to be about 7 years and that for neoprene- or hypalon-coated fabric about 10 years with proper maintenance. The weight of the fabric is about 4 oz/ft^2 (10 N/m^2) and the thickness is about 0.02 to 0.4 in (0.5 to 1 mm). Fiberglass coated with Teflon Fluorocarbon resin has a life of about 25 years. It is translucent and has self-cleansing properties. For longer life, a membrane made of metal such as thin stainless steel may be used if economic considerations permit.

Jointing. Three different methods of jointing are generally employed for fabrics: high-frequency welding or heat sealing, cementing, and stitching. Stitched joints, though cheaper, have a lower efficiency than welded or glued joints. Normally a multiple-sewn seam or a double-folded seam is used (Fig. 8-3).

8-2.2 Inflation System

Constant pumping is necessary for single-wall air structures and the reliability of pressurization equipment is essential. Small structures require hand- or foot-operated pumps or bellows. For large structures, high-capacity turbine blowers may be employed. Standby fans are essential to cater for the possibility of a breakdown, and a separate source of supply should be maintained in case of a power failure, with facility for automatic changeover for both fans and supply. The requirement of constant pumping is not as big a handicap as it may first appear, since many modern buildings depend completely on an uninterrupted operation of mechanical or electrical equipment, such as elevators and air-conditioning equipment.

8-3 CONSTRUCTION DETAILS

The construction details such as the anchor system for the fabric, as well as the supporting cables, form as important a design problem here as in the conventional cable roofs. The safety of these structures is almost directly linked to the safety of the anchor system. Also important are details of erection and maintenance. All these aspects are briefly touched upon in this section.

8-3.1 Anchorage

The pressure differential across the membrane causes uplift forces on the whole structure, and these must be resisted by firmly anchoring the structure to the ground. The membrane forces must be transmitted evenly from the skin to the anchorage and loads resisted with an adequate factor of safety. When cables are used, anchorage forces are concentrated at cable points, often requiring substantially large anchors. In a ground anchorage system, the membrane is

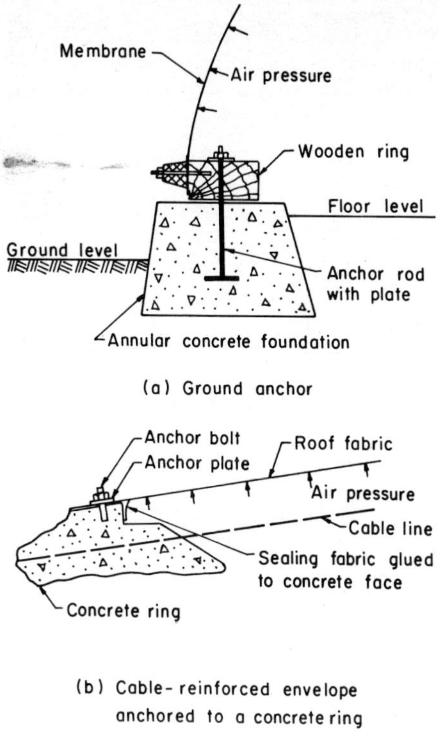

FIGURE 8-4 Typical details for anchoring membrane fabric.

attached to the concrete foundation at regular intervals. For satisfactory results, the anchors are attached to the reinforcement rods of the foundation before the concrete is poured. A seal must be provided along the perimeter to prevent loss of air pressure. The membrane may also be supported into a ring beam at an elevation above ground level. The ring beam is supported on a wall or on a number of columns, in a manner similar to that seen for many cable roofs. The U.S. Pavilion shown in Fig. 8-1 is supported thus. Anchor systems for cables are adequately covered in Chaps. 1 and 7 and need no repetition here. Other underground anchorages such as spiral screws, helical plates, and spreading and driven arrowhead anchors may also be used, as shown in Fig. 8-4.

8-3.2 Doors and Windows

Various types of doors are used to prevent pressure loss at an entrance—a door with a central pivot, a revolving door, an air

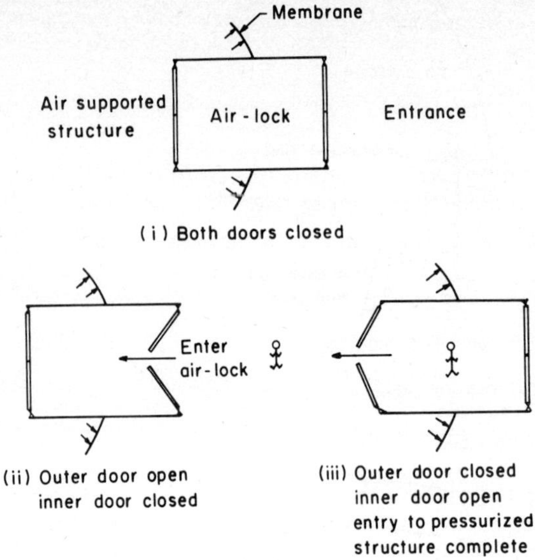

FIGURE 8-5

curtain, or an airlock (Fig. 8-5). A slip-pocket, the simplest door, is generally employed for domestic structures, such as covers for swimming pools. An airlock consists of an airtight tunnel with doors at either end that is designed so that one set of doors is always closed automatically. These provide a satisfactory answer to the need for vehicular access. Air curtains have been developed for heavy personnel traffic. To compensate for the pressure loss, fans are located outside the entrance and operate when the door opens. To minimize the use of artificial light during the daytime, translucent windows of fabric can be incorporated in the roof. The fabric is reinforced, where necessary, to avoid stress concentration.

8-3.3 Erection

The first step in erecting the structure is the completion of the supporting and anchoring system. This is followed by the erection of the entrances, and the membrane material is attached to the fixing points. The inflation equipments are installed and the structure inflated. To avoid damage to the membrane, inflation or deflation is done during periods of low wind speed. A structure covering an area of 1000 m^2 could be inflated in 30 to 45 min, depending upon the fan capacity. On an existing foundation, a task force of one engineer and four unskilled persons may take only 4 to 5 days to erect a structure covering 1000 m^2 [22]. Dismantling is a simpler and quicker operation for such structures, and deflation can be achieved in 15 to 20 min. Packaging of the structure can be accomplished in less than a day for this structure by the same team of five men.

8-3.4 Maintenance

Regular checking of the building is necessary so that the safety is not impaired. Maintenance of equipments, periodic checking of anchorages, operation of the airlock/emergency exit, and inspection of the membrane for abrasive wear and tear should be carried out as a routine. Repair of small tears in fabric can be achieved in situ by gluing a patch on the torn fabric. In an industrial climate, the membrane may be washed yearly by hosing for aesthetic reasons as well as for good interior lighting.

8-4 ANALYSIS AND DESIGN

8-4.1 Loading

The dead load of such roofs is usually small enough to be neglected except when cables are also used. In addition, the important loads are due to wind, snow, or live load. Live loads would normally occur on account of maintenance teams. Other than this there does not seem any reason for live load to occur. The shapes of these roofs are such that snow would tend to slide off the surface, unless the roof has a flat curvature or extreme climatic conditions exist. In areas of heavy snowfall, snow-load requirements often dictate the inflation pressure, and an arrangement to heat and melt away the snow may well be necessary. Both live and snow loads will try to reduce the tension that is generated in the membrane and the cables by internal pressure, and would thus try to deflate the roof. Whereas, for a general discussion of wind loads the reader is referred to Sec. 4-5.1, it is worth considering here the effect wind can have on an air-supported roof. The wind distribution may be such as to cause pressure, suction, or a combination of the two on the surface. While suction will try to increase the membrane and cable tension, pressure will try to deflate it. It follows, therefore, that the maximum tension in the material and anchor forces are likely under the combined effect of internal air pressure and wind suction. On the other hand, the system may try to slacken under the combination of live load/snow load and external wind pressure. The requirement of internal air pressure needs to be checked against this possibility.

8-4.2 Membrane Stresses

For a doubly curved membrane under internal pressure p_a normal to the surface, it can easily be proved by the membrane theory that

$$\frac{T_m}{r_m} + \frac{T_c}{r_c} = p_a$$

if r_c = radius of curvature in circumferential direction
r_m = radius of curvature in meridional direction
T_c = membrane force in circumferential direction
T_m = membrane force in meridional direction

In a dome, $r_c = r_m = r$; $T_c = T_m = T$; hence, $T = p_a r/2$. For a cylindrical surface, $r_m = \infty$, $r_c = r$, and $T_c = p_a r$.

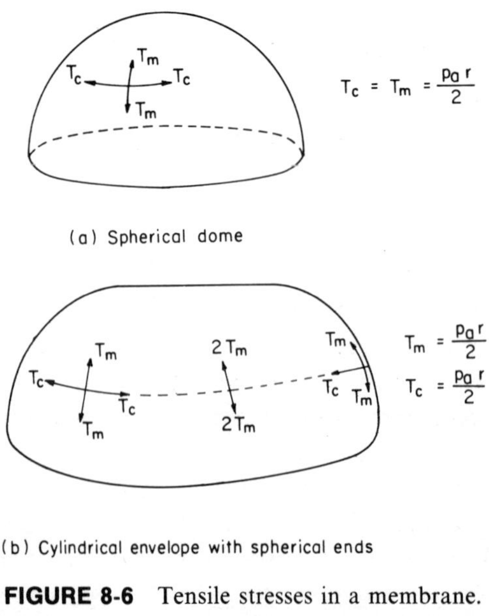

FIGURE 8-6 Tensile stresses in a membrane.

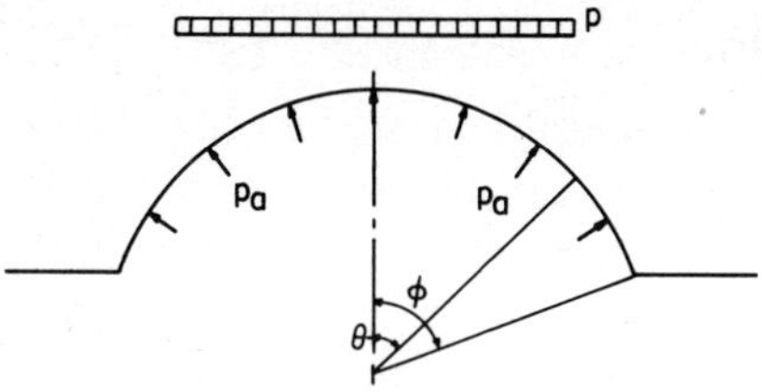

FIGURE 8-7 Membrane under uniform load.

In a spherical form, the membrane is stressed equally at every point in all directions, but its application may not always be practical. For a rectangular plan, the cylindrical form with or without spherical ends is very popular for uses such as warehouses, factories, indoor sports, and places of assembly. For a cylinder with spherical ends, the longitudinal membrane stress in the cylinder is the same as the stress in the spherical end (Fig. 8-6).

Expressions for calculating membrane forces under distributed loads such as snow load, are given below (Fig. 8-7).

1. *Uniform load on a spherical surface.* If the self-weight of the membrane is neglected, it can be easily derived that

$$T = \frac{p_a r}{2} - \frac{pr}{2}\left(\frac{\sin\theta}{\sin\phi}\right)^2$$

2. *Uniform load on a cylindrical surface with spherical ends.*

$$T_c = p_a r - pr\,\frac{\sin\theta}{\sin\phi}$$

$$T_m = \frac{p_a r}{2}$$

For the live load, θ may be taken equal to ϕ. However, for snow load, the value of $\theta(<\phi)$ should be appropriately chosen to account for the steepness of the surface at the ends (Fig. 8-6). The analysis is based on the classic membrane theory applicable to shell structures and assumes linear behavior. Due to large deformations of the membrane and nonlinear behavior of the fabric, a nonlinear theory should be used for a rigorous solution. However, this is by no means straightforward. As such, simpler methods of analysis are employed generally and model studies often considered useful. Analysis of cable nets supporting the membrane can be carried out as described in Chaps. 2 to 4.

8-4.3 Capacity of Inflation Equipment

Calculation of fan capacity is largely an empirical process because of the difficulty of estimating the magnitude of air losses. The major points of leakage are through door openings, openings for ventilation, and around joints of the membrane with anchor/supporting structure. The capacity of an inflation system should be designed to cope with critical conditions that may occur during initial inflation, peak loading periods, and emergency situations that may occur from a tear in the membrane or during periods when doors have to be kept open excessively. Air should be supplied uniformly around the periphery of the building by a peripheral ducting system having numerous outlets that discharge upwards (Fig. 8-8).

8-5 SAFETY AND SERVICEABILITY

An air-supported roof with a sound structural design is normally as safe as any other well-designed structure. However, a few other possibilities that may cause damage need consideration. These are failure of pumps or power, fire, flying debris, etc. Also important for consideration is the effect of raised pressure on the persons inside the building.

Pump or Power Failure. In the event of the failure of the air pumping system, the roof will begin to deflate. However, it will take a fairly long time for a large envelope to collapse as the air can leak out only slowly and the weight of the membrane is very small. The

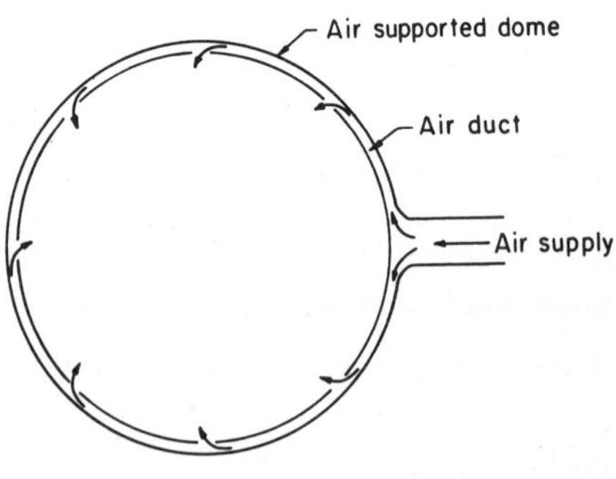

FIGURE 8-8 Peripheral air supply.

failure is not sudden, as in the case of some other conventional buildings. The people living inside are therefore given ample time for evacuation.

Fire Hazard. Fire tests indicate that the coated fabric *does not support combustion*. If a hole is created in the skin by fire, an airstream is created through the hole on account of the pressure differential and tends to blow out the fire. Since the skin is thin, the fire intensity is small and as such it takes comparatively little energy to blow the fire out. Fire tests conducted by Schulz [78] indicate that it is difficult to inflame the envelope due to pressure differential of the air and it took about 12 min to burn a hole in a PVC-coated fabric used for the experiment. Due to the pressure differential, the flame was put out and the hole was enlarged only insignificantly. Results of a full-scale fire test on the U.S. Pavilion at Expo '70 [35] indicated that the smoke generation was slow and that there would be ample time to evacuate all occupants. When the material is PVC-coated fiberglass, it was observed during the test that the PVC melted and then caught fire, but there was hardly any damage to the base fabric of the fiberglass.

Effect of Elevated Pressure on Man. To maintain the shape of the structure, only a moderate pressure is generally required for single-walled air-supported roof structures. Since man has been able to safely withstand pressures in a wide range, no health hazard appears to be presented by a continuous stay in an air-supported structure.

APPENDIX

Computer Program for Static Analysis of a Cable System

GENERAL

The program given here uses FORTRAN IV and can analyze a pin-jointed structure comprising tension (cable) and compression members with axial forces only, loaded at the node points, and supported on unyielding supports. The procedure of analysis used is the same as described in Sec. 3-2.1 and employs numerical scheme 2 (Newton-Raphson method), as given in Sec. 3-3.2. It is possible to analyze both plane and space assemblies for:

1. A given configuration, structure properties, and member pretension, loaded at node points (IPT = 0), to determine member forces and deformations

2. Determination of the initial geometry, given an assumed set of coordinates, structure properties and member pretension and assuming no loads (IPT = 1)

For the second type (IPT = 1) of problem (illustrated in Example 3-2, Sec. 3-2.2) the program is arranged to establish the equilibrium configuration for the structure with given data in the first case of loading (which is really a no-load case). The output for the first loading case, along with other relevant data, is stored as input for subsequent cases of loading, for which the procedure of analysis is identical to the one used for problems of the first type (IPT = 0). Provision is made to handle slackness in cable elements as described in Sec. 3-2.5. A flow diagram for the program is given in Fig. A-1.

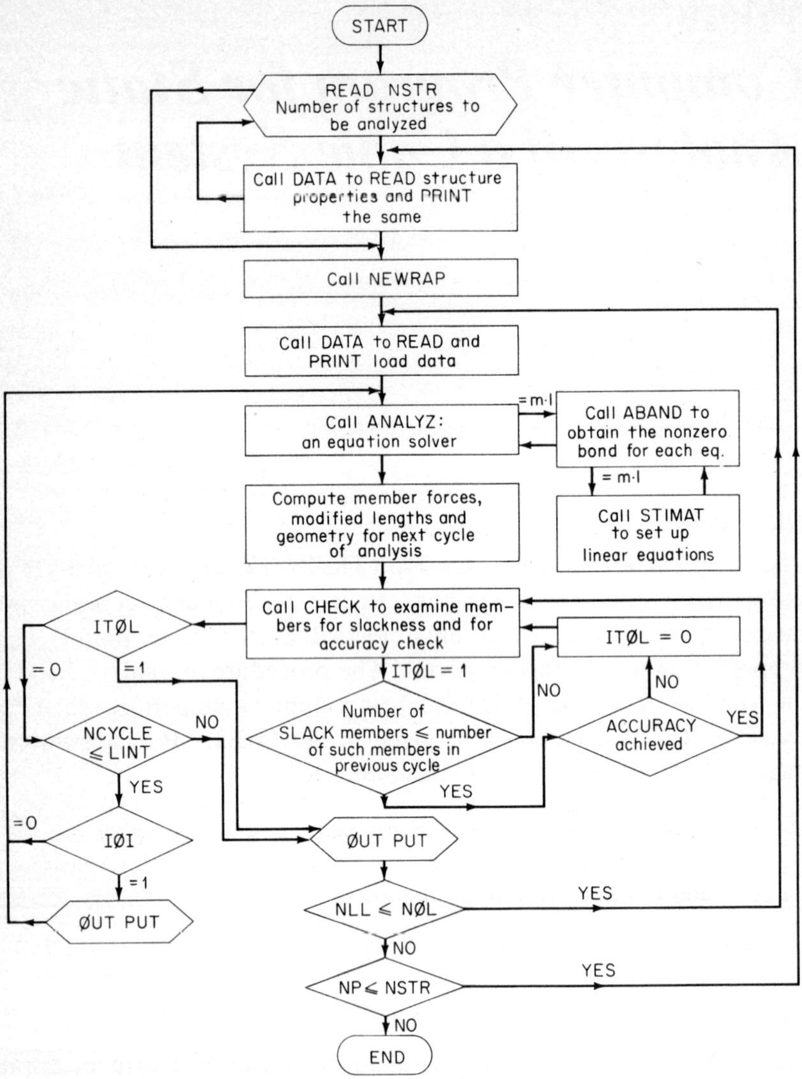

FIGURE A-1 Flow diagram.

DATA

The data required to be input and the corresponding notation used in the program are as follows:

Structure

NSTR = number of structures to be analyzed in a single run

IPT = indicates class of problem to be analyzed

=0/1 (See description in preceding paragraph.)
NDF = number of degrees of freedom per node
= 2, for plane structures
= 3, for space structures
NOL = number of loading cases per structure
NOJ = total number of joints
NOSJ = number of support joints
JTI = joint-type indicator
= 0/1: Free to move in space/support joint (restrained)
SJC = joint coordinates (For system of axes, see Fig. 3-1.)
NOCS = number of members
IJOM = numbers of joints incident on member ends (2 per member)
SMF = pretension force (negative for struts)
ICPR = indicates whether for a particular member prestress is to be maintained as constant through the first loading case.
= 0, not constant   for IPT = 0, ICPR = 0
= 1, constant   for IPT = 1, ICPR = 1 or 0

(See Example 3-2. ICPR = 0 for members 1 to 16 and ICPR = 1 for members 17 to 40.)

YM = Young's modulus
AOC = cross-sectional area
MTI = member-type indicator
= 0, cable element—cannot take compressive force
= 1, strut

Loads and Output

PJL = joint loads (consistent in direction with coordinate axes)
TEMP = $\alpha \times t$ for each member, where α is the coefficient of linear expansion and t is the temperature change
IOI = indicates whether output is to be printed in each cycle of iteration or only final output for each loading case.
= 0, only final output
= 1, output for each cycle
LNIT = limit desired to be placed on the maximum number of cycles of iteration for a loading case
TOL = percentage tolerance permitted on deflections in an accuracy check to terminate iterations
DPI = absolute value of deflection, the values less than which will be ignored in accuracy check

DIMENSIONS

The dimensions of storage locations required depend upon the size of problem and are therefore given in terms of the following letters, except where a numeral is used. As is well understood, however, the dimensions as given on an actual program have to be in terms of numerals.

i = number of members (NOCS)
j = number of joints (NOJ)
k = the maximum number of members incident on any particular joint, i.e., for structures in Figs. 3-11, 3-13, and 3-16: this number would be 3, 4, 4, respectively.
l = number of degrees of freedom per node (NDF)
m = NOFJ
$n = \frac{1}{2} m \times b$
b = nonzero bandwidth of an equation

It is convenient to pick an approximate (but overestimated) value of n, by taking the maximum value of b.

EXAMPLE

The cable truss analyzed in Sec. 3-4.1 (Fig. 3-4a) is modified slightly (see Fig. A-2) and chosen for illustrating the data and output obtained from the program. The modification is that the two cables are connected directly to each other at the center point, instead of being connected through a vertical hanger. The truss, assumed to have a known geometry under the initial prestress (IPT = 0), is analyzed for two cases of applied load; namely, cases VII and VIII (see Table 3-6).

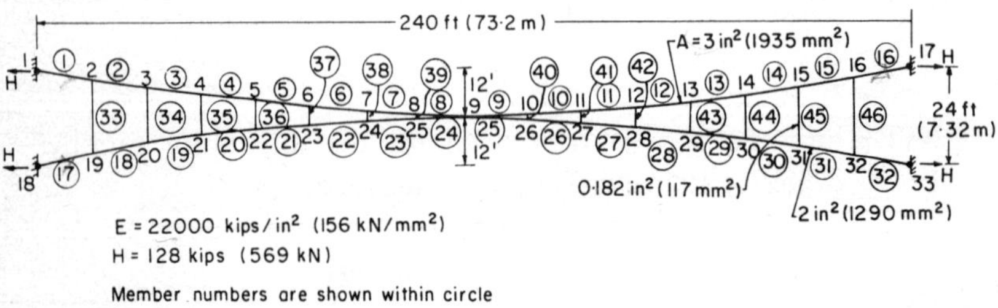

FIGURE A-2 Truss used to illustrate program input and output details.

LISTING OF INSTRUCTIONS, DATA, AND OUTPUT

```
PROGRAM INSTRUCTIONS **
      COMMON  NDF,NOL,IPT,NOJ,NOSJ,JTI,SJC,NOCS,IJOM,SMF,ICPR,YM,AOC,
     1 MTI,PJL,TEMP,DJD,IOI,LNIT,TOL,DPI,NP,NOFJ,JNFJ,JNIC,NINJ,IMOJ,
     2 RJC,RMF,RLS,SLS,YMS,GRT
      DIMENSION JTI(J),SJC(J,L),IJOM(I,2),SMF(I),ICPR(I),
     1 YM(I),AOC(I),MTI(I),PJL(M,L),TEMP(I),DJD(J,L),JNFJ(M),
     2 JNIC(M),NINJ(M),IMOJ(J,K),RJC(J,L),RMF(I),RLS(I),SLS(I),
     3 YMS(I)
      READ 11,NSTR
      DO 1 NP=1,NSTR
      NLL=0
      CALL DATA(NLL)
      CALL NEWRAP
    1 CONTINUE
      PRINT 12
   11 FORMAT(I4)
   12 FORMAT(1X,34H STOPPED EXECUTION IN MAIN PROGRAM)
      STOP
      END


      SUBROUTINE DATA(NLL)
      READS,GENERATES AND PRINTS DATA


      ** REPEAT COMMON DIMENSION BLOCK **


      DIMENSION P(L)
    1 FORMAT(12I6)
    2 FORMAT(6F12.0)
    3 FORMAT(3F12.0,4I6)
    4 FORMAT(I6,3F12.0)
      IF(NLL.NE.0)GO TO 1000
      READ 1,IPT,NDF,NOL,NOJ,NOSJ,NOCS
      DO 5 J=1,NOJ
    5 READ 4,JTI(4),(SJC(J,K),K=1,NDF)
      DO 6 M=1,NOCS
    6 READ 3,SMF(M),YMS(M),AOC(M),(IJOM(M,I),I=1,2),ICPR(M),MTI(M)
      GO TO 2000
C     THE FOLLOWING PARAGRAPH IS CALLED REPEATEDLY BY SUBROUTINE NEWRAP
C     TO READ AND PRINT DATA FOR EACH LOADING CASE OF A PARTICULAR
C     STRUCTURE WHEREAS THE REST OF THE DATA READ ABOVE IS REQUIRED
C     ONLY ONCE PER STRUCTURE
C     LOADS TO BE READ FOR UNSUPPORTED(OR FREE)JOINTS ONLY
 1000 READ 1,LNIT,IOI
      READ 2,TOL,DPI
      DO 12 J=1,NOFJ
      READ 4,JN,(P(M),M=1,NDF)
      K=JNFJ(JN)
      DO 11 M=1,NDF
   11 PJL(K,M)=P(M)
   12 CONTINUE
      READ 2,(TEMP(M),M=1,NOCS)
C     PRINTS OUT THE LOADING AND TOLERANCE VALUES ETC.
      PRINT 13,NLL
   13 FORMAT(////1X,22H LOADING CASE NUMBER =,I2)
      PRINT 14
   14 FORMAT(//1X,10H JOINT NO.,19X,6H LOADS/)
      DO 15 J=1,NOFJ
      JK=JNIC(J)
      JL=JNFJ(JK)
   15 PRINT 16,JK,(PJL(JL,K),K=1,NDF)
   16 FORMAT(4X,I3,6X,3F12.6)
      PRINT 17,LNIT,TOL,DPI
   17 FORMAT(/1X,27H LIMITING NUMBER OF CYCLES=,I2,/1X,
     1 46H PERCENTAGE TOLERANCE ON ERROR IN DEFLECTIONS=,E12.4,/1X,
     2 57H ORDER OF MAGNITUDE OF DEFLECTION PERMITTED TO BE IGNORED,
     3 19H IN ACCURACY CHECK=,E12.4)
      PRINT 18
   18 FORMAT(/1X,15H ALPHA*T VALUES)
      PRINT 19,(TEMP(M),M=1,NOCS)
   19 FORMAT(1X,6E12.4)
      RETURN
C     GENERATES SOME DATA NECESSARY
```

```
 2000   NOFJ=NOJ-NOSJ
C       GENERATES THE LIST FOR MEMBERS INCIDENT ON EACH JOINT
        DO 21 J=1,NOJ
        I=0
        NINJ(J)=0
        DO 21 K=1,NOCS
        JS=IJOM(K,1)
        JE=IJOM(K,2)
        IF(J.NE.JS.AND.J.NE.JE)GO TO 21
        I=I+1
        NINJ(J)=NINJ(J)+1
        IMOJ(J,I)=K
   21   CONTINUE
C       ARRANGES NUMBERS PERTAINING TO FREE JOINTS  IN A CONTINUOUS AND
C       SERIAL ARRAY RESPECTIVELY
        K=1
        DO 22 I=1,NOJ
        JNFJ(I)=0
        IF(JTI(I).EQ.1)GO TO 22
        JNFJ(I)=K
        JNIC(K)=I
        K=K+1
   22   CONTINUE
C       COMPUTES MEMBER LENGTHS
        DO 24 MN=1,NOCS
        ASUM=0.
        JS=IJOM(MN,1)
        JE=IJOM(MN,2)
        DO 23 N=1,NDF
        RJD=SJC(JS,N)-SJC(JE,N)
   23   ASUM=ASUM+RJD*RJD
        SLS(MN)=SQRT(ASUM)
   24   CONTINUE
        DO 25 M=1,NOCS
   25   YM(M)=YMS(M)
C       PRINTS OUT THE DATA FOR EACH STRUCTURE
        PRINT 31,NP
   31   FORMAT(////1X,20H DATA FOR STRUCTURE ,I3,3X,17HREAD OR GENERATED)
        PRINT 32,NDF,NOL,NOJ,NOFJ,NOSJ,NOCS
   32   FORMAT(///1X,27H NO. OF DEGREES OF FREEDOM=,I3,/1X,
     1 22H NO. OF LOADING CASES=,I3,/1X,15H NO. OF JOINTS=,I3,3X,
     2 20H NO. OF FREE JOINTS=,I3,3X,23H NO. OF SUPPORT JOINTS=,I3,/1X,
     3 23H NO. OF CABLE SEGMENTS=,I3)
        PRINT 33
   33   FORMAT(//1X,6H JOINT,6X,4H JTI,5X,2H X,14X,2H Y,14X,2H Z/)
        DO 34 J=1,NOJ
   34   PRINT 35,J,JTI(J),(SJC(J,K),K=1,NDF)
   35   FORMAT(1X,I3,10X,I2,3(4X,E12.4))
        PRINT36
   36   FORMAT(//1X,7H MEMBER,9X,4H AOC,13X,4H YMS,12X,4H SMF,12X,
     1 4H SLS,10X,5H IJOM,3X,5H ICPR,2X,4H MTI/)
        DO 37 M=1,NOCS
   37   PRINT 38,M,AOC(M),YM(M),SMF(M),SLS(M),(IJOM(M,I),I=1,2),ICPR)M*,
     1  MTI(M)
   38   FORMAT(1X,I6,3X,4(4X,E12.4),4X,2I4,4X,I2,4X,I2)
        RETURN
        END


        SUBROUTINE NEWRAP
C       CARRIES OUT ANALYSIS BY NEWTON-RAPHSON PROCEDURE

        ** REPEAT COMMON DIMENSION BLOCK **

        DIMENSION A(M.L),TRI(N),X(M.L),MB(M.L),CMF(I),NS(I)
        DIMENSION DJS(L),DJE(L)
        LIMIT=N
        N=NDF*NOFJ
        DO 300 NLL=1,NOL
C       INITIALIZES FORCE,LENGTH,MEMBER STIFFNESS,COORDINATES AND
C       DEFLECTION VALUE FOR EACH LOADING CASE
        DO 1 J=1,NOJ
        DO 1 K=1,NDF
        RJC(J,K)=SJC(J,K)
    1   DJD(J,K)=0.
        DO 2 M=1,NOCS
        RLS(M)=SLS(M)
```

```
      RMF(M)=SMF(M)
      CMF(M)=SMF(M)
C     FOR CMF SEE SUBROUTINE CHECK
    2 YM(M)=YMS(M)
      CALL DATA(NLL)
C     ALTER MEMBER FORCES TO ACCOUNT FOR TEMPERATURE CHANGES
      DO 11 M=1,NOCS
      RMF(M)=RMF(M)-TEMP(M)*YM(M)*AOC(M)
   11 CMF(M)=RMF(M)
C     INITIALIZES NUMBER OF SLACK MEMBERS(II)TO ZERO
      II=0
C     CALLS SUBROUTINE TO ANALYZE THE STRUCTURE FOR DEFLECTIONS UNDER
C     GIVEN LOADING OR RESIDUAL DUE TO NONLINEARITY
      NCYCLE=1
  100 CALL ANALYZ(A,TRI,X,MB,N,LIMIT,NOJ,JTI,NDF,NLL,LTRI)
C     COMPUTES THE NEW STIFFNESS OF THE SYSTEM FOR THE NEXT CYCLE
C     OF ANALYSIS
      I=0
      DO 22 J=1,NOJ
      IF(JTI(J).EQ.1)GO TO 22
      DO 21 K=1,NDF
      I=I+1
   21 DJD(J,K)=X(I)
   22 CONTINUE
      DO 26 MN=1,NOCS
      JS=IJOM(MN,1)
      JE=IJOM(MN,2)
      DO 23 K=1,NDF
      DJS(K)=DJD(JS,K)
   23 DJE(K)=DJD(JE,K)
      XA=AOC(MN)
      CM=CMF(MN)
      CALL FORLEN(MN,DJS,DJE,JS,JE,XA,CM,SFL,SFF)

      RLS(MN)=SFL
   24 IF(NLL.GT.1)GO TO 25
      IF(ICPR(MN).EQ.1)GO TO 26
   25 RMF(MN)=SFF
      CMF(MN)=CM
   26 CONTINUE
      DO 27 J=1,NOJ
      DO 27 K=1,NDF
   27 RJC(J,K)=RJC(J,K)+DJD(J,K)
C     SUBROUTINE TO CHECK FOR ACCURACY AND MEMBER SLACKNESS
      CALL CHECK(ITOL,NCYCLE,CMF,NS,II,NLL)
C     ITOL INDICATES WHETHER OR NOT BOTH REQUIREMENTS ARE SATISFIED
C     ITOL=1 INDICATES AFFIRMATION
      NCYCLE=NCYCLE+1
      IF(ITOL.EQ.1.OR.NCYCLE.GT.LNIT)GO TO 200
      IF(IOI)100,100,200
  200 CALL OUTPUT(NCYCLE,NS,II,NLL,LTRI)
      IF(ITOL.EQ.1)GO TO 300
      IF(NCYCLE-LNIT)100,100,31
   31 PRINT 32
   32 FORMAT(/1X,50H DESIRED ACCURACY NOT ACHIEVED IN PERMITTED NUMBER,
     1 14H OF ITERATIONS/)
  300 CONTINUE
      RETURN
      END



      SUBROUTINE ANALYZ(A,TRI,X,MB,N,LIMIT,NOJ,JTI,NDF,NLL,LTRI)
C     SOLVES LINEAR ALGEBRAIC EQUATIONS BY GAUSS ELIMINATION PROCEDURE
C     TAKING INTO ACCOUNT THE NON-ZERO BAND WIDTH AND SENDS DEFLECTION
C     VALUES TO SUBROUTINE NEWRAP
      DIMENSION A(1),TRI(1),MB(1),X(1),JTI(1)
      MB(1)=1
      I=0
      DO 15 JJ=1,NOJ
      IF(JTI(JJ).EQ.1)GO TO 15
      DO 14 KK=1,NDF
      I=I+1
      CALL ABAND(A,KLOW,KHIGH,CON,N,JJ,KK,NLL)
      CON=-CON
      KL=KLOW
      KH=KHIGH+1-KLOW
```

```
      KU=I-1
      IF(KU)7,7,1
    1 K1=0
      IF(KU-KL)7,2,2
    2 DO 6 K=KL,KU
      K1=K1+1
      LL=MB(K)
      LU=MB(K+1)-1
      L1=K1
      DO 5 L=LL,LU
      L1=L1+1
      A(L1)=A(L1)+TRI(L)*A(K1)
      IF(L1-KH)5,5,3
    3 IF(A(L1))4,5,4
    4 KH=L1
    5 CONTINUE
    6 CON=CON+A(K1)*X(K)
    7 K1=I+1-KLOW
      KL=K1+1
      L=MB(I)-1
      IF(A(K1))10,8,10
    8 ERR=I
      PRINT 9,ERR
    9 FORMAT(1X,12H EQUATION NO,F4.0,2X,27H HAS A ZERO IN THE DIAGONAL)
   10 DO 11 K=KL,KH
      L=L+1
   11 TRI(L)=-A(K)/A(K1)
      IF(I.LT.N)MB(I+1)=L+1
      X(I)=-CON/A(K1)
      IF(L-LIMIT)14,12,12
   12 ERR=I
      PRINT 13,ERR
   13 FORMAT(////1X,33H ARRAY TRI IS FULL IN EQUATION NO,F4.0,/1X,
     1 47H INCREASE THE DIMENSION ON TRI AND CHANGE LIMIT)
   14 CONTINUE
   15 CONTINUE
      LTRI=L
      ERR=0.
      N2=N-1
      DO 16 I=1,N2
      JA=N-I
      JL=MB(JA)
      JU=MB(JA+1)-1
      J2=JA
      DO 16 J=JL,JU
      J2=J2+1
   16 X(JA)=X(JA)+TRI(J)*X(J2)


      RETURN
      END

      SUBROUTINE ABAND(E,KLOW,KHIGH,CON,N,JJ,KK,NLL)
C     SENDS THE NON-ZERO BAND IN EQUATION AND ITS WIDTH TO ANALYZ
C     CALLS STIMAT FOR THE EQUATIONS
      DIMENSION E(M,L)
      CALL STIMAT(E,CON,JJ,KK,N,NLL)
      DO 2 K=1,N
      IF(E(K))1,2,1
    1 KLOW=K
      GO TO 3
    2 CONTINUE
    3 DO 5 K=1,N
      J=N+1-K
      IF(E(J))4,5,4
    4 KHIGH=J
      GO TO 6
    5 CONTINUE
    6 DO 7 L=KLOW,KHIGH
      ISHIFT=L+1-KLOW
    7 E(ISHIFT)=E(L)
      LL=KHIGH-KLOW+2
      DO 8 K=LL,N
    8 E(K)=0.
      RETURN
      END
```

```
      SUBROUTINE STIMAT(E,CON,JNN,K,N,NLL)
C     SENDS  STIFFNESS MATRIX ROW BY ROW TO ABAND

      ** REPEAT COMMON DIMENSION BLOCK **

      DIMENSION E(M,L),MALO(J),DR(L),MR(L),MC(L),AS(L),ASM(L)
      CON=0.
      NM=NINJ(JNN)
      DO 1 I=1,N
    1 E(I)=0.
      DO 2 L=1,NDF
    2 ASM(L)=0.
      DO 3 L=1,NOJ
      IF(JTI(L).EQ.1)GO TO 3
      JM=JNFJ(L)
      MALO(L)=NDF*(JM-1)+1
    3 CONTINUE
      DO 18 I=1,NM
      MNJ=IMOJ(JNN,I)
      YE=YM(MNJ)
      JS=IJOM(MNJ,1)
      JE=IJOM(MNJ,2)
      IF(JNN.EQ.JS)JNR=JE
      IF(JNN.EQ.JE)JNR=JS
      DO11 L=1,NDF
      DR(L)=RJC(JNR,L)-RJC(JNN,L)
      MR(L)=MALO(JNN)+L-1
      IF(JTI(JNR).EQ.1)GO TO 11
      MC(L)=MALO(JNR)+L-1
   11 CONTINUE
      SF=RMF(MNJ)
      SL=RLS(MNJ)
      A=SF/SL
      IF(NLL.EQ.1.AND.ICPR(MNJ).EQ.1)GO TO 12
      B=YE*AOC(MNJ)-SF
      GO TO 13
   12 B=-SF
   13 SL3=SL*SL*SL
      B=B/SL3
      DO 16 L=1,NDF
      IF(L.EQ.K)GO TO 14
      AS(L)=B*DR(K)*DR(L)
      GO TO 15
   14 AS(L)=A+B*DR(K)*DR(K)
   15 IF(JTI(JNR).EQ.1)GO TO 16
      ML=MC(L)
      E(ML)=AS(L)
   16 CONTINUE
      DO 17 L=1,NDF
   17 ASM(L)=ASM(L)-AS(L)
      CON=CON+A*DR(K)
   18 CONTINUE
      DO 19 L=1,NDF
      ML=MR(L)
   19 E(ML)=ASM(L)
      JM=JNFJ(JNN)
      CON=CON+PJL(JM,K)
      CON=-CON
      ML=MR(K)
      IF(E(ML))21,20,21
   20 E(ML)=1.
      CON=0.
   21 RETURN
      END


      SUBROUTINE FORLEN(MN,DJS,DJE,JS,JE,XA,CM,SFL,SFF)

      ** REPEAT COMMON DIMENSION BLOCK **

      DIMENSION DJS(L),DJE(L)
      SL=RLS(MN)
      SS=SL*SL
      C1=0.
      C2=0.
      DO 1 N=1,NDF
      A=RJC(JS,N)-RJC(JE,N)
```

```
      B=DJS(N)-DJE(N)
      C1=C1+B*B
    1 C2=C2+A*B
      C=C1+2.*C2
      D=SL+SQRT(SS+C)
      DL=C/D
      SFL=SL+DL
      SFF=RMF(MN)+XA*YM(MN)*DL/SL
      CM=CM+XA*YMS(MN)*DL/SL
      RETURN
      END

      SUBROUTINE CHECK(ITOL,NCYCLE,CMF,NS,II,NLL)

      ** REPEAT COMMON DIMENSION BLOCK **

      DIMENSION NS(1),CMF(1)
C     CHECK FOR SLACKNESS IN CABLE SEGMENTS
      ITOL=1
      I=0
      IF(NCYCLE.GT.1)GO TO 6
      PRINT 5
    5 FORMAT(//1X,33H DETAILS OF MEMBER BECOMING SLACK)
    6 PRINT 7,NCYCLE
    7 FORMAT(//1X,21H ITERATIVE CYCLE NO.=,I2)
      PRINT 8
    8 FORMAT(/1X,11H MEMBER NO.,4X,21H COMPUTED TOTAL FORCE)
      DO 3 M=1,NOCS
C     IGNORES CHECK IF THE MEMBER IS A STRUT
      IF(MTI(M).EQ.1)GO TO 3
C     IF MEMBER IS SLACK,SET ITS STIFFNESS AND FORCE TO ZERO
      IF(CMF(M))1,1,2
    1 YM(M)=0.
      RMF(M)=0.
      PRINT 100,M,CMF(M)
C     CMF IS THE MEMBER FORCE COMPUTED IGNORING THE LIMITATION OF A
C     CABLE ELEMENT TO SUSTAIN COMPRESSION. THIS VALUE IS OUTPUT ONLY
C     FOR SLACK MEMBERS
      I=I+1
      NS(I)=M
      GO TO 3
    2 YM(M)=YMS(M)
    3 CONTINUE
C     IF NUMBER OF SLACK MEMBERS IN THIS AND THE PREVIOUS CYCLE OF
C     ITERATION ARE EQUAL,CHECK FOR DEFLECTION TOLERANCES. OTHERWISE GO
C     BACK FOR ANOTHER CYCLE OF ANALYSIS TO NEWRAP
      IF(I-II)4,10,4
    4 II=I
      ITOL=0
      RETURN
C     CHECK FOR DEFLECTION TOLERANCES
   10 GRT=0.
      DO 13 J=1,NOJ
      IF(JTI(J).EQ.1)GO TO 13
      DO 12 K=1,NDF
      DJ=ABS(DJD(J,K))
      RJ=ABS(RJC(J,K)-SJC(J,K))
      IF(RJ.EQ.0.)GO TO 12
      IF(RJ.LT.DPI)GO TO 12
      RT=DJ/RJ
      IF(RT.LT.TOL/100.)GO TO 11
      ITOL=0
   11 IF(RT.GE.GRT)GRT=RT
   12 CONTINUE
   13 CONTINUE
      GRT=100.*GRT
  100 FORMAT(1X,I4,10X,E12.4)
      RETURN
      END

      SUBROUTINE OUTPUT(NCYCLE,NS,II,NLL,LTRI)

      ** REPEAT COMMON DIMENSION BLOCK **
```

```
      DIMENSION NS(I),DJ(L)
      NC=NCYCLE-1
      IF(IPT.EQ.1.AND.NLL.EQ.1.AND.NC.EQ.1)GO TO 1
      GO TO 3
    1 PRINT 2
    2 FORMAT(/1X,49H DEFLECTION VALUES FOR THIS LOADING CASE ARE THE ,
     1 55HCORRECTIONS TO THE INITIALLY ASSUMED GEOMETRY, IN ORDER,/1X,
     2 56HTHAT EQUILIBRIUM CAN BE SATISFIED FOR THE GIVEN CONSTANT,
     3 35H MEMBER FORCES IN SPECIFIED MEMBERS/)
    3 PRINT 4,NC
    4 FORMAT(////1X,21H ITERATIVE CYCLE NO.=,I2)
      PRINT 5
    5 FORMAT(27X,12H DEFLECTIONS/1X,
     1 10H JOINT NO.,3X,2H U,15X,2H V,15X,2H W/)
      DO 7 J=1,NOJ
      DO 6 K=1,NDF
    6 DJ(K)=RJC(J,K)-SJC(J,K)
    7 PRINT 8,J,(DJ(K),K=1,NDF)
    8 FORMAT(4X,I3,3X,3(3X,E12.4))
      PRINT 9
    9 FORMAT(//1X,10HMEMBER NO.,8X,11HTOTAL FORCE,7X,
     1 15HCHANGE IN FORCE/)
      DO 10 M=1,NOCS
      FORCE=RMF(M)-SMF(M)
   10 PRINT 11,M,RMF(M),FORCE
   11 FORMAT(4X,I3,8X,E16.6,2X,E16.6)
      PRINT 12,GRT
   12 FORMAT(/1X,50H LARGEST PERCENTAGE VALUE OF,DEFLECTION CORRECTION,
     1 22H /DEFLECTION OBTAINED=,E16.6)
      PRINT 13,LTRI
   13 FORMAT(/1X,26H LENGTH OF ARRAY TRI USED=,I5/)
      IF(II.EQ.0)GO TO 15
      PRINT 14,(NS(M),M=1,II)
   14 FORMAT(/1X,25H MEMBERS GOING SLACK ARE,,25I4/)
   15 RETURN
      END
```

```
** DATA INPUT **
   1
     0      2       2     33      4      46
     1  0.0            0.0
     0  15.0           2.8112
     0  30.            5.2496
     0  45.            7.312
     0  60.            8.9984
     0  75.            10.3130
     0  90.            11.2496
     0  105.           11.8112
     0  120.           12.
     0  135.           11.8112
     0  150.           11.2496
     0  165.           10.3130
     0  180.           8.9984
     0  195.           7.3120
     0  210.           5.2496
     0  225.           2.8118
     1  240.           0.0
     1  0.0            24.
     0  15.            21.1888
     0  30.            19.7504
     0  45.            16.6888
     0  60.            15.0016
     0  75.            13.6870
     0  90.            12.7504
     0  105.           12.1888
     0  135.           12.1888
     0  150.           12.7504
     0  165.           13.6870
     0  180.           15.0016
     0  195.           16.6880
     0  210.           19.7504
     0  225.           21.1888
     1  240.           24.
122.0892     22000.        3.0                  1     2     0     1
121.5752     22000.        3.0                  2     3     0     1
121.1289     22000.        3.0                  3     4     0     1
```

```
120.7559     22000.     3.0              4    5    0    1
120.4599     22000.     3.0              5    6    0    1
120.2336     22000.     3.0              6    7    0    1
120.0840     22000.     3.0              7    8    0    1
120.0095     22000.     3.0              8    9    0    1
120.0095     22000.     3.0              9   10    0    1
120.0040     22000.     3.0             10   11    0    1
120.2336     22000.     3.0             11   12    0    1
120.4599     22000.     3.0             12   13    0    1
120.7559     22000.     3.0             13   14    0    1
121.1289     22000.     3.0             14   15    0    1
121.5752     22000.     3.0             15   16    0    1
122.0892     22000.     3.0             16   17    0    1
122.0892     22000.     2.0             18   19    0    1
121.5504     22000.     2.0             19   20    0    1
121.4753     22000.     2.0             20   21    0    1
120.7559     22000.     2.0             21   22    0    1
120.4599     22000.     2.0             22   23    0    1
120.2336     22000.     2.0             23   24    0    1
120.0840     22000.     2.0             24   25    0    1
120.0095     22000.     2.0             25    9    0    1
120.0095     22000.     2.0              9   26    0    1
120.0840     22000.     2.0             26   27    0    1
120.2336     22000.     2.0             27   28    0    1
120.4599     22000.     2.0             28   29    0    1
120.7559     22000.     2.0             29   30    0    1
121.4753     22000.     2.0             30   31    0    1
121.5504     22000.     2.0             31   32    0    1
122.0892     22000.     2.0             32   33    0    1
3.2          22000.     0.182            2   19    0    0
3.2          22000.     0.182            3   20    0    0
3.2          22000.     0.182            4   21    0    0
3.2          22000.     0.182            5   22    0    0
3.2          22000.     0.182            6   23    0    0
3.2          22000.     0.182            7   24    0    0
3.2          22000.     0.182            8   25    0    0
3.2          22000.     0.182           10   26    0    0
3.2          22000.     0.182           11   27    0    0
3.2          22000.     0.182           12   28    0    0
3.2          22000.     0.182           13   29    0    0
3.2          22000.     0.182           14   30    0    0
3.2          22000.     0.182           15   31    0    0
3.2          22000.     0.182           16   32    0    0
      6       0
1.0          0.001
      2  0.0          7.5
      3  0.0          7.5
      4  0.0          7.5
      5  0.0          7.5
      6  0.0          7.5
      7  0.0          7.5
      8  0.0          7.5
      9  0.0          5.25
     10  0.0          3.0
     11  0.0          3.0
     12  0.0          3.0
[illegible] [illegible] [illegible]
     14  0.0          3.0
     15  0.0          3.0
     16  0.0          3.0
     19  0.0          0.0
     20  0.0          0.0
     21  0.0          0.0
     22  0.0          0.0
     23  0.0          0.0
     24  0.0          0.0
     25  0.0          0.0
     26  0.0          0.0
     27  0.0          0.0
     28  0.0          0.0
     29  0.0          0.0
     30  0.0          0.0
     31  0.0          0.0
     32  0.0          0.0
0.0          0.0        0.0        0.0        0.0        0.0
0.0          0.0        0.0        0.0        0.0        0.0
```

```
     0.0         0.0         0.0         0.0         0.0         0.0
     0.0         0.0         0.0         0.0         0.0         0.0
     0.0         0.0         0.0         0.0         0.0         0.0
     0.0         0.0         0.0         0.0         0.0         0.0
     0.0         0.0         0.0         0.0         0.0         0.0
     0.0         0.0         0.0         0.0
         6        0
     1.0          0.001
          2  0.0         6.0
          3  0.0         6.0
          4  0.0         6.0
          5  0.0         6.0
          6  0.0         6.0
          7  0.0         6.0
          8  0.0         6.0
          9  0.0         3.0
         10  0.0         0.0
         11  0.0         0.0
         12  0.0         0.0
         13  0.0         0.0
         14  0.0         0.0
         15  0.0         0.0
         16  0.0         0.0
         19  0.0         0.0
         20  0.0         0.0
         21  0.0         0.0
         22  0.0         0.0
         23  0.0         0.0
         24  0.0         0.0
         25  0.0         0.0
         26  0.0         0.0
         27  0.0         0.0
         28  0.0         0.0
         29  0.0         0.0
         30  0.0         0.0
         31  0.0         0.0
         32  0.0         0.0
     0.0         0.0         0.0         0.0         0.0         0.0
     0.0         0.0         0.0         0.0         0.0         0.0
     0.0         0.0         0.0         0.0         0.0         0.0
     0.0         0.0         0.0         0.0         0.0         0.0
     0.0         0.0         0.0         0.0         0.0         0.0
     0.0         0.0         0.0         0.0         0.0         0.0
     0.0         0.0         0.0         0.0         0.0         0.0
     0.0         0.0         0.0         0.0
** OUTPUT **
  DATA FOR STRUCTURE   1   READ OR GENERATED
  NO. OF DEGREES OF FREEDOM=  2
  NO. OF LOADING CASES=  2
  NO. OF JOINTS= 33    NO. OF FREE JOINTS= 29     NO. OF SUPPORT JOINTS=  4
  NO. OF CABLE SEGMENTS= 46
  JOINT        JTI      X               Y               Z
   1            1     0.0             0.0
   2            0     0.1500E 02      0.2811E 01
   3            0     0.3000E 02      0.5250E 01
   4            0     0.4500E 02      0.7312E 01
   5            0     0.6000E 02      0.8998E 01
   6            0     0.7500E 02      0.1031E 02
   7            0     0.9000E 02      0.1125E 02
   8            0     0.1050E 03      0.1181E 02
   9            0     0.1200E 03      0.1200E 02
  10            0     0.1350E 03      0.1181E 02
  11            0     0.1500E 03      0.1125E 02
  12            0     0.1650E 03      0.1031E 02
  13            0     0.1800E 03      0.8998E 01
  14            0     0.1950E 03      0.7312E 01
  15            0     0.2100E 03      0.5250E 01
  16            0     0.2250E 03      0.2812E 01
  17            1     0.2400E 03      0.0
  18            1     0.0             0.2400E 02
  19            0     0.1500E 02      0.2119E 02
  20            0     0.3000E 02      0.1975E 02
  21            0     0.4500E 02      0.1669E 02
  22            0     0.6000E 02      0.1500E 02
  23            0     0.7500E 02      0.1369E 02
  24            0     0.9000E 02      0.1275E 02
  25            0     0.1050E 03      0.1219E 02
  26            0     0.1350E 03      0.1219E 02
  27            0     0.1500E 03      0.1275E 02
  28            0     0.1650E 03      0.1369E 02
  29            0     0.1800E 03      0.1500E 02
  30            0     0.1950E 03      0.1669E 02
  31            0     0.2100E 03      0.1975E 02
  32            0     0.2250E 03      0.2119E 02
  33            1     0.2400E 03      0.2400E 02
```

MEMBER	AOC	YMS	SMF	SLS	IJON		ICPR	MTI
1	0.3000E 01	0.2200E 05	0.1221E 03	0.1526E 02	1	2	0	1
2	0.3000E 01	0.2200E 05	0.1216E 03	0.1520E 02	2	3	0	1
3	0.3000E 01	0.2200E 05	0.1211E 03	0.1514E 02	3	4	0	1
4	0.3000E 01	0.2200E 05	0.1208E 03	0.1509E 02	4	5	0	1
5	0.3000E 01	0.2200E 05	0.1205E 03	0.1506E 02	5	6	0	1
6	0.3000E 01	0.2200E 05	0.1202E 03	0.1503E 02	6	7	0	1
7	0.3000E 01	0.2200E 05	0.1201E 03	0.1501E 02	7	8	0	1
8	0.3000E 01	0.2200E 05	0.1200E 03	0.1500E 02	8	9	0	1
9	0.3000E 01	0.2200E 05	0.1200E 03	0.1500E 02	9	10	0	1
10	0.3000E 01	0.2200E 05	0.1200E 03	0.1501E 02	10	11	0	1
11	0.3000E 01	0.2200E 05	0.1202E 03	0.1503E 02	11	12	0	1
12	0.3000E 01	0.2200E 05	0.1205E 03	0.1506E 02	12	13	0	1
13	0.3000E 01	0.2200E 05	0.1208E 03	0.1509E 02	13	14	0	1
14	0.3000E 01	0.2200E 05	0.1211E 03	0.1514E 02	14	15	0	1
15	0.3000E 01	0.2200E 05	0.1216E 03	0.1520E 02	15	16	0	1
16	0.3000E 01	0.2200E 05	0.1221E 03	0.1526E 02	16	17	0	1
17	0.2000E 01	0.2200E 05	0.1221E 03	0.1526E 02	18	19	0	1
18	0.2000E 01	0.2200E 05	0.1216E 03	0.1507E 02	19	20	0	1
19	0.2000E 01	0.2200E 05	0.1215E 03	0.1531E 02	20	21	0	1
20	0.2000E 01	0.2200E 05	0.1208E 03	0.1509E 02	21	22	0	1
21	0.2000E 01	0.2200E 05	0.1205E 03	0.1506E 02	22	23	0	1
22	0.2000E 01	0.2200E 05	0.1202E 03	0.1503E 02	23	24	0	1
23	0.2000E 01	0.2200E 05	0.1201E 03	0.1501E 02	24	25	0	1
24	0.2000E 01	0.2200E 05	0.1200E 03	0.1500E 02	25	9	0	1
25	0.2000E 01	0.2200E 05	0.1200E 03	0.1500E 02	9	26	0	1
26	0.2000E 01	0.2200E 05	0.1201E 03	0.1501E 02	26	27	0	1
27	0.2000E 01	0.2200E 05	0.1202E 03	0.1503E 02	27	28	0	1
28	0.2000E 01	0.2200E 05	0.1205E 03	0.1506E 02	28	29	0	1
29	0.2000E 01	0.2200E 05	0.1208E 03	0.1509E 02	29	30	0	1
30	0.2000E 01	0.2200E 05	0.1215E 03	0.1531E 02	30	31	0	1
31	0.2000E 01	0.2200E 05	0.1216E 03	0.1507E 02	31	32	0	1
32	0.2000E 01	0.2200E 05	0.1221E 03	0.1526E 02	32	33	0	1
33	0.1820E 00	0.2200E 05	0.3200E 01	0.1838E 02	2	19	0	0
34	0.1820E 00	0.2200E 05	0.3200E 01	0.1450E 02	3	20	0	0
35	0.1820E 00	0.2200E 05	0.3200E 01	0.9377E 01	4	21	0	0
36	0.1820E 00	0.2200E 05	0.3200E 01	0.6003E 01	5	22	0	0
37	0.1820E 00	0.2200E 05	0.3200E 01	0.3374E 01	6	23	0	0
38	0.1820E 00	0.2200E 05	0.3200E 01	0.1501E 01	7	24	0	0
39	0.1820E 00	0.2200E 05	0.3200E 01	0.3776E 00	8	25	0	0
40	0.1820E 00	0.2200E 05	0.3200E 01	0.3776E 00	10	26	0	0
41	0.1820E 00	0.2200E 05	0.3200E 01	0.1501E 01	11	27	0	0
42	0.1820E 00	0.2200E 05	0.3200E 01	0.3374E 01	12	28	0	0
43	0.1820E 00	0.2200E 05	0.3200E 01	0.6003E 01	13	29	0	0
44	0.1820E 00	0.2200E 05	0.3200E 01	0.9376E 01	14	30	0	0
45	0.1820E 00	0.2200E 05	0.3200E 01	0.1450E 02	15	31	0	0
46	0.1820E 00	0.2200E 05	0.3200E 01	0.1838E 02	16	32	0	0

LOADING CASE NUMBER = 1

JOINT NO.		LOADS
2	0.0	7.500000
3	0.0	7.500000
4	0.0	7.500000
5	0.0	7.500000
6	0.0	7.500000
7	0.0	7.500000
8	0.0	7.500000
9	0.0	5.250000
10	0.0	3.000000
11	0.0	3.000000
12	0.0	3.000000
13	0.0	3.000000
14	0.0	3.000000
15	0.0	3.000000
16	0.0	3.000000
19	0.0	0.0
20	0.0	0.0
21	0.0	0.0
22	0.0	0.0
23	0.0	0.0
24	0.0	0.0
25	0.0	0.0
26	0.0	0.0
27	0.0	0.0
28	0.0	0.0
29	0.0	0.0
30	0.0	0.0
31	0.0	0.0
32	0.0	0.0

LIMITING NUMBER OF CYCLES= 6

PERCENTAGE TOLERANCE ON ERROR IN DEFLECTIONS= 0.1000E 01

ORDER OF MAGNITUDE OF DEFLECTION PERMITTED TO BE IGNORED IN ACCURACY CHECK= 0.1000E-02

ALPHA*T VALUES

0.0	0.0	0.0	0.0	0.0	0.0
0.0	0.0	0.0	0.0	0.0	0.0
0.0	0.0	0.0	0.0	0.0	0.0
0.0	0.0	0.0	0.0	0.0	0.0
0.0	0.0	0.0	0.0	0.0	0.0
0.0	0.0	0.0	0.0	0.0	0.0
0.0	0.0	0.0	0.0	0.0	0.0
0.0	0.0	0.0	0.0		

```
DETAILS OF MEMBER BECOMING SLACK
ITERATIVE CYCLE NO.= 1
MEMBER NO.        COMPUTED TOTAL FORCE
 34             -0.1273E 01
 45             -0.2486E 01
ITERATIVE CYCLE NO.= 2
MEMBER NO.        COMPUTED TOTAL FORCE
 34             -0.1319E 03
 45             -0.1739E 03
ITERATIVE CYCLE NO.= 3
MEMBER NO.        COMPUTED TOTAL FORCE
 34             -0.1904E 03
 45             -0.1942E 03
ITERATIVE CYCLE NO.= 4
MEMBER NO.        COMPUTED TOTAL FORCE
 34             -0.2269E 03
 45             -0.1979E 03
ITERATIVE CYCLE NO.= 5
MEMBER NO.        COMPUTED TOTAL FORCE
 34             -0.2334E 03
 45             -0.1979E 03
ITERATIVE CYCLE NO.= 6
MEMBER NO.        COMPUTED TOTAL FORCE
 34             -0.2338E 03
 45             -0.1979E 03
ITERATIVE CYCLE NO.= 6

                                 DEFLECTIONS
JOINT NO.      U                     V                      W
    1          0.0                  0.0
    2        -0.7224E-01            0.5063E 00
    3        -0.1120E 00            0.9071E 00
    4        -0.1306E 00            0.1237E 01
    5        -0.1274E 00            0.1461E 01
    6        -0.1090E 00            0.1585E 01
    7        -0.8122E-01            0.1613E 01
    8        -0.4926E-01            0.1540E 01
    9        -0.1872E-01            0.1365E 01
   10        -0.4959E-02            0.1149E 01
   11         0.4120E-02            0.9445E 00
   12         0.9186E-02            0.7530E 00
   13         0.1102E-01            0.5764E 00
   14         0.1004E-01            0.4113E 00
   15         0.2640E-02            0.1946E 00
   16         0.4700E-02            0.1247E 00
   17         0.0                   0.0
   18         0.0                   0.0
   19         0.4865E-01            0.4938E 00
   20        -0.3365E-01            0.6401E-01
   21         0.1221E 00            0.1228E 01
   22         0.1112E 00            0.1452E 01
   23         0.8737E-01            0.1577E 01
   24         0.5501E-01            0.1605E 01
   25         0.1822E-01            0.1534E 01
   26        -0.3445E-01            0.1148E 01
   27        -0.4517E-01            0.9432E 00
   28        -0.5148E-01            0.7513E 00
   29        -0.5415E-01            0.5740E 00
   30        -0.5392E-01            0.4099E 00
   31         0.8974E-01           -0.5291E 00
   32        -0.4608E-02            0.1225E 00
   33         0.0                   0.0
MEMBER NO.          TOTAL FORCE              CHANGE IN FORCE
    1               0.255362E 03             0.133273E 03
    2               0.253727E 03             0.132151E 03
    3               0.252435E 03             0.131306E 03
    4               0.251274E 03             0.130518E 03
    5               0.250387E 03             0.129927E 03
    6               0.249737E 03             0.129503E 03
    7               0.249321E 03             0.129237E 03
    8               0.249121E 03             0.129111E 03
    9               0.199405E 03             0.793958E 02
   10               0.199734E 03             0.797305E 02
   11               0.200097E 03             0.798635E 02
   12               0.200551E 03             0.800909E 02
   13               0.201102E 03             0.803458E 02
   14               0.201900E 03             0.807714E 02
```

```
  15            0.202373E 03      0.807975E 02
  16            0.203397E 03      0.813083E 02
  17            0.207614E 02     -0.101328E 03
  18            0.206840E 02     -0.100866E 03
  19            0.206840E 02     -0.100791E 03
  20            0.206362E 02     -0.100120E 03
  21            0.206183E 02     -0.998416E 02
  22            0.206138E 02     -0.996198E 02
  23            0.206287E 02     -0.994553E 02
  24            0.206850E 02     -0.993246E 02
  25            0.704672E 02     -0.495423E 02
  26            0.703452E 02     -0.497388E 02
  27            0.703522E 02     -0.498814E 02
  28            0.704341E 02     -0.500258E 02
  29            0.705737E 02     -0.501822E 02
  30            0.708812E 02     -0.505941E 02
  31            0.708811E 02     -0.506693E 02
  32            0.713108E 02     -0.507784E 02
  33            0.589591E 00     -0.261041E 01
  34            0.0              -0.320000E 01
  35            0.566582E 00     -0.263342E 01
  36            0.371177E 00     -0.282882E 01
  37            0.384928E 00     -0.281507E 01
  38            0.378261E 00     -0.282174E 01
  39            0.384645E 00     -0.281535E 01
  40            0.181424E 01     -0.138576E 01
  41            0.181708E 01     -0.138292E 01
  42            0.183601E 01     -0.136399E 01
  43            0.180025E 01     -0.139975E 01
  44            0.271657E 01     -0.483428E 00
  45            0.0              -0.320000E 01
  46            0.273536E 01     -0.464636E 00
LARGEST PERCENTAGE VALUE OF,DEFLECTION CORRECTION /DEFLECTION OBTAINED=
LENGTH OF ARRAY TRI USED= 1258                                  0.217128E-01
MEMBERS GOING SLACK ARE,  34  45
DESIRED ACCURACY NOT ACHIEVED IN PERMITTED NUMBER OF ITERATIONS
LOADING CASE NUMBER = 2
JOINT NO.                     LOADS
    2           0.0        6.000000
    3           0.0        6.000000
    4           0.0        6.000000
    5           0.0        6.000000
    6           0.0        6.000000
    7           0.0        6.000000
    8           0.0        6.000000
    9           0.0        3.000000
   10           0.0        0.0
   11           0.0        0.0
   12           0.0        0.0
   13           0.0        0.0
   14           0.0        0.0
   15           0.0        0.0
   16           0.0        0.0
   19           0.0        0.0
   20           0.0        0.0
   21           0.0        0.0
   22           0.0        0.0
   23           0.0        0.0
   24           0.0        0.0
   25           0.0        0.0
   26           0.0        0.0
   27           0.0        0.0
   28           0.0        0.0
   29           0.0        0.0
   30           0.0        0.0
   31           0.0        0.0
   32           0.0        0.0
 LIMITING NUMBER OF CYCLES= 6
 PERCENTAGE TOLERANCE ON ERROR IN DEFLECTIONS=  0.1000E 01

 ORDER OF MAGNITUDE OF DEFLECTION PERMITTED TO BE IGNORED IN ACCURACY CHECK=
 ALPHA*T VALUES                                                     0.1000E-02
  0.0        0.0        0.0        0.0        0.0        0.0
  0.0        0.0        0.0        0.0        0.0        0.0
  0.0        0.0        0.0        0.0        0.0        0.0
  0.0        0.0        0.0        0.0        0.0        0.0
  0.0        0.0        0.0        0.0        0.0        0.0
```

```
0.0         0.0         0.0         0.0         0.0         0.0
0.0         0.0         0.0         0.0         0.0         0.0
0.0         0.0         0.0         0.0
DETAILS OF MEMBER BECOMING SLACK
ITERATIVE CYCLE NO.= 1
MEMBER NO.      COMPUTED TOTAL FORCE
 34           -0.2113E 01
 45           -0.3269E 01
ITERATIVE CYCLE NO.= 2
MEMBER NO.      COMPUTED TOTAL FORCE
 34           -0.1767E 03
 45           -0.1388E 03
ITERATIVE CYCLE NO.= 3
MEMBER NO.      COMPUTED TOTAL FORCE
 34           -0.2102E 03
 45           -0.1640E 03
ITERATIVE CYCLE NO.= 4
MEMBER NO.      COMPUTED TOTAL FORCE
 34           -0.2274E 03
 45           -0.1648E 03
ITERATIVE CYCLE NO.= 5
MEMBER NO.      COMPUTED TOTAL FORCE
 34           -0.2279E 03
 45           -0.1648E 03
ITERATIVE CYCLE NO.= 6
MEMBER NO.      COMPUTED TOTAL FORCE
 34           -0.2279E 03
 45           -0.1648E 03
ITERATIVE CYCLE NO.= 6
                              DEFLECTIONS
JOINT NO.     U                  V                      W
    1        0.0                0.0
    2      -0.6553E-01         0.4346E 00
    3      -0.9770E-01         0.7472E 00
    4      -0.1152E 00         0.1015E 01
    5      -0.1110E 00         0.1161E 01
    6      -0.9378E-01         0.1206E 01
    7      -0.6950E-01         0.1152E 01
    8      -0.4356E-01         0.9979E 00
    9      -0.2167E-01         0.7400E 00
   10      -0.2238E-01         0.4602E 00
   11      -0.2707E-01         0.2341E 00
   12      -0.3333E-01         0.6194E-01
   13      -0.3864E-01        -0.5604E-01
   14      -0.4074E-01        -0.1227E 00
   15      -0.5937E-01        -0.2919E 00
   16      -0.2298E-01        -0.9393E-01
   17        0.0                0.0
   18        0.0                0.0
   19       0.4456E-01         0.4247E 00
   20      -0.4074E-01        -0.8022E-01
   21       0.1120E 00         0.1007E 01
   22       0.9984E-01         0.1154E 01
   23       0.7602E-01         0.1200E 01
   24       0.4506E-01         0.1147E 01
   25       0.1105E-01         0.9937E 00
   26      -0.2493E-01         0.4602E 00
   27      -0.2237E-01         0.2341E 00
   28      -0.1683E-01         0.6182E-01
   29      -0.1117E-01        -0.5630E-01
   30      -0.8469E-02        -0.1195E 00
   31       0.1252E 00        -0.8976E 00
   32       0.2100E-01        -0.8763E-01
   33        0.0                0.0
MEMBER NO.          TOTAL FORCE          CHANGE IN FORCE
    1             0.217109E 03          0.950200E 02
    2             0.215704E 03          0.941285E 02
    3             0.214700E 03          0.935709E 02
    4             0.213700E 03          0.929441E 02
    5             0.212966E 03          0.925058E 02
    6             0.212427E 03          0.921939E 02
    7             0.212084E 03          0.920000E 02
    8             0.211903E 03          0.918939E 02
    9             0.143900E 03          0.238903E 02
   10             0.144048E 03          0.240444E 02
   11             0.144233E 03          0.239996E 02
   12             0.144480E 03          0.240204E 02
```

13	0.144790E 03	0.240344E 02
14	0.145382E 03	0.242532E 02
15	0.145382E 03	0.238067E 02
16	0.146118E 03	0.240290E 02
17	0.399363E 02	-0.821529E 02
18	0.397834E 02	-0.817670E 02
19	0.397833E 02	-0.816920E 02
20	0.396841E 02	-0.810718E 02
21	0.396426E 02	-0.808173E 02
22	0.396243E 02	-0.806093E 02
23	0.396385E 02	-0.804455E 02
24	0.397157E 02	-0.802938E 02
25	0.107772E 03	-0.122374E 02
26	0.107776E 03	-0.123077E 02
27	0.107899E 03	-0.123348E 02
28	0.108116E 03	-0.123437E 02
29	0.108417E 03	-0.123385E 02
30	0.109025E 03	-0.124500E 02
31	0.109025E 03	-0.125253E 02
32	0.109815E 03	-0.122739E 02
33	0.111690E 01	-0.208310E 01
34	0.0	-0.320000E 01
35	0.108259E 01	-0.211741E 01
36	0.711735E 00	-0.248826E 01
37	0.735000E 00	-0.246500E 01
38	0.722882E 00	-0.247712E 01
39	0.725844E 00	-0.247416E 01
40	0.306330E 01	-0.136699E 00
41	0.308007E 01	-0.119933E 00
42	0.310427E 01	-0.957260E-01
43	0.306869E 01	-0.131310E 00
44	0.461003E 01	0.141003E 01
45	0.0	-0.320000E 01
46	0.459303E 01	0.139303E 01

LARGEST PERCENTAGE VALUE OF,DEFLECTION CORRECTION /DEFLECTION OBTAINED= 0.121534E-01
LENGTH OF ARRAY TRI USED= 1258
MEMBERS GOING SLACK ARE, 34 45
STOPPED EXECUTION IN MAIN PROGRAM

Bibliography

REFERENCES

1. Abu Sitta, S., and Elhaskar, I.: The Dynamic Response of Tension Roofs to Turbulent Winds, paper presented at International Conference on Tension Roof Structures, Polytechnic of Central London, April 8–10, 1974.

2. Agarwal, T. P.: An Experimental Study of Behaviour of Hyperbolic Paraboloid Cable Roof, master's dissertation, University of Roorkee, India, 1968.

3. "Air Structures—a Survey," Her Majesty's Stationery Office, London, 1971.

4. Baron, F., and Venkatesan, M. S.: Nonlinear Analysis of Cable and Truss Structures, *Proc. ASCE,*[1] Struct. Div., vol. 97, no. ST2, February 1971.

5. Braga, F., and Care, A.: Study on Cable Networks Subject to Loads However Distributed, paper presented at International Conference on Tension Roof Structures, Polytechnic of Central London, April 8–10, 1974.

6. Bubner, E.: Net Structures—a Survey, paper presented at International Conference on Tension Roof Structures, Polytechnic of Central London, April 8–10, 1974.

7. Buchholdt, H. A., and McMillan, B. R.: Iterative Methods for the Solution of Pretensioned Cable Structures and Pinjointed Assemblies Having Significant Geometrical Displacements, *Proceedings of the IASS*[2] *Pacific Symposium Part II on Tension Structures and Space Frames*, Tokyo and Kyoto, October 1971.

[1] American Society of Civil Engineers
[2] International Association of Shell and Spatial Structures

8. Buchholdt, H. A., and Chana, J. S.: The Influence of Design Parameters on the Natural Frequencies of Cable Beams, paper presented at International Conference on Tension Roof Structures, Polytechnic of Central London, April 8–10, 1974.

9. Buchholdt, H. A., Das, N. K., and Al-Hilli, A. J.: A Gradiant Method for the Analysis of Cable Structures with Flexible Boundaries, paper presented at International Conference on Tension Roof Structures, Polytechnic of Central London, April 8–10, 1974.

10. Buchholdt, H. A., and Dixon, R.: Cable Roof for a Small Water Reservoir, paper presented at International Conference on Tension Roof Structures, Polytechnic of Central London, April 8–10, 1974.

11. Buchholdt, H. A., and Vadgama, N.: Tension Anchors for Cable Structures, paper presented at International Conference on Tension Roof Structures, Polytechnic of Central London, April 8–10, 1974.

12. Burley, E., and Harvey, R. C.: Behavior of Tension Structures subjected to Uniformly Distributed Cable Loadings, paper presented at International Conference on Tension Roof Structures, Polytechnic of Central London, April 8–10, 1974.

13. Butler, A. A.: Long Span Cable Roof Structures, *Proc. Inst. Civ. Eng., (London)*, vol. 52, November 1972.

14. *Cable Construction in Contemporary Architecture*, Booklet 2264A, Bethlehem Steel Corp., Bethlehem, Pa.

15. Cable Suspended Roof Construction State-of-the-Art, *Proc. ASCE, J. Struct. Div.*, vol. 97, no. ST6, June 1971.

16. Christiano, P., Seely, G. R., and Stefan, H.: Transient Wind Loads on Circular Concave Cable Roofs, *Proc. ASCE, J. Struct. Div.*, vol. 100, no. ST11, November 1974.

17. Complex Concrete Ring Supports Cable Roof, *Constr. Methods Equip.*, May 1960.

18. Davenport, A. G.: The Application of Statistical Concepts to the Wind Loading of Structures, *Proc. Inst. Civ. Eng., (London)*, vol. 19, 1961.

19. Davenport, A. G.: Note on the Distribution of Largest Values of Random Function with Application to Gust Loading, *Proc. Inst. Civ. Eng., (London)*, vol. 28, 1964.

20. Davenport, A. G.: Buffeting of Structures by Gusts, *Proceedings of the 1963 International Symposium on the Effects of Wind on Structures*, Her Majesty's Stationery Office, London (England), 1965.

21. Davenport, A. G.: Gust Loading Factors, *Proc. ASCE, J. Struct. Div.*, vol. 93, no. ST3, June 1967.

22. Dent, R. N.: *Principles of Pneumatic Architecture*, The Architectural Press, London, 1971.

23. Elhaskar, I.: Dynamic Response of Tension Roofs to Turbulent Wind, masters' thesis, The University of Western Ontario, London, June 1974.

24. Eras, G., and Elze, H.: "Zur Berechnung Und Statisch Vorteilhaften Formagebung von Seilnetzwerken," Hanging Roofs, North Holland Publishing Company, Amsterdam, 1963.

25. Esquillan, N., and Saillard, Y.(ed.): "Hanging Roofs," *Proceedings of the IASS Colloquium on Hanging Roofs, Continuous Metallic Shell roofs and superficial lattice Roofs* (Paris 1962), North Holland Publishing Company, Amsterdam, 1963.

26. Finzi, L., and Maier, G.: *La Tensostructura del Palasport di Genova*, Facolta Di Architecttura, Politecnico Milano Pubblicazione, No. 24, 1963–64.

27. Franchi, A., and Scirocco, F.: On the Dynamics of Plane Pretensioned Cable Structures, Theoretical and Experimental Research, *Proceedings of the IASS Pacific Symposium Part II on Tension Structures and Space Frames*, Tokyo and Kyoto, October 1971.

28. Gabriel, K.: Structural Design of Cable-suspended Structures, paper presented at International Conference on Tension Roof Structures, Polytechnic of Central London, April 8–10, 1974.

29. Gallo-Curico, A., and Piccarreta, F.: Experimental Tests on a Hanging Roof Model of the New Palasport in Milan, paper presented at International Conference on Tension Roof Structures, Polytechnic of Central London, April 8–10, 1974.

30. Gaylord, E. H., Jr., and Gaylord, C. N.: *Structural Engineering Handbook*, McGraw-Hill, New York, 1968.

31. Gill, V. L., and Petersen Juhl, G.: The Bicycle Wheel, *Consult. Eng.*, December 1974.

32. Greenberg, D. P.: Inelastic Analysis of Suspension Roof Structures, *Proc. ASCE, J. Struct. Div.*, vol. 96, no. ST5, May 1970.

33. Hangleiter, V.: Problems of Accuracy with Prestressed Cable-Net-Structures—Report on Observations and Tests with the Cable-nets for the Olympic Roofs, paper presented at International Conference on Tension Roof Structures, Polytechnic of Central London, April 8–10, 1974.

34. Howard, S.: Prestressing Prevents Flutter of Cable Roofs, *Architectural Record*, August 1959.

35. Isono, Y., and Nakahara, Y.: Fabrication and Construction of U.S. Pavilion Air Structures, *Proceedings of the IASS Pacific Symposium Part II on Tension Structure and Space Frames*, Tokyo and Kyoto, October 1971.

36. Jennings, A.: Discussion to (O'Brien 64), *Proc. ASCE, J. Struct. Div.*, vol. 91, no. ST1, February 1965.

37. Jensen, J. J.: *Eine Statische Und Dynamische Untersuchung Der Seil-Und-Membrantragwerke*, Division of Structural Mechanics, The Norwegian Institute of Technology, University of Trondheim, Report No. 70-1, September 1970.

38. Jensen, J. J.: An Investigation of Static and Dynamic Behavior of Suspension Structures, *Proceedings of the IASS Pacific Symposium—Part II on Tension Structures and Space Frames*, Tokyo and Kyoto, October 1971.

39. Jensen, J. J.: Das Dynamisch Verhalten, Eines Vorgespannten Kabelnetzes, Division of Structural Mechanics, The Norwegian Institute of Technology, University of Trondheim, Report No. 72-3, October 1972.

40. Jensen, J. J.: Dynamics of Tension Roof Structures, paper presented at International Conference on Tension Roof Structures, Polytechnic of Central London, April 8–10, 1974.

41. Kar, A. K., and Okazaki, C. Y.: Convergence in Highly Non-linear Cable Net Problems, *Proc. ASCE, J. Struct. Div.*, vol. 99, no. ST3, March 1973.

42. Krishna, P.: Analysis of Funicular Suspension Systems, Ph.D. thesis, University of London, 1964.

43. Krishna, P.: Experiments on the Wire Model of a Suspended Cable Structure, *J. Inst. Eng. (India)*, vol. XLVIII, no. 5, pt. CI3, January 1968.

44. Krishna, P.: Discussion to "Numerical Solution of Nonlinear Structures by T. J. Poskitt," *Proc. ASCE, J. Struct. Div.*, vol. 94, no. ST6, June 1968.

45. Krishna, P.: Theoretical Analysis of Pretensioned Cable Networks, *Proc. IASS Pacific Symposium—Part II on Tension Structures and Space Frames*, Tokyo and Kyoto, October 1971.

46. Krishna, P., and Agarwal, T. P.: Study of Suspended Roof Model, *Proc. ASCE, J. Struct. Div.*, vol. 97, no. ST6, June 1971.

47. Krishna, P., and Agarwal, T. P.: Approximate Analysis of Cable Structures, paper presented at International Conference on Tension Roof Structures, Polytechnic of Central London, April 8–10, 1974.

48. Krishna, P., Chandra, B., and Gupta, V. K.: Vibration Characteristics of Cable Suspended System, *Bull. Indian Soc. Earthquake Technol.*, December 1973.

49. Krishna, P., Chandra, B., and Gupta, V. K.: Vibration Characteristics of Cable Systems, paper presented at International Conference on Tension Roof Structures, Polytechnic of Central London, April 8–10, 1974.

50. Krishna, P., and Natrajan, P. R.: Analysis of Doubly Curved Suspended Cable Roof Networks, *Bull. IASS*, no. 34, June 1968.

51. Krishna, P., and Sparkes, S. R.: Analysis of Pretensioned Cable Systems, *Proc. Inst. Civ. Eng. (London)*, vol. 39, January 1968.

52. Leonhardt, F., Egger, H., and Haug, E.: The German Pavilion at Expo '67, Montreal—A Prestressed Cable Network Structure (in German), *Der Stahlbau*, vol. 37, no. 4, April 1968.

53. Leonhardt, F., and Schlaich, J.: Structural Design of Roof over the Sports Arenas for the 1972 Olympic Games: Some Problems of Prestressed Cable Nets, *Struct. Eng.*, vol. 50, no. 3, March 1972.

54. Leonhardt, F., and Schlaich, J.: Cable Suspended Roof for Munich Olympic, *Civ. Eng.*, vol. 52, no. 7, July 1972.

55. Leonhardt, F., and Schlaich, J.: *Vorgespannte Seilnetz Konstruktionen Das Olympiadach in Munchen*, SFB64/Mitteilungen 19/1973, Universitat Stuttgart, 1973.

56. Lin, T. Y., and Young, B. W.: Two Large Shells of Post-tensioned Precast Concrete, *Civ. Eng.*, vol. 35, no. 7, July 1965.

57. Magnusson, K.: Cable Roofs—Points of View Regarding Economics and Insurance, paper presented at International Conference on Tension Roof Structures, Polytechnic of Central London, April 8–10, 1974.

58. Maier, G., and De Donato, O.: Elastic Analysis of Plane Pretensioned Systems, *Proceedings of IASS Pacific Symposium Part II on Tension Structures and Space Frames*, Tokyo and Kyoto, October 1971.

59. Makowski, Z. S., Nooshin, H., and Butterworth, J. W.: *Report on Checking of the Design of the Cable Roof for Farahabad Sports Stadium*, Space Structures Research Centre, University of Surrey, England, August 1971.

60. *Manual for Structural Applications of Steel Cables for Buildings*, 1973, American Iron and Steel Institute.

61. Mayr, G.: *Untersuchungen an Netzlitzen-knoten and endbeschlagen*, SFB Report 7/73, Wermer-Verlag, Düsseldorf, 1974.

62. Merritt, F. S.: Curved Roof on Cable Spans Big Arena, *Eng. News-Rec.*, vol. 150, no. 6, Feb. 5, 1953.

63. Meyerhof, G. G.: The Ultimate Bearing Capacity of Foundations, *Geotechnique*, London, vol. 2, 1951.

64. Meyerhof, G. G., and Adams, J. I.: The Ultimate Uplift Capacity of Foundations, *Can. Geotech. J.*, vol. 5, no. 4, 1969.

65. Meyerhof, G. G.: Uplift Resistance of Inclined Anchors and Piles, *Proceedings of the 8th International Conference on Soil Mechanics and Foundation Engineering*, Moscow, 1973.

66. Mollmann, H.: Analysis of Plane Prestressed Cable Structures, *Proc. ASCE, J. Struct. Div.*, vol. 96, no. ST10, October 1970.

67. Mollmann, H.: Analytical Solution for a Cable Net over a Rectangular Plan, *Proceedings of the IASS Pacific Symposium—Part II on Tension Structures and Space Frames*, Tokyo and Kyoto, October 1971.

68. Mollmann, H.: Analysis of Prestressed Cable Systems Supported by Elastic Boundary Structures, paper presented at International Conference on Tension Roof Structures, Polytechnic of Central London, April 8–10, 1974.

69. Mollmann, H.: *Analysis of Hanging Roofs by Means of the Displacement Method*, Polyteknisk Forlag, Lyngby, Denmark, 1974.

70. Natrajan, P. R.: Comparative Study of Methods of Analysis for Suspended Cable Roof Systems, masters' thesis, University of Roorkee, India, 1967.

71. Nooshin, H., and Butterworth, J. W.: Experimental Study of a Prestressed Cable Roof, paper presented at International Conference on Tension Roof Structures, Polytechnic of Central London, April 8–10, 1974.

72. Norris, Charles H., et al.: *Structural Design for Dynamic Loads*, McGraw-Hill, New York, 1959.

73. O'Brien, W. T., and Francis, A. J.: Cable Movements under Two Dimensional Loads, *Proc. ASCE, J. Struct. Div.*, vol. 90, no. ST3, June 1964.

74. O'Brien, W. T.: Static Behaviour of Pretensioned Cable Trusses, dissertation submitted for the Diploma of Imperial College, London, July 1968.

75. Otto, Frei, and Schleyer, F. B.: *Tensile Structures*, M.I.T. Press, 1969.

76. Our Two Largest Airports—Dulles International Airport, *Prog. Archit.*, vol. 44, no. 8, August 1963.

77. Pearson, R., and Dickson, M.: A Study in Structural Performance and Protection in Fire of a Cable Roof Design, paper presented at the International Conference on Tension Roof Structures, Polytechnic of Central London, April 8–10, 1974.

78. *Proceedings of the First International Colloquium on Pneumatic Structures*, IASS, University of Stuttgart, Germany, May 1967.

79. Pugsley, A. G.: On the Natural Frequency of Suspension Chains, *Quart. J. Appl. Math.*, vol. 2, 1949.

80. Pugsley, A. G.: *Theory of Suspension Bridges*, Arnold, 1957.

81. Ranjan, G.: The Pull-out Capacity of Rigid Piles in Sand, First Iranian Congress of Civil Engineering and Engineering Mechanics, Pahlavi University, Shiraz, Iran, 1972.

82. Rice, P.: Notes on the Design of Cable Roofs, *Proceedings of the IASS Pacific Symposium—Part II on Tension Structures and Space Frames*, Tokyo and Kyoto, October 1971.

83. Sandi, H., and Hagiescu, S.: Influence of the Boundary Structure on the

Behaviour of Cable Girders, paper presented at the International Conference on Tension Roof Structures, Polytechnic of Central London, April 8–10, 1974.

84. Schlaich, J.: Membrane-Skin and Cable-Net Cooling Towers, paper presented at the International Conference on Tension Roof Structures, Polytechnic of Central London, April 8–10, 1974.

85. Schupack, M.: Cable Supported Roof Cuts Cost, *Civ. Eng.*, vol. 28, no. 4, April 1958.

86. Selection of Wire Ropes for Engineering Applications, British Ropes Group, Publication no. 589, June 1967.

87. Severud, F. N.: Materials Combined to Advantage—Concrete in Compression, Steel in Tension, *Civ. Eng.*, vol. 24, no. 3, March 1954.

88. Siev, A.: Stability of Prestressed Suspended Roofs, D.Sc. thesis Technion-Israel Institute of Technology, Haifa, Israel, 1961.

89. Siev, A., and Eidelman, J.: Stress Analysis of Prestressed Cable Roofs, *Proc. ASCE, J. Struct. Div.*, vol. 90, no. ST4, August 1964.

90. Smith, R. C.: *Materials of Construction*, 2d ed., McGraw-Hill, New York, 1973.

91. Space Gain Solves Uplift Problem in 404 ft. Cable Roof, *Eng. News-Rec.*, vol. 176, no. 23, June 9, 1966.

92. Sports Complex Features Cable Supported Roof, *Civ. Eng.*, vol. 37, no. 1, January 1967.

93. Tezcan, S. S., Cherry, S., and Mahapatra, B. C.: Dynamic Analysis of Cable Structures, *Proceedings of the IASS Pacific Symposium Part II on Tension Structures and Space Frames*, Tokyo and Kyoto, October 1971.

94. Thorburn, S.: "Large Diameter Piles Founded on Bedrock," Large Bored Piles, *Proceedings of Symposium*, Institute of Civil Engineers, 1966.

95. Thornton, C. H., and Birnstiel, C.: Three Dimensional Suspension Structures, *Proc. ASCE., J. Struct. Div.*, vol. 93, no. ST2, April 1967.

96. Tottenham, H., and Khan, M. A.: Elastic Non-linear Analysis of Prestressed Cable Networks using Equivalent Continuum Theory, *Proceedings of the IASS Pacific Symposium Part II on Tension Structures and Space Frames*, Tokyo and Kyoto, October 1971.

97. Tottenham, H., and Williams, P. G.: Cable Net: Continuous System Analysis, *Proc. ASCE, J. Eng. Mech. Div.*, vol. 96, no. EM3, June 1970.

98. Tsuboi, Y., and Kawaguchi, M.: Design Problem of Suspension Roof Structure—Tokyo Olympic Swimming Pools, Report of the Institute of Industrial Science, Tokyo University, 15-2, 111-164, November 1965. (Published in German in *Der Stahlbau*, Heft 3, March 1966.)

99. Uhlenbeck, G.: *Diffraction of Shock Waves Around Various Obstacles*, Engineering Research Institute, University of Michigan, Ann Arbor, March 1950.

100. White, K., Happold, E., and Dickson, M.: Notes on the Theoretical Costs of Cable Structures, paper presented at International Conference on Tension Roof Structures, Polytechnic of Central London, April 8–10, 1974.

101. Wyatt, T. A.: Secondary Stresses in Parallel Wire Suspension Cable, *Proc. ASCE, J. Struct. Div.*, vol. 86, no. ST7, July 1960.

102. Yokoo, Y., Nakamura, T., Terabayashi, T., and Mukai, H.: Nonlinear Behaviour of Counter Stressed Dual Plane Cable Frames, *Proceedings of the IASS Pacific Symposium Part II on Tension Structures and Space Frames*, Tokyo and Kyoto, October 1971.

X 103. Yokoo, Y., Nakamura, T., Heki, K., and Kawamata, S. (ed): Tension Structures and Space Frames, *Proceedings of the IASS Pacific Symposium Part II on Tension Structures and Space Frames*, Tokyo and Kyoto, October 1971, Tokyo, 1972.

ADDITIONAL REFERENCES

Argyris, J. H., and Scharpf, D. W.: Large Deflection Analysis of Prestressed Network, *Proc. ASCE, J. Struct. Div.*, vol. 98, no. ST3, March 1972.

———, Aicher, W., and Fluh, H.: Experimental Analysis of Oscillations and Determination of Damping Values of Cables and Cable Net Structures, paper presented at International Conference of Tension Roof Structures, Polytechnic of Central London, April 8–10, 1974.

———, Dunne, P., and Angelopoulus, T.: Dynamic Analysis of Tension Roof Structures, paper presented at International Conference on Tension Roof Structures, Polytechnic of Central London, April 8–10, 1974.

Avent, R. R.: Nonlinear Field Analysis of Structural Nets, *Proc. ASCE, J. Struct. Div.*, vol. 95, no. ST5, May 1969.

Bandel, H. K.: Orthogonal Nets in the Form of Hyperbolic Paraboloids Subjected to Vertical Loading and Temperature Change (in German), *Der Bauingenieur*, vol. 34, no. 10, 1959.

———: The Hanging Cable Under Three-dimensional Loading and Temperature Change (in German), *Der Bauingenieur*, vol. 37, no. 4, 1962.

Batish, G. N.: Membrane Analysis of Cable Roofs, thesis presented for degree of Ph.D. at University of Michigan, Ann Arbor, 1966.

Bethlehem Wire Rope for Bridges, Towers, Aerial Tramways, Catalog 2277, Bethlehem Steel Corp., Bethlehem, Pa.

Bowman, W.: 1958 World's Fair Is a Structural Show, *Eng. News-Rec.*, vol. 159, no. 13, Sept. 26, 1957.

Buchanan, G. R.: Two-dimensional Cable Analysis, *Proc. ASCE, J. Struct. Div.*, vol. 96, no. ST7, July 1970.

Buchholdt, H. A.: The Behaviour of Small Prestressed Cable Roof Subjected to U.D.L., *Space Structures*, University of Surrey, England, August 1966.

———: A Non-linear Deformation Theory Applied to Two-dimensional Pretensioned Cable Assemblies, *Proc. Inst. Civ. Eng. (London)*, vol. 42, January 1969.

———: Tension Structures, *Struct. Eng.*, vol. 48, no. 2, February 1970.

———: Pretensioned Cable Girders, *Proc. Inst. Civ. Eng. (London)*, vol. 45, March 1970.

———: The Behaviour of Prestressed Circular Cable Grids, *Proceedings of the IASS Pacific Symposium—Part II on Tension Structures and Space Frames*, Tokyo and Kyoto, October 1971.

———: A Non-linear Vector Method for the Analysis of Vertically and Laterally Loaded Cables and Cable Assemblies, *Proc. Inst. Civ. Eng. (London)*, vol. 55, March 1973.

———: Analysis and Design of Cable Structures, *J. Inst. Eng. (India)*, vol. 53, pt. CI5, May 1973.

———, Gill, V., and McMillan, B.: The Design and Testing of a Cable Beam Structure for Prefabrication, *Proceedings of the IABSE[3] Ninth Congress*, Amsterdam, 1972.

Building Engineering, Part 2, Parabolic Cable Roof, *Architectural Forum*, vol. 98, no. 61, June 1953.

Cable, Concrete-Pylon Supported Roofs, *Prog. Archit.*, vol. 41, no. 3, March 1960.

Cable Hung Roof Goes to the Races, *Eng. News-Rec.*, vol. 161, no. 10, Sept. 4, 1958.

Cable Hung Olympic Roof Spans 300 ft., *Eng. News-Rec.*, vol. 163, no. 5, July 30, 1959.

Cables Supported Steel Plate Roof, *Eng. News-Rec.*, vol. 175, no. 19, Nov. 4, 1965.

Cables Support 130 ft. Hangar Roof, *Eng. News-Rec.*, vol. 157, no. 26, Dec. 27, 1956.

Chaudhari, B. S.: Some Aspects of Dynamic Behaviour of Cable Networks, Ph.D. dissertation, University of Pennsylvania, Philadelphia, 1969.

Clark, S. H.: Cable Supported Roof for Olympic Arena, *Civ. Eng.*, vol. 29, no. 9, September 1959.

Conrad, Roland: *Frei Otto Tension Structures*, Praeger, New York, 1970.

[3] International Association of Bridge and Structural Engineering.

Davis, R. M. (ed.): Space Structures, Blackwell, Oxford, England, 1967.

Day, A. S., and Bunce, J. W.: Analysis of Cable Networks, *Civ. Eng. Public Works Rev.*, April 1970.

Dean, D. L., and Ugrate, C. P.: Analysis of Structural Nets, *Publications, IABSE*, vol. 23, Zurich, 1963.

Design Award, *Prog. Archit.*, vol. 39, no. 1, January 1958.

Dirlewanger, H., Geisler, E., and Knorzer, I.: Architectural Psychology and the Design of Wide-Span Surface Structures, paper presented at International Conference on Tension Roof Structures, Polytechnic of Central London, April 8–10, 1974.

Ferretti, A. S., and Zingali, A.: A Large Span Hanging Roof: The "Palasport" in Milan, Preliminary Report, *Proceedings of the IABSE Ninth Congress*, Amsterdam, 1972.

Firt, V.: Two-membrane Suspended Roofs, paper presented at International Conference on Tension Roof Structures, Polytechnic of Central London, April 8–10, 1974.

Fischer, F., Jr.: Cable Supported Warehouse Roof Provides Large Open Area, *Civ. Eng.*, vol. 36, no. 9, September 1966.

Francis, A. J.: Single Cables Subjected to Loads, *Civil Eng. Trans. Inst. Engrs., Australia*, vol. CE-7, no. 2, October 1965.

———: Analysis of Suspension Cable Behaviour, *The Engineer*, vol. 219, no. 5709, June 25, 1965.

Fresh Forms and New Directions from a Special Kind of Problem Solving, *Architectural Record*, vol. 143, no. 5, May 6, 1968.

Gensert, R. M., et. al.: *Cable Roof Structures*, Booklet 2318, Bethlehem Steel Corp., Bethlehem, Pa., 1968.

Greenberg, D. P.: Cable Suspended Roofs, *Third Progress Report to the American Iron and Steel Institute*, Department of Structural Engineering, Cornell University, August 1968.

———: An Equivalent Stiffness Method for Suspension Roof Analysis, *Proceedings of the IABSE Ninth Congress*, Amsterdam, 1972.

Grigorian, M.: Free Vibrations of Supported Cable Nets, *Proc. Inst. Civ. Eng. (London)*, vol. 45, 1970.

Gupta, V. K.: Analysis and Design of Cable Roof Systems, master's dissertation, University of Roorkee, India, 1971.

Haug, E., and Powell, G. H.: Analytical Shape Finding for Cable Nets, *Proceed-*

ings of the IASS Pacific Symposium Part II on Tension Structures and Space Frames, Tokyo and Kyoto, October 1971.

Jawerth, D.: Prestressed Suspension Systems Composed of Cables of Opposite Curvature with Inclined Suspenders (German), *Der Stahlbau*, vol. 28, no. 5, May 1959.

———, and Schulz, H.: On the Natural Frequency and Aerodynamic Stability of Suspended Roofs (in German), *Der Stahlbau*, vol. 35, no. 1, January 1966.

Jennings, A.: The Free Cable, *The Engineer*, vol. 214, no. 5579, Dec. 28, 1962.

Jonatowski, J. J., and Birnsteil, C.: Inelastic Stiffened Suspension Space Structures, *Proc. ASCE, J. Struct. Div.*, vol. 96, no. ST6, June 1970.

Knudson, W. C.: Response of Cable Net Structures Under Dynamic Loads, *Proceedings of the IASS Pacific Symposium Part II on Tension Structures and Space Frames*, Tokyo and Kyoto, October 1971.

Konishi, Y., Moritani, T., and Takemura, Y.: Structural Design of Prestressed Cable Roof, *Proceedings of the IASS Pacific Symposium Part II on Tension Structures and Space Frames*, Tokyo and Kyoto, October 1971.

Krishna, P.: Modern Trends in Suspended Structural Systems, Symposium on Changing Concepts of Human Habitations, CBRI, Roorkee, India, December 1965.

———: Structural Performance of Suspended Cable Roofs, *Cement and Concrete*, Sahu Cements Publication, India, January 1968.

———, and Gupta, V. K.: Dynamic Response of Cable Structures, *Earthquake Engineering—Jai Krishna Sixtieth Anniversary Commemoration Volume*, Sarita Prakashan, Meerut, India, 1974.

———, and Kohli, A. K.: Design of a Doubly Curved Suspended Cable Roof with Prestressed Concrete Beams, *Indian Concr. J.*, vol. 42, no. 11, November 1968.

———, and Sawhney, P. S.: Design of Cable Roofs, *Proceedings of the Symposium on Modern Trends in Civil Engineering*, Department of Civil Engineering, University of Roorkee, India, November 1972.

———, and Sparkes, S. R.: An Influence Coefficient Method of Analysis for Cable Systems, *Proc. Inst. Civ. Engrs. (London)*, vol. 40, September 1968.

Leonard, J. W.: Nonlinear Dynamics of Curved Cable Elements, Technical Note, *Proc. ASCE, J. Eng. Mech. Div.*, vol. 99, no. EM3, June 1973.

———: Incremental Response of 3-D Cable Net Works, Technical Note, *Proc. ASCE, J. Eng. Mech. Div.*, vol. 99, no. EM3, June 1973.

Liudkovsky, I. G. (ed.): Suspended Roofs with Circular Plan (in Russian), Moscow, 1962.

Markland, E.: Deflection of a Cable Due to a Single Point Load, *Philosophical Magazine*, vol. 42, no. 23, London, September 1951.

Michalos, J., and Birnstiel, C.: Movements of a Cable due to Changes in Loading, *Proc. ASCE, J. Struct. Div.*, vol. 86, no. ST12, December 1960.

Modjtadhedi, D., and Shore, S.: The Non-linear Steady State Response of Flexible Cable Network, *Proceedings of the IASS Pacific Symposium Part II on Tension Structures and Space Frames*, Tokyo and Kyoto, October 1971.

Mojowiecki, M.: Tension Structures Jawerth System, *Acier stahl*, April 1971.

Mollmann, H.: A Study in the Theory of Suspension Structures, Akademisk Forlag, Copenhagen, Denmark, 1965.

Mollmann, H., and Mortensen, P. L.: The Analysis of Prestressed Suspended Roofs (Cable Nets), *Space Structures*, ed. by R. M. Davis, Blackwell, Oxford, England, 1967.

Morris, N. F.: Dynamic Response of Cable Networks, *Proc. ASCE, J. Struct. Div.*, vol. 100, no. ST10, October 1974.

Murray, T. M., and Willems, N.: Analysis of Inelastic Suspension Structures, *Proc. ASCE, J. Struct. Div.*, vol. 97, no. ST12, December 1971.

Noesgen, J.: A Contribution of the Load Bearing Behaviour of Prestressed Cable Net Structures, paper presented at International Conference on Tension Roof Structures, Polytechnic of Central London, April 8–10, 1974.

O'Brien, T. W.: General Solution of Suspended Cable Problems, *Proc. ASCE, J. Struct. Div.*, vol. 93, no. ST1, February 1967.

Otto, F.: *The Hanging Roof*, Ullstein Verlag, Berlin, West Germany, 1954.

Pirner, M.: Statical Calculation of Prestressed Suspended Cable Roofs (in Czech), *Inz. Stavby*, no. 11, Prague, 1962.

———: Natural Vibrations of Prestressed Nets having the Form of Translatory Surfaces (in Czech), *Stavebnicky Cas.*, vol. XI, no. 10, Bratislava, 1963.

———: Berechnung von vorgespannten Seilnetzen, *Bauplanung-Bautechnik*, vol. 19, heft 11, November 1965.

———: Doubly Curved, Prestressed Cable Nets, Under Static and Dynamic Loads (in Polish), *Arch. Inz. Ladowy*, vol. 11, no. 1, Warsaw, 1965.

———: Theory of Statically and Dynamically Loaded Prestressed Networks (in Czech), *Sb. Vys. Sk. Dopravnej Ziline*, no. 4, 1965.

Prestressing Stabilizes Unusual Cable Roof, *Eng. News-Rec.*, vol. 164, no. 17, April 28, 1960.

Roy, B. N.: Stiffness Matrix Analysis of Prestressed Cable Systems, master's dissertation, University of Roorkee, India, 1969.

Saafan, S. A.: Theoretical Analysis of Suspension Roofs, *Proc. ASCE, J. Struct. Div.*, vol. 96, no. ST2, February, 1970.

Safronie, R.: Dynamic Stability of Prestressed Suspension Roofs, doctoral thesis, Institute of Civil Engineering, Bucharest, 1969.

———: The Mechanical Stability of Cable Systems, *Rev. Roum. Sci. Tech. Ser. Mec. Appl.*, April 1971.

———: The Mechanical Behaviour of Cable Systems, *Rev. Roum. Sci. Tech. Ser. Mec. Appl.*, April 1972.

Savvy Smooths Erection of 427 Ft Cable Roof, *Eng. News-Rec.*, vol. 177, no. 2, July 14, 1966.

Sayar, K. F.: On the Statical Analysis of a Hanging Roof as a Prestressed Cable Network, *Proceedings of the IASS Pacific Symposium Part II on Tension Structures and Space Frames*, Tokyo and Kyoto, October 1971.

———: On the Statics of a Tension Roof Structure, paper presented at International Conference on Tension Roof Structures, Polytechnic of Central London, April 8–10, 1974.

Scalzi, J. B., and McGrath, W. K.: Mechanical Properties of Structural Cables, *Proc. ASCE, J. Struct. Div.*, vol. 97, no. ST12, December 1971.

———, Podolny, W., Jr., and Teng, W. C.: *Design Fundamentals of Cable Roof Structures*, Booklet ADUSS 55-3580-01, United States Steel Corp., Pittsburgh, Pa., October 1969.

Seely, G. R., Stefan, H., and Christiano, P.: Static Wind Loads on Circular Concave Cable Roofs, *Proc. ASCE, J. Struct. Div.*, vol. 100, no. ST8, August 1974.

Shore, S., and Bathish, G. N.: Membrane Analysis of Cable Roofs, Space Structures, Blackwell, Oxford, England, 1967.

———, and Chaudhari, B.: *Dynamic Response of Cable Systems*, report by Civil Engineering Division, University of Pennsylvania, sponsored by American Iron and Steel Institute, July 1968.

Siev, A.: A General Analysis of Prestressed Nets, *Publications, IABSE*, Zurich, vol. 23, 1963.

Simpson, W. E.: Cable Suspended Roof for San Antonio Assembly Building, *Civ. Eng.*, vol. 30, no. 11, November 1960.

Singh, B. P.: Membrane Analogy for Linear and Nonlinear Vibrations of Cable Networks, Ph.D. thesis, Indian Institute of Technology, Kanpur, India, October 1972.

Sinicyn, A. P.: The Dynamics of Flexible Roof of Large Sag (in German), *Hangedacher*, ed. by I. M. Rabinovic, Bauverlag, Wiesbaden, West Germany, 1966.

Soler, A. I., and Afshari, H.: On the Analysis of Cable Network Vibrations Using Galerkin's Method, *J. Appl. Mech.*, vol. 37, no. 3, 1970; *Transactions ASME*, vol. 92, series E, 1970.

Steel Cables Suspended 220 Foot Diameter Roof from Concrete Cantilevers, *Eng. News-Rec.*, vol. 172, no. 24, June 11, 1964.

Sterner, J. A.: Fabrication and Erection of Cable Supported Roof, *Civ. Eng.*, vol. 37, no. 10, October 1967.

Suspended Steel Blanket Roof, *Architectural Forum*, vol. 104, no. 4, April 1956.

Thornton, C. H., and Cuoco, D. A.: Innovative Approaches to the Erection of Tension Roof Structures, paper presented at International Conference on Tension Roof Structures, Polytechnic of Central London, April 8–10, 1974.

Thurnherr, F.: Erection of the Olympic Tents at Munich Using VSL Post-Tensioning Equipment, paper presented at International Conference on Tension Roof Structures, Polytechnic of Central London, April 8–10, 1974.

Tsuboi, Y., and Kawaguchi, M.: *Design Problems of Suspension Roof Structure—Tokyo Olympic Swimming Pools*, IASS Bulletins 27 and 28, 1966.

Vasarhelyi, M. A.: A 340,000 Cubic Metre Capacity Water Reservoir in Durban, Natal, Sixth F.I.P. Congress, Prague, June 1970.

Virdi, K. S.: Design of Suspended Cable Roofs, master's dissertation, University of Roorkee, India, 1969.

Wakabayashi, M., Nakamura, T., Minami, K., and Shibata, M.: An Approach to Structural Design of a Prestressed Cable Net Structure, *Proceedings of the IASS Pacific Symposium Part II on Tension Structures and Space Frames*, Tokyo and Kyoto, October 1971.

Wide-Span Cable Roof for Whorley Reservoir of Tees Valley and Cleveland Water Board, *Water Water Eng.*, January 1972, pp. 18–19.

Williamson, R. A., Stevenson, H. F., and Vandarpas, P. W.: Computer Design of a Tower and Cable System, *Proc. ASCE, J. Struct. Div.*, vol. 92, no. ST12, December 1966.

Zetlin, L.: Basic Design Principles of Cable Roofs including Statics, Dynamics and Systems, *Cable Roof Structures*, Booklet 2318, Bethlehem Steel Corp., Bethlehem, Pa.

———: Steel Cable Creates Novel Structural Space System, *AISC Engineering J.*, January 1964.

Zingali, A.: Remarks on the Dynamic Characteristics of a Large Net Hanging Roof Structure, paper presented at International Conference on Tension Roof Structures, Polytechnic of Central London, April 8–10, 1974.

Index